Hands-On Mathematical Optimization with Python

This practical guide to optimization combines mathematical theory with hands-on coding examples to explore how Python can be used to model problems and obtain the best possible solutions. Presenting a balance of theory and practical applications, it is the ideal resource for upper-undergraduate and graduate students in applied mathematics, data science, business, industrial engineering, and operations research programs, as well as practitioners in related fields.

Beginning with an introduction to the concept of optimization, this text presents the key ingredients of an optimization problem and the choices one needs to make when modeling a real-life problem mathematically. Topics covered range from linear and network optimization to convex optimization and optimizations under uncertainty.

The book's Python code snippets, and the accompanying suite of more than 50 Jupyter notebooks on the author's GitHub, allow students to put the theory into practice and solve problems inspired by real-life challenges, while numerous exercises sharpen students' understanding of the methods discussed.

KRZYSZTOF POSTEK is Senior Optimization Data Scientist with the Boston Consulting Group in Amsterdam. He received his PhD in Operations Research in 2017 from Tilburg University. After his postdoc at the Technion – Israel Institute of Technology, he spent several years as a faculty member at Erasmus University Rotterdam and Delft University of Technology. His research interests revolve mostly around optimization under uncertainty.

ALESSANDRO ZOCCA is Assistant Professor in the Department of Mathematics at the Vrije Universiteit Amsterdam. He received his PhD in Mathematics from the University of Eindhoven in 2015. He was a postdoctoral researcher first at CWI Amsterdam, and then at the California Institute of Technology, supported by a NWO Rubicon grant. His work lies in the area of applied probability, learning, and optimization, drawing motivation in particular from applications to power systems reliability.

JOAQUIM A.S. GROMICHO acts as Science and Education Officer for ORTEC and is full professor of Business Analytics at the University of Amsterdam. He received his PhD in Optimization in 1995 from the Erasmus University Rotterdam, before spending two years as Assistant Professor at the University of Lisbon. He serves the Dutch Statistics and OR Society as editor in chief of *STAtOR*, a magazine on applications and impact, and the steering committee of the EURO Practitioner's Forum.

JEFFREY C. KANTOR earned his PhD in Chemical Engineering from Princeton University in 1981. After a postdoc at the University of Tel Aviv, he joined the Chemical Engineering Department at the University of Notre Dame. His research interests focused on the theory and application of nonlinear control theory and techniques to chemical and biological processes. His awards have included an NSF Presidential Young Investigator Award, a Camille and Henry Dreyfus Research Scholar Award, and is a Fellow of the American Association for the Advancement of Science. He enjoyed modeling for optimization and contributed to the Pyomo community.

Hands-On Mathematical Optimization with Python

Krzysztof Postek
Boston Consulting Group, Amsterdam

Alessandro Zocca
Vrije Universiteit Amsterdam

Joaquim A. S. Gromicho
University of Amsterdam & ORTEC

Jeffrey C. Kantor[†]
University of Notre Dame, Indiana

CAMBRIDGE
UNIVERSITY PRESS

Shaftesbury Road, Cambridge CB2 8EA, United Kingdom

One Liberty Plaza, 20th Floor, New York, NY 10006, USA

477 Williamstown Road, Port Melbourne, VIC 3207, Australia

314–321, 3rd Floor, Plot 3, Splendor Forum, Jasola District Centre, New Delhi – 110025, India

103 Penang Road, #05–06/07, Visioncrest Commercial, Singapore 238467

Cambridge University Press is part of Cambridge University Press & Assessment,
a department of the University of Cambridge.

We share the University's mission to contribute to society through the pursuit of
education, learning and research at the highest international levels of excellence.

www.cambridge.org
Information on this title: www.cambridge.org/highereducation/isbn9781009493505

DOI: 10.1017/9781009493512

First published 2025

A catalogue record for this publication is available from the British Library

Library of Congress Cataloging-in-Publication Data
Names: Postek, Krzysztof, 1989- author.
Title: Hands-on mathematical optimization with Python / Krzysztof Postek
Boston Consulting Group, Amsterdam [and three others].
Description: Cambridge, United Kingdom ; New York, NY, USA : Cambridge
University Press, 2024. | Includes bibliographical references and index.
Identifiers: LCCN 2024015052 (print) | LCCN 2024015053 (ebook) |
ISBN 9781009493505 (paperback) | ISBN 9781009493512 (epub)
Subjects: LCSH: Mathematical optimization. | Python (Computer program language)
Classification: LCC QA402.5 .P68 2024 (print) | LCC QA402.5 (ebook) |
DDC 519.60285/5133–dc23/eng/20240412
LC record available at https://lccn.loc.gov/2024015052
LC ebook record available at https://lccn.loc.gov/2024015053

ISBN 978-1-009-49350-5 Paperback

Additional resources for this publication at www.cambridge.org/highereducation/ISBN/9781009493505/resources
and mobook.github.io/MO-book/intro.html

In dedication to Jeffrey's delight in learning, his enthusiasm for teaching, and the generosity that he shared with his family, colleagues, and students.

Contents

Figures

Tables

Preface

Are you a student of engineering, business, economics, operations research, or applied mathematics who needs to use optimization in some way? Are you a data science professional who wants to find the best attainable solution for real-world decision problems? Then you will need to learn mathematical optimization.

While machine learning is the right tool for classification and prediction tasks such as predicting the likelihood of customer churn or forecasting product prices, mathematical optimization is the go-to technology to make the best possible decisions on how to allocate the budget to prevent customer churn, how much and when to buy or produce given the price forecasts, or which path to take when traveling or which sequence of tasks to perform. For that reason, mastering mathematical optimization is fundamental to developing modern decision-making solutions.

This book is your comprehensive, hands-on guide to mastering mathematical optimization. We believe in learning by doing, which is why we use a hands-on approach by providing examples using Python throughout the book. Through our experience, we discovered that such an approach also greatly helps students' understanding of the underlying mathematical concepts.

Not assuming a prior education in optimization or operations research, we provide:

- *thorough coverage of the mathematical fundamentals* of optimization modeling;
- *illustrative* $50+$ *Jupyter notebooks* with example applications of the methods presented;
- *opportunity to test your knowledge* on a comprehensive base of $80+$ exercises with solutions to selected exercises at the end of the book.

In this way, we help you design, prove mathematical correctness, and industrialize optimization models right from the start. Given some basic knowledge of Python or eagerness to acquire it, you will learn how to implement your models in Python using the Pyomo package, feed data into those models, select the suitable solver (software that determines the best solutions), and find the optimal solutions to your complex real-world problems with ease.

The code scripts in this book are also freely available in an online companion [Pos+23]. Start unlocking the power of mathematical optimization today!

Book Structure

The book is divided into two parts, ensuring a coherent progression tailored to different academic levels and professional requirements. Part One encompasses the first five chapters, which introduce concepts and materials frequently encountered in popular textbooks on mathematical optimization but seldom presented together. This part is deliberately designed to align with the foundational requirements of undergraduate programs, offering a comprehensive hands-on approach that prepares students for advanced studies and impactful professions. Commencing with Chapter 6, Part Two marks a transition to topics that are less commonly addressed in conventional literature. This is specifically crafted for graduate-level courses and ambitious professionals aiming to elevate their expertise and impact. It presents cutting-edge methodologies while maintaining the hands-on approach of the first part.

The book also includes a linear algebra primer in Appendix A to accommodate the necessary linear algebra terminology and notation.

Each chapter's methods are illustrated with several Python code scripts, and exercises are available to test the understanding of the concepts introduced. At the end of the book, you will find the solutions to one exercise from each chapter. The solutions for the remaining exercises are available to instructors only on the publisher's website at www.cambridge.org/highereducation/ISBN/9781009493505/resources.

This book is suitable for two one-semester courses offered at undergraduate and graduate levels, depending on the student's background. Courses that may adopt this textbook are very diverse; an extensive but not exhaustive list includes: Operations Research, Supply Chain and Logistics Management, Process Operations, Process Control, Operations Management, Advanced Analytics, Prescriptive Analytics, Business Analytics, Chemical Process Control, Combinatorial and Integer Optimization, Stochastic Optimization, Linear Optimization, Systems Engineering, Operations Planning, Optimization for Sustainability, Nonlinear Optimization, Robust Optimization, and Mathematical Optimization.

Prerequisites

Only elementary knowledge from the following branches of mathematics is required to follow the text:

- *Linear algebra*: matrices and vectors, though the essential concepts the reader needs to know are reviewed in Appendix A.
- *Calculus*: first and second derivatives of single-variate and multi-variate functions.
- *Probability*: expectation, probability, the cumulative distribution function, and the probability density function.

Additionally, it is helpful if the reader has a basic familiarity with programming in Python, but this knowledge can be easily acquired by studying this book.

Acknowledgments

The book arose from courses that we have taught at our respective universities and we are greatly indebted to our own teachers and our students who helped shape this material. During the book writing process, we also benefited from the proofreading, suggestions, examples, and exercises of numerous people. We want to particularly thank (in alphabetical order):

Alex Dowling, Ana Isabel Barros, Britt van Veggel, Caroline Jagtenberg, Dick den Hertog, Filipe Brandão, Francisco Saldanha da Gama, Gerrit Timmer, Goos Kant, Jan Brinkhuis, Jasper van Doorn, Jelke van Hoorn, Justin Starreveld, Laurens Bliek, Leon Lan, Maria Teresa Melo, Nicolau Santos, Niels van der Laan, Ronald Buitenhek, Rosario Paradiso and Wouter Kool.

In closing, we must acknowledge the invaluable support received from those not directly involved in our academic endeavors but whose influence permeates our work. Our heartfelt gratitude extends to our partners and family members: Heidi, Marlies, Angel, Joana, Tina-Luíza, and Diane. Their unwavering support and encouragement have been our constant source of strength and inspiration.

1

Mathematical Optimization

Mathematical optimization models are mathematical means to find the best possible solutions to real-life optimization problems. They consist of three parts: decision variables that describe possible solutions, constraints that define conditions that these solutions need to satisfy, and an objective function that assigns a value to each solution, expressing how "good" it is.

Once you input the relevant data and parameters into the model, your goal is to find a solution that meets all constraints and has the best value, like a navigation system that finds the suitable route for your vehicle and gets you to your destination as fast as possible.

In this chapter, our goal is to introduce you to these concepts and how they are implemented, describing the interactions between different theoretical (model) and software (programming language, modeling package, and solver) components. To keep things concrete, we will center the discussion around a practical example that will be introduced in Section 1.1.

In Section 1.2, we will abstract away from this single example to give the general mathematical notation used to model problems, making an important disclaimer in Section 1.3 that although models are well-defined mathematical objects, the very process of building them is a combination of art and science. Section 1.4 will explain the general "rules of the game" of solving problems, including the various software choices that can be made. In particular, we explain our choice for the Pyomo library as the modeling language for the code snippets in this book.

In Section 1.5 we give an overview of the basic Pyomo components and show how to implement and solve the starting optimization problem. Finally, in Section 1.6, we will give an overall view of what this book consists of.

1.1 A Motivating Example: Production Planning

A small company has developed two versions of a new product. Each version of the product is made from the same raw material that costs $10 per gram and requires two different types of specialized labor.

U is the higher-priced version of the product. U sells for $270 per unit and requires 10 g of raw material, 1 hour of labor type A, and 2 hours of labor type B. Due to the higher price, the market demand for U is limited to 40 units per week. V is the lower-priced version of the product that sells for $210 per unit with unlimited demand and requires 9 g of raw material, 1 hour of labor type A, and 1 hour of labor type B. These data are summarized in Table 1.1.

The availability of labor and the raw materials inventory limits weekly production in the company. The raw material must be ordered in advance and has a short shelf life. Any raw

Table 1.1 *Summary of the problem data and parameters.*

Product version	Raw material required	Labor A required	Labor B required	Market demand	Price
U	10 g	1 hour	2 hours	$\leq$40 units	$270
V	9 g	1 hour	1 hour	unlimited	$210

Table 1.2 *Summary of the cost and availability of raw material and labor.*

Resource	Amount available	Cost
Raw material	no limits	$10 / g
Labor A	80 hours	$50 / hour
Labor B	100 hours	$40 / hour

Table 1.3 *Summary of the decision variables and their bounds.*

Decision variable	Description	Lower bound	Upper bound
x_M	amount of raw material used	0	—
x_A	amount of Labor A used	0	80
x_B	amount of Labor B used	0	100
y_U	number of U units to produce	0	40
y_V	number of V units to produce	0	—

material left over at the end of the week is discarded. Table 1.2 details the cost and availability of raw materials and labor.

The company wants to maximize its gross profit. How much raw material should be ordered in advance for each week? How many units of U and V should the company produce each week?

The above is an example of an optimization problem for which we will now create a mathematical model. To do this, the starting point is to list decision variables relevant to the problem. **Decision variables** are quantities that can be modified to achieve a desired outcome. While some decision variables introduced at this stage may become redundant later, the goal at this point is to create a comprehensive list of variables that will be useful in expressing the problem's objective and constraints.

In Table 1.3, we list the decision variables with symbols, descriptions, and any lower and upper bounds known from the problem data.

The next step is to formulate an **objective function** describing how we will measure the value of candidate solutions to the problem. In this case, the value of the solution is measured by the profit, which we want to maximize:

$$\max \text{ profit.}$$

Profit, in turn, is equal to the difference between revenue and cost of operations:

$$\text{profit} = \text{revenue} - \text{cost.}$$

Revenue and cost are linear **expressions** that can be written in terms of the decision variables:

$$\text{revenue} = 270y_U + 210y_V$$
$$\text{cost} = 10x_M + 50x_A + 40x_B.$$

As shown here, an expression is an algebraic combination of variables that can be referred to by name. Expressions are helpful when the same combination of variables appears in multiple places in a model or when it is desirable to break up longer expressions into smaller units. Here, we have created expressions for revenue and cost that simplify the objective function.

The decision variables y_U, y_V, x_M, x_A, x_B must satisfy the specific conditions described in the problem statement. **Constraints** are mathematical relationships between decision variables or expressions, formulated either as equalities or inequalities. In this problem, for each resource, there is a linear constraint that limits overall production:

$$10y_U + 9y_V \leq x_M \qquad \text{(raw material)}$$
$$1y_U + 1y_V \leq x_A \qquad \text{(labor A)}$$
$$2y_U + 1y_V \leq x_B \qquad \text{(labor B)}.$$

We are now ready to formulate the full mathematical optimization problem in the following way that we shall stick to throughout the book. First, we state the objective function to be maximized (or minimized), then we list all the constraints, and last, we report all the decision variables and their bounds:

$$\max \quad 270y_U + 210y_V - 10x_M - 50x_A - 40x_B \qquad (1.1)$$
$$\text{such that} = \text{s.t.} \quad 10y_U + 9y_V \leq x_M$$
$$1y_U + 1y_V \leq x_A$$
$$2y_U + 1y_V \leq x_B$$
$$0 \leq x_M$$
$$0 \leq x_A \leq 80$$
$$0 \leq x_B \leq 100$$
$$0 \leq y_U \leq 40$$
$$0 \leq y_V.$$

This completes the mathematical description of this example of a production planning problem. An **optimal solution** of the problem is any vector of decision variables that satisfies the constraints and achieves the maximum/minimum objective.

However, even for a simple problem like this, it is not immediately clear what the optimal solution is. This is precisely where mathematical optimization algorithms come into play – they are generic procedures that can find the optimal solutions to problems as long as these can be formulated in a standardized fashion as above.

In the following sections, we will dive into all the issues related to modeling and solving optimization problems. If you are interested, however, directly in the implementation and solution of the above problem in Python, we refer you to Section 1.5.

Remark 1.1 In many textbooks, it is customary to list the decision variables under the symbol max to distinguish between them and any other parameters (which could also be expressed as symbols or letters) in the problem formulation. However, throughout this book, we stick to the convention that decision variables are only those for which the domain is explicitly stated as part of the problem formulation.

1.2 A General Mathematical Optimization Model

While the previous section introduced the components of an optimization based on an example, we now abstract from all examples and give a general mathematical formulation. Our intent is to illustrate exactly what makes mathematical optimization a powerful tool: Once translated into a mathematical model, optimization problems from various application domains "look the same." This similarity is what enables the creation of universal algorithms and solvers capable of solving these models regardless of the specific field from which the original problem emerged.

As already said, a mathematical optimization model consists of three components:

- **Decision variables**, which correspond to actions or choices that we have to make in our decision problem: how many products to make for each product type, or in other problems, whether to open a new manufacturing facility, which supply routes to use or for which prices we should sell our products.
- An **objective function**, which is used to evaluate a specific solution (i.e., a specific choice of values for the decision variables introduced earlier). For the objective function, a "goal" should be specified, either *maximize* or *minimize* it – in the production planning example, we aimed to maximize the profit. In other applications, the objective can be to minimize operational costs or to maximize the number of satisfied customers.
- **Constraints** that restrict the possible values of our decision variables, that is, conditions that must be satisfied – in the production planning example, the number of products is constrained by the availability of raw materials and labor. Other common examples of constraints are the requirement that a maximum allowed budget is met, that all demands of important customers are met, or that warehouse capacities are not exceeded. Constraints define the **feasible set** of our model, that is, the set of all solutions that meet the constraints.

The goal is to find a **global optimum**: a feasible selection of the decision variables' values leading to the best objective function value. Although quite simple, the example of Section 1.1 already includes all the key ingredients of a mathematical optimization problem.

Mathematically, we can describe optimization problems as follows. Given an objective function $f \colon X \to \mathbb{R}$ to be minimized, with $X \subseteq \mathbb{R}^n$ being the set of **feasible set** of candidate solutions, we seek to find $x \in X$ satisfying the following condition $f(y) \geq f(x) \quad \forall y \in X$, that is, the solution we find is at least as good as all other possible solutions. Similarly, we can define a maximization problem by changing the last condition to $f(y) \leq f(x) \quad \forall y \in X$. In both cases, we refer to such solutions as optimal solutions. The general way to formulate a minimization problem is

$$\min \quad f(x)$$
$$\text{s.t.} \quad x \in X,$$

and similarly for a maximization problem. Different types of functions f and sets X lead to different types of optimization problems and different solution techniques. We refer the reader to [Kit11] for a nice overview of highlights in the history of optimization and to [KIG14] for an overview of techniques.

1.3 Modeling is Both Art and Science

Although the mathematical formulation of the previous section could give the impression that a clear and well-defined model always exists for a given problem, this is actually not true, and usually, there is no "single and best" model. In fact, modeling as a process of translating a problem into a mathematically solvable object is a mixture of art and science, during which three fundamental questions must be asked:

- *What to model?* First, the modeler must translate a problem from the real world to an abstracted mathematical representation. Not every aspect of the real world can or should be taken into account by the model, so there are many choices to be made in this first step, which typically have a significant impact on the model and solution approach.
- *How to model?* There can be multiple equivalent model formulations. Conceptually, equivalent models solve the same optimization problem, but the applicable solution approaches or computational complexity may differ.
- *How to interpret the model solution?* The solution must be evaluated and translated back to the original real-world problem after solving the model.

Addressing these questions should be treated as a continuous (or cyclical) process, not as sequential steps. For example, if the final solution turns out to be impractical, we need to adjust the model. If specific desired properties cannot be modeled efficiently, perhaps we should reevaluate what to include in the model. A mathematical model is a tool, not a goal (well, except for mathematicians to study). Models will "always be flawed" and our challenge lies in refining them to the point where they become practically useful.

1.4 Modeling and Solving Problems With Pyomo

Having put all the necessary disclaimers on what models and modeling are, we are ready to start facing the crucial questions: How do we obtain the best possible solution and what is the value of such a solution?

One possible approach is to check if your problem falls within a class of well-known problems for which efficient algorithms already exist. If so, you can implement one of such algorithms, making the necessary modifications to fit your specific context. Alternatively, you may develop an algorithm on your own, but then you also need to argue why the solution found is (near-)optimal. Should either of these strategies appear feasible, by all means, go ahead with the implementation, as customized algorithms are often good candidates for quick execution times.

However, in general, it is easiest to avoid deploying an algorithm on your own and instead find the solution through software known as a **solvers**. A solver usually also guarantees optimality, that is, a certificate that no other solution is better. While for most general optimization problems this guarantee is only local, that is, no solution "close" to the one found can beat it, for most of the models discussed in this book, the guarantee will be global too, that is, the found solution will be just as good or better than any other solution, as we explain later in Chapter 5.

The actual amount of time it takes to solve the problem will vary depending on the type of problem, so this book is divided into chapters that cover different kinds of problems. For example, the problems discussed in Chapter 3 will be extremely difficult to solve, while problems of comparable size that fall under Chapter 4 can be solved in no time. Complexity theory is a branch of mathematics that attempts to analyze and differentiate the difficulty of various problems. In this book, we will hint at the practical implications of the complexity of different problems.

While a computer is definitely able to run a solution algorithm to a given optimization problem, it is not (yet) able to model the problem by itself – it is the human who learns to understand the relationships between different decisions and the constraints and (possibly multiple) objectives to be met. Optimization modeling is thus a marriage of human ingenuity and the processing powers of computers. For that reason, it is essential that the programming language/tool in which we formulate optimization problems most closely resembles the way we would think about it. In this book, we adopt a Python-based interface, the `Pyomo` modeling package.

Pyomo is an algebraic modeling language for mathematical optimization that is integrated into the Python programming environment. It enables users to create optimization models consisting of decision variables, expressions, objective functions, and constraints. Pyomo provides tools to transform models and then solve them using a variety of open-source and commercial solvers.

Figure 1.1 illustrates the relationship between the user, the modeling package, and the solver. The user utilizes Pyomo to formulate an optimization problem in a way that is most convenient for a human to conceptualize. Pyomo, in turn, translates this optimization model into the most convenient form for a specific solver. The solver then finds an optimal solution to the problem and sends it back to Pyomo, which presents it to the user.

Figure 1.1 highlights a crucial benefit of Pyomo: The same model can be solved by different solvers without having to be rewritten in the solver-specific Python language. This matters a

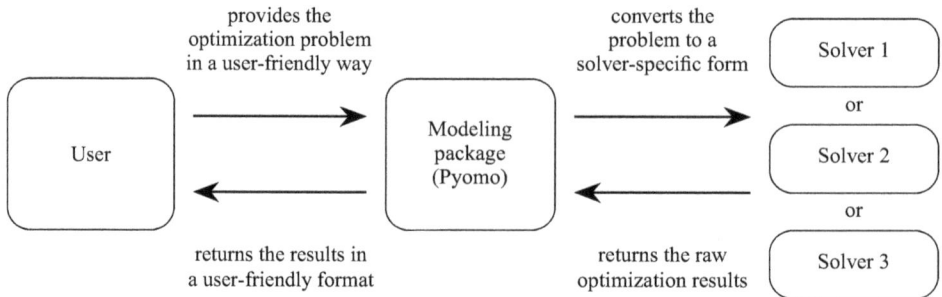

Figure 1.1 Relationship between the user, modeling package, and the ability to choose a solver.

lot in practice, as you may initially use an open-source solver and then discover that as your problem instance grows in size, you may need to switch to a more powerful commercial solver.

Solvers can be compared with various numerical linear algebra or machine learning libraries in Python – some of them excel at certain tasks, while others do better on other types of tasks. There are a variety of commercial and open-source solvers available, and it is essential to comprehend the major differences between them and to interpret the solution output they generate. Throughout the book, we will also provide instructions on how to do this.

1.5 Getting Started With Pyomo: Back to Production Planning

Throughout the book, we include `Python` code snippets[1] for most of the examples we introduce, showing how each optimization model can be implemented using the `Pyomo` library. This section aims to familiarize the reader with the essential components of this modeling language. Although informative, this section is not an exhaustive guide, and readers seeking a comprehensive understanding of Pyomo are encouraged to consult [Har+17] and the complete Pyomo documentation, available at [Pyo23].

We introduce the basic components of the `Pyomo` library while defining the Pyomo model corresponding to the motivating example about optimal production planning presented in Section 1.1.

Parameter values from that mathematical model are included directly in the Pyomo model for simplicity. This method works well for problems with a small number of decision variables and constraints, but limits the reusability of the model. In later chapters, we will demonstrate the use of sets and indexed parameters, variables, and constraints that are essential for more generic "data-driven" applications.

Additionally, we make use of Python decorators to designate Pyomo expressions, objectives, and constraints. While decorators may be unfamiliar to some Python users, or even current Pyomo users, they significantly improve the readability of the Pyomo model. This feature is relatively new and is available in recent versions of Pyomo.

Preliminary Step: Installing Pyomo and Solvers

We start by verifying the installation of Pyomo and the designated solvers.

In this book, whenever possible, we use open-source solvers that are freely available. The differences between open-source and commercial solvers are similar to other software domains. While open-source solvers are free and enable you to play more with the embedded algorithms, commercial solvers typically scale better to large problems and usually come with better support. Generally, no single solver is superior for all problems, so it is useful to formulate your problem in a way that allows you to experiment with multiple solvers.

In the first four chapters of this book, we consistently use HiGHS, see [Hal+], as the solver. HiGHS is an open-source, high-performance solver for linear and mixed-integer optimization. Another open-source alternative with the same capabilities is the COIN-OR CBC solver, see [For+]. In later chapters, we will occasionally need solvers with different capabilities, which will be introduced in due course.

---

[1] The complete code for all examples can be found in a dedicated online collection of Jupyter notebooks [Pos+23] and its corresponding GitHub repository.

The following code snippet imports the Pyomo library and the HiGHS solver. If this code runs on Google Colab, it first installs them, while if it is run locally, we tacitly assume that Python, Pyomo, and the HiGHS solver have been previously installed. It is then set to use HiGHS as a solver via the appsi module, and a test is performed to verify that it is available. For later use, the solver interface is stored in a global object `SOLVER`.

```python
import sys

if 'google.colab' in sys.modules:
    %pip install pyomo >/dev/null 2>/dev/null
    %pip install highspy >/dev/null 2>/dev/null

solver = 'appsi_highs'

import pyomo.environ as pyo
SOLVER = pyo.SolverFactory(solver)

assert SOLVER.available(), f"Solver {solver} is not available."
```

Note that there are other suitable solvers, both open-source, such as GLPK, IPopt, and CBC, on one hand, and commercial, such as CPLEX, Gurobi, Xpress, and Mosek, on the other hand.

Step 1. Import Pyomo

The first step in a new Pyomo model is to import the required components into the Python environment. The module `pyomo.environ` provides the components most commonly used to build Pyomo models. Although this module was imported in the previous code snippet, we mention it again here to emphasize our standardized conventions. Throughout this book, a uniform standard is maintained for importing `pyomo.environ` consistently using the alias `pyo`.

```python
import pyomo.environ as pyo
```

Most of the examples that we present in this book, in addition to Pyomo, use several Python packages that are supposed to be installed and imported before being used. We refer to these packages using common aliases as follows:

```python
import matplotlib.pyplot as plt
import networkx as nx
import numpy as np
import pandas as pd
```

Step 2. The ConcreteModel Object

Pyomo models can be named using any standard Python variable name. In the following code, an instance of `ConcreteModel` is created and stored in a Python variable named `model`. It is best to use a short name since it will appear as a prefix for every Pyomo variable and constraint. `ConcreteModel` accepts an optional string argument to title subsequent reports.

`pyo.ConcreteModel()` is used to create a model object when the problem data is known at the time of construction. Alternatively, `pyo.AbstractModel()` can create models where the problem data will be provided later to create specific model instances. However, this is normally not needed when using the "data-driven" approach demonstrated in this book.

The `.display()` method is a convenient means of displaying the current contents of a Pyomo model. When developing new models, this is a valuable tool to verify that the model is being constructed as intended. At this stage, the major components of the model are empty.

```
model = pyo.ConcreteModel("Production Planning: Version 1")
model.display()
```

Step 3. Decision Variables

Decision variables are created with the Pyomo `pyo.Var()` class. Decision variables can be assigned to any valid Python identifier. We assign decision variables to the model instance using the Python "dot" notation. Variable names are chosen to reflect their names in the mathematical model.

`pyo.Var()` accepts optional keyword arguments. The most commonly used keyword arguments are:

- "domain" specifies a set of values for a decision variable. By default, the domain is the set of all real numbers. Other commonly used domains are `pyo.Binary`, `pyo.NonNegativeReals` and `pyo.NonNegativeIntegers`.
- "bounds" is an optional keyword argument to specify a tuple containing lower and upper bounds values. Specifying any known and fixed bounds on the decision variables is good modeling practice. `None` can be used as a placeholder if one of the two bounds is unknown. Specifying the bounds as `(0, None)` is equivalent to specifying the domain as `pyo.NonNegativeReals`.

The use of optional keywords is shown in the following code. Displaying the model shows that the value of the decision variables is not yet known.

```
model.x_M = pyo.Var(bounds=(0, None))
model.x_A = pyo.Var(bounds=(0, 80))
model.x_B = pyo.Var(bounds=(0, 100))
model.y_U = pyo.Var(bounds=(0, 40))
model.y_V = pyo.Var(bounds=(0, None))
```

Step 4. Expressions

Pyomo expressions are mathematical formulas that involve decision variables. The following cell creates expressions for revenue and cost that are assigned to `model.cost` and `model.revenue`, respectively.

```
model.cost = 10 * model.x_M + 50 * model.x_A + 40 * model.x_B
model.revenue = 270 * model.y_U + 210 * model.y_V
```

Step 5. Objective

The objective is to maximize gross profit, where gross profit is defined as the difference between revenue and cost. There are two alternative ways this objective could be specified in Pyomo, both illustrated in the code below.

```
# define the objective function directly using the Objective class
model.profit = pyo.Objective(expr=model.revenue - model.expense,
↪    sense=pyo.maximize)


# define the objective function using decorators
@model.Objective(sense=pyo.maximize)
def profit(m):
    return m.revenue - m.cost
```

The first method is to use `pyo.Objective()` where the expression to be optimized is assigned with the `expr` keyword and the type of objective is assigned with the `sense` keyword. For clarity, it is good practice to always include the `sense` keyword argument.

Recent Pyomo releases provide a second method that uses Python decorators, see [Smi+], to specify an objective. Python decorators modify the behavior of the function defined in the next line. In this case, the decorator `@model.Objective()` modifies the behavior of `profit()` so that it returns an expression for the profit to Pyomo. The keyword `sense` sets the type of objective, which can either be to maximize or minimize the value returned by the objective function. After being decorated, the function `profit()` takes the Pyomo model as its first argument and adds its name to the model attributes.

In effect, Pyomo decorators are tags that insert functions into a Pyomo model to serve as expressions, objectives, or constraints. Decorators can improve the readability and maintainability of more complex models. They also simplify the syntax for creating other Pyomo object expressions, constraints, and other optimization-related elements.

Step 6. Constraints

Constraints are logical relationships between expressions that define the range of feasible solutions in an optimization problem. Logical relationships can be equality ($=$), less-than ($\leq$), or greater-than ($\geq$).

Constraints can be created with `pyo.Constraint()`. The constraint is passed as a keyword argument `expr` to `pyo.Constraint()`. For this application, the constraints could be expressed as follows:

```
model.raw_materials = pyo.Constraint(expr=10 * model.y_U + 9 * model.y_V <=
↪    model.x_M)
model.labor_A = pyo.Constraint(expr=1 * model.y_U + 1 * model.y_V <=
↪    model.x_A)
model.labor_B = pyo.Constraint(expr=2 * model.y_U + 1 * model.y_V <=
↪    model.x_B)
```

Alternatively, the `@model.Constraint()` decorator "tags" the output of the following function as a constraint. For the present example, the constraints are expressed with decorators below. In this book and its companion notebooks, we use decorators whenever possible to improve the readability and maintainability of Pyomo models.

```
@model.Constraint()
def raw_materials(m):
    return 10 * m.y_U + 9 * m.y_V <= m.x_M


@model.Constraint()
def labor_A(m):
    return 1 * m.y_U + 1 * m.y_V <= m.x_A


@model.Constraint()
def labor_B(m):
    return 2 * m.y_U + 1 * m.y_V <= m.x_B
```

Step 7. Solve the Model

Now that the model is fully specified, the next step is calculating a solution. This is done by applying the solver object we created in Step 1 with `SolverFactory` to the model, as shown in the following code. The optional keyword `tee=True` causes the solver to print a detailed output, which can be helpful in debugging problems that arise when developing a new model.

```
results = SOLVER.solve(model, tee=True)
```

Step 8. Reporting the Solution

The final step in most applications is to report the solution in a suitable format. We demonstrate simple tabular and graphic reports using the `Pandas` library for this example. Later chapters will show other ways to report and visualize the solutions.

Pyomo provides several functions to create model reports that contain solution values. The `pprint()` method can be applied to the entire model or individual components of the model. Note that alongside `pprint()`, `display()` also exists: The former works well even before solving the model, while the latter assumes that a solution has been found. If the model has already been solved, it becomes a matter of taste which one to use.

```
# display the whole model
model.pprint()
# display a component of the model
model.profit.pprint()
```

After a solution to a Pyomo model has been successfully computed, the values of the objective, expressions, and decision variables can be accessed with `pyo.value()`. When combined with Python f-strings, [Pyt23], `pyo.value()` provides a convenient means to create formatted reports. Pyomo provides a shortcut notation for accessing the solution.

After a solution has been computed, a function with the same name as the decision variable is created that will report the value of the solution.

```python
# access the value of a variable or expression
pyo.value(model.profit)

# print the value of variables or expressions using f-strings
print(f" Profit = {pyo.value(model.profit): 9.2f}")
print(f"Revenue = {pyo.value(model.revenue): 9.2f}")
print(f"   Cost = {pyo.value(model.cost): 9.2f}")

# shortcut to access the value of the solution
print("x_A =", model.x_A())
print("x_B =", model.x_B())
print("x_M =", model.x_M())
```

Alternatively, one may create a report with Pandas. Pandas is an open-source library for working with data in Python and is widely used in the data science community. Here, we use a Pandas `Series()` object to hold and display solution data.

```python
# create pandas series for production and raw materials
production = pd.Series(
    {
        "U": pyo.value(model.y_U),
        "V": pyo.value(model.y_V),
    }
)

raw_materials = pd.Series(
    {
        "A": pyo.value(model.x_A),
        "B": pyo.value(model.x_B),
        "M": pyo.value(model.x_M),
    }
)

# Create a 1x2 grid of subplots and configure global settings
fig, ax = plt.subplots(1, 2, figsize=(8, 3.5))
plt.rcParams["font.size"] = 12
colors = plt.cm.tab20c.colors
color_sets = [[colors[0], colors[4]], [colors[16], colors[8], colors[12]]]
datasets = [production, raw_materials]
titles = ["Production", "Raw Materials"]

# Plot data on subplots
for i, (data, title, color_set) in enumerate(zip(datasets, titles,
    color_sets)):
    data.plot(ax=ax[i], kind="barh", title=title, alpha=0.7, color=color_set)
    ax[i].set_xlabel("Units")
    ax[i].invert_yaxis()
plt.tight_layout()
plt.show()
```

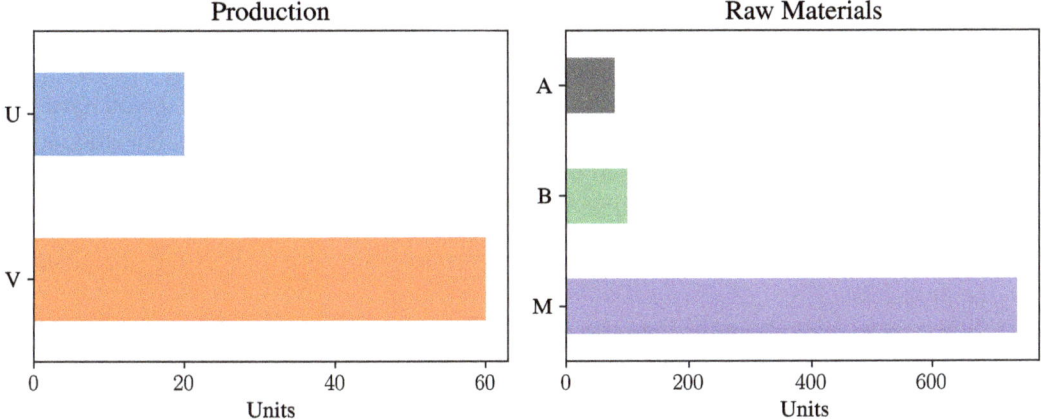

Figure 1.2 Visualization of the optimal solution and the corresponding values of the required labor A and B and market demand.

We can then visualize them using the Matplotlib library, for instance, with bar charts (see Figure 1.2).

1.6 What to Expect From This Book

This book covers both the basic theoretical and applied (numerical) aspects of modeling optimization problems. We find this combination to be extremely important.

First, because optimization is a domain of mathematics that has grown mostly with specific applications in mind, presenting it without the applied angle misses the point. Second, understanding the mathematics underlying the optimization tools provides the user with the necessary skills to choose the right approach, interpret the solutions, and create their own optimization tools.

For these reasons, we will gradually build up our mathematical understanding and ability to model optimization problems, from the simplest to the most complicated ones. Next, we will show how to formulate them using Pyomo, solve them using suitable solvers, and interpret the obtained solutions.

To combine these two smoothly, at the end of each chapter we present a complete example in which a real-life situation is translated into a model, indicating the modeling decisions we make each time. In addition, we will show how to interpret the output provided by the solvers to the optimization models we built.

Interpreting the output is the reverse of modeling: going from abstraction back to reality. Suppose that a model to find a path includes individual variables that state whether each possible segment is used. These variables allow us to identify the used segments and link them in the correct sequence by matching the end point of one with the starting point of the other.

Throughout the book, to keep the notation concise, we often use vector and matrix notation. For that reason, we have included the necessary (basic) knowledge of linear algebra in Appendix A.

Exercises

Exercise 1.1 You have just booked a weekend away and selected the bag to carry your belongings. It is time to decide what to bring with you, and on top of your bed, you have just laid out all potentially relevant items. The whole set seems to fit in your bag, but since some walking is involved, you have stipulated a maximum weight you are willing to carry. You just read Chapter 1 and decided to formulate your luggage decision as an optimization problem.

(a) Can you describe your decisions about what to take with you along the lines of Section 1.2?
(b) Now you wonder about the data that you need to express your constraints and objective. Can you obtain all the data you need? What do you need to do?
(c) You are now able to describe the potential items to carry in an objective way, including data. This is what it looks like:

item	gram	utility
slippers	150	10
pajamas	350	9
pair of trousers	250	6
blue T-shirt	100	6
red T-shirt	120	7
thermal bottle	500	6
dental hygiene kit	120	9
this book	550	8
the latest Dan Brown	850	8
binoculars	350	6
underwear	100	10
socks	50	9

You have decided not to carry more than one and a half kilograms (you do not care about the volume considerations). Using these data, how would you model the optimization of your luggage composition to maximize your total utility?
(d) Knowing that you can add a domain to your decision variables and make them assume the "yes/no" behavior you need by stating `domain=pyo.Binary`, write the code that you would need to reach your decision.
(e) Whether you take this book is a no-brainer: of course you do! How does that modify your problem?

2

Linear Optimization

2.1 Formulation

The simplest and most scalable type of optimization problem is one in which the objective function and constraints are formulated using the simplest type of functions – linear functions. We refer to this class of problems as **linear optimization (LO)** problems.

The following example illustrates a first linear optimization problem.

Example 2.1 (Building microchips – Problem formulation) The company BIM (Best International Machines) produces two types of microchips, logic chips (1 g silicon, 1 g plastic, 4 g copper) and memory chips (1 g germanium, 1 g plastic, 2 g copper). Each of the logic chips can be sold for a \$12 profit, and each of the memory chips for a \$9 profit. The current stock of raw materials is as follows: 1000 g silicon, 1500 g germanium, 1750 g plastic, and 4800 g copper. How many microchips of each type should be produced to maximize profit while respecting the availability of raw material stock?

Let $x \geq 0$ denote the number of logic chips to be produced and $y \geq 0$ the number of memory chips. In the problem described above, the goal is to maximize the total profit. Since the profit is \$12 for each logic chip and \$9 for each memory chip, the total profit to maximize is equal to

$$12x + 9y.$$

While maximizing this quantity, we have to respect some constraints. We know that we cannot use more raw materials than those available in stock. For copper, this means that the joint usage for logic chips, which is equal to $4x$ g (4 g per chip for each of the x chips), and for memory chips, which is equal to $2y$ g (2 g per chip for each of the y chips), cannot exceed the maximum availability of 4800 g of copper:

$$4x + 2y \leq 4800.$$

Similarly, we can deduce the condition for silicon, which involves only logic chips (memory chips do not require this material),

$$x \leq 1000,$$

the condition for germanium, which involves only memory chips (logic chips do not require this material),

$$y \leq 1500,$$

15

and the condition for plastic, which involves both types of chips,

$$x + y \leq 1750.$$

Putting all the components together, we obtain an optimization problem of the following form:

$$
\begin{aligned}
\max \quad & 12x + 9y \\
\text{s.t.} \quad & x \leq 1000 && \text{(silicon)} \\
& y \leq 1500 && \text{(germanium)} \\
& x + y \leq 1750 && \text{(plastic)} \\
& 4x + 2y \leq 4800 && \text{(copper)} \\
& x, y \geq 0.
\end{aligned}
\tag{2.1}
$$

The problem described in Example 2.1 is relatively small, featuring only $n = 2$ decision variables and $m = 4$ constraints. However, it is easy to imagine that adding more products and constraints would significantly complicate matters. In such cases, explicitly listing each constraint and fully expanding all expressions could obfuscate the overall structure, making it challenging to discern the key aspects of the problem.

In fact, it is much more common to formulate, analyze, and compare LO problems using vectors and matrices. This format not only aligns more closely with computational implementation but also greatly facilitates the identification of the similarities between various LO problems, regardless of whether they are about chip production or food manufacturing. If you are new to or need to refresh on how equations and inequalities can be formulated using vectors and matrices, we refer you to Appendix A. We will now present a generalized form of an LO problem using vectors and matrices and demonstrate how it encompasses the specific instance described in Example 2.1.

Example 2.2 (Building microchips – Matrix formulation) As a first step towards building a vector–matrix formulation of our problem, we rename the decision variables x and y as x_1 and x_2, obtaining:

$$
\begin{aligned}
\max \quad & 12x_1 + 9x_2 \\
\text{s.t.} \quad & x_1 \leq 1000 && \text{(silicon)} \\
& x_2 \leq 1500 && \text{(germanium)} \\
& x_1 + x_2 \leq 1750 && \text{(plastic)} \\
& 4x_1 + 2x_2 \leq 4800 && \text{(copper)} \\
& x_1, x_2 \geq 0.
\end{aligned}
\tag{2.2}
$$

With this notation, we denote the vector of decision variables by $x = \begin{pmatrix} x_1 \\ x_2 \end{pmatrix}$, where x_1 and x_2 are just the respective components.

We now rewrite the objective function using the vector form. For this, we define the vector

$$c = \begin{pmatrix} 12 \\ 9 \end{pmatrix}$$

so that the objective

$$\max \ c^\top x = \max \ 12x_1 + 9x_2.$$

For the constraints, we define the problem coefficients as

$$A = \begin{bmatrix} 1 & 0 \\ 0 & 1 \\ 1 & 1 \\ 4 & 2 \end{bmatrix}, \quad \text{and} \quad b = \begin{pmatrix} 1000 \\ 1500 \\ 1750 \\ 4800 \end{pmatrix}.$$

The system of inequalities $Ax \le b$, when read row-by-row, correctly replicates all the constraints:

$$Ax = \begin{bmatrix} x_1 \\ x_2 \\ x_1 + x_2 \\ 4x_1 + 2x_2 \end{bmatrix} \le \begin{pmatrix} 1000 \\ 1500 \\ 1750 \\ 4800 \end{pmatrix} = b \quad \Longleftrightarrow \quad \begin{cases} x_1 \le 1000 \\ x_2 \le 1500 \\ x_1 + x_2 \le 1750 \\ 4x_1 + 2x_2 \le 4800. \end{cases}$$

In this way, our optimization problem becomes:

$$\max \quad c^\top x \tag{2.3}$$
$$\text{s.t.} \quad Ax \le b$$
$$x \ge 0.$$

A significant advantage of expressing problems in concise matrix form is not only a reduction in space but also the clarity it provides in revealing the similarities across problems from different domains. This uniformity also allows us to use a single tool (a solver) to solve all of them. Furthermore, when unnecessary details and equations are abstracted into a single matrix–vector expression, the theoretical analysis of linear optimization problems is streamlined, facilitating the exploration of the following key questions:

- Is there an optimal solution?
- Is there only one or more solutions?
- How do we prove the optimality of a solution?

To answer such questions in one go for an entire class of optimization problems, it is customary to define a so-called **canonical form** of the optimization problem, which specifies (i) whether the objective is a maximization or minimization, (ii) if the constraints are inequalities or equalities, and (iii) what signs the variables take. Once we commit to a specific canonical form, we can derive all sorts of useful results and properties, which will hold for all problems in the considered class anyway (because any problem can be transformed to a given form), without the need to consider cases like "if the problem is a maximization then ..., and if it is a minimization then...."

In this book, we adhere to the standard convention of representing LO problems with an objective of minimization, all constraints being of the type $\ge$, and all variables being nonnegative. In other words, we work with the following general formulation for LO problems:

$$\min \quad \boldsymbol{c}^\top \boldsymbol{x} \tag{2.4}$$
$$\text{s.t.} \quad A\boldsymbol{x} \geq \boldsymbol{b}$$
$$\boldsymbol{x} \geq 0,$$

where n (decision) variables are grouped in a vector $\boldsymbol{x} \in \mathbb{R}^n$, $\boldsymbol{c} \in \mathbb{R}^n$ are the objective coefficients, and m linear constraints are specified by the matrix $A \in \mathbb{R}^{m \times n}$ and the vector $\boldsymbol{b} \in \mathbb{R}^m$.

Of course, LO problems could also (i) be maximization problems, (ii) involve equality constraints and constraints of the form $\leq$, and (iii) have unbounded or nonpositive decision variables x_i. In fact, any LO problem with such characteristics can be easily converted to the canonical form (2.4) by adding/removing variables and/or multiplying specific inequalities by -1. To illustrate this point, we shall convert our example problem to this canonical form, building upon the formulation we already had.

Example 2.3 (Building microchips – Matrix formulation to canonical minimization form) Recall the problem from (2.2) and its matrix formulation (2.3). We begin transforming the LO problem mentioned above into its canonical form, starting with its objective. Maximization of a given quantity is equivalent to minimization of the opposite of that quantity, therefore setting

$$\bar{\boldsymbol{c}} = \begin{pmatrix} -12 \\ -9 \end{pmatrix}$$

will be enough because minimizing

$$\bar{\boldsymbol{c}}^\top \boldsymbol{x} = -12x_1 - 9x_2$$

is the same as maximizing $12x_1 + 9x_2$.

For the constraints, we can also multiply all the left-hand and right-hand side coefficients by -1 to build the following matrix and vector:

$$\overline{A} = \begin{bmatrix} -1 & 0 \\ 0 & -1 \\ -1 & -1 \\ -4 & -2 \end{bmatrix}, \quad \text{and} \quad \bar{\boldsymbol{b}} = \begin{pmatrix} -1000 \\ -1500 \\ -1750 \\ -4800 \end{pmatrix}.$$

One can easily check that the system of inequalities $\overline{A}\boldsymbol{x} \geq \bar{\boldsymbol{b}}$, when read row-by-row, indeed yields all the problem constraints:

$$\overline{A}\boldsymbol{x} = \begin{bmatrix} -x_1 \\ -x_2 \\ -x_1 - x_2 \\ -4x_1 - 2x_2 \end{bmatrix} \geq \begin{pmatrix} -1000 \\ -1500 \\ -1750 \\ -4800 \end{pmatrix} = \bar{\boldsymbol{b}} \quad \Longleftrightarrow \quad \begin{cases} x_1 \leq 1000 \\ x_2 \leq 1500 \\ x_1 + x_2 \leq 1750 \\ 4x_1 + 2x_2 \leq 4800. \end{cases}$$

We have thus reformulated our original problem to its canonical form as

$$\min \quad \bar{\boldsymbol{c}}^\top \boldsymbol{x}$$
$$\text{s.t.} \quad \overline{A}\boldsymbol{x} \geq \bar{\boldsymbol{b}}$$
$$\boldsymbol{x} \geq 0.$$

Having put some effort into formulating our first problem, we are now ready to actually analyze this problem and the set of its feasible/optimal solutions.

Example 2.4 (Building microchips – Problem analysis) To build some intuition for the feasible set, we can create a small list of feasible solutions that are easy to guess and calculate the corresponding value of the objective function, see Table 2.1. They can be obtained by calculating points in which multiple constraints are satisfied with equality.

Table 2.1 *List of all corner points of the feasible set and the corresponding value of the objective function.*

x_1	x_2	value
0.0	0.0	0.0
0.0	1500.0	13,500.0
250.0	1500.0	16,500.0
650.0	1100.0	17,700.0
1000.0	0.0	12,000.0
1000.0	400.0	15,600.0

Using the fact that we have a two-dimensional problem, we can then visualize the entire feasible set (see Figure 2.1).

The optimization problem in (2.2) can be implemented in Pyomo as follows:

```
m = pyo.ConcreteModel("BIM production planning")

# Decision variables and their domains
m.x1 = pyo.Var(domain=pyo.NonNegativeReals)
m.x2 = pyo.Var(domain=pyo.NonNegativeReals)

# Objective function
m.profit = pyo.Objective(expr=12 * m.x1 + 9 * m.x2, sense=pyo.maximize)

# Constraints
m.silicon = pyo.Constraint(expr=m.x1 <= 1000)
m.germanium = pyo.Constraint(expr=m.x2 <= 1500)
m.plastic = pyo.Constraint(expr=m.x1 + m.x2 <= 1750)
m.copper = pyo.Constraint(expr=4 * m.x1 + 2 * m.x2 <= 4800)

SOLVER.solve(m)
```

Note that the constraints imposing the nonnegativity of the two variables, that is, $x_1, x_2 \geq 0$, are not implemented explicitly as Pyomo constraints, but are included instead in the variable declaration with the `domain=NonNegativeReals` argument.

If instead we want to implement the same optimization problem but leverage its canonical form, we can do so using the `numpy` library as follows:

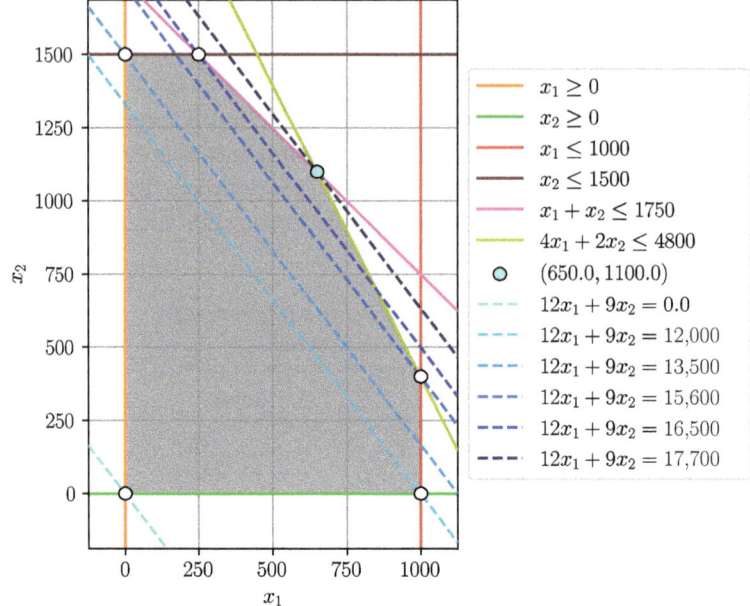

Figure 2.1 The feasible set (in gray), enclosed by the linear constraints (solid lines) and the isolines (parallel dashed lines) corresponding to the objective function.

```python
m = pyo.ConcreteModel("BIM BIM production planning in matrix form")

# Define the number of variables and constraints
n_vars = 2
n_constraints = 4

# Decision variables and their domain
m.x = pyo.Var(range(n_vars), domain=pyo.NonNegativeReals)

# Define the vectors and matrices
c = np.array([-12, -9])
A = np.array([[-1, 0], [0, -1], [-1, -1], [-4, -2]])
b = np.array([-1000, -1500, -1750, -4800])

# Objective function
m.profit = pyo.Objective(
    expr=sum(c[i] * m.x[i] for i in range(n_vars)), sense=pyo.minimize
)

# Constraints
m.constraints = pyo.ConstraintList()
for i in range(n_constraints):
    m.constraints.add(expr=sum(A[i, j] * m.x[j] for j in range(n_vars)) >=
    ↪  b[i])

SOLVER.solve(m)
```

The above example was rather straightforward to model since the choice of the decision variables was obvious and the constraints were easy to formulate. Additionally, it was easy to find the optimal solution by hand, and it was evident to the "naked eye" that it was indeed optimal.

Surprisingly, even much larger and seemingly more complicated problems can be modeled using linear constraints only. For such problems, however, we are often not able to find the solution by hand, and one typically cannot judge "by eye" that a particular solution is an optimal one. To move on to working confidently with real-life problems, we need to gain more knowledge about LO.

In the following sections, we shall expand on what we have learned so far. First, we will extend our capabilities to modeling various situations using linear constraints. Second, we will provide a formal definition of the search for the optimality certificate of a given solution. In the end, we will explain intuitively how numerical algorithms for LO problems work.

2.2 Modeling Techniques

For some expressions, such as those in Example 2.1, it is clear that we can express them using linear functions. However, it is more difficult to quickly recognize the same for many relevant objective functions and constraints from real-world applications. At the same time, since LO problems are by far the easiest problems to solve, a problem should be expressed using linear constraints as long as possible. To this end, we will present a range of useful LO modeling techniques.

2.2.1 Absolute Values

It is often useful to use absolute values to formulate an optimization model. However, their presence makes the problem much harder to solve, since the absolute value is a nonlinear function. In this subsection, we show that in most cases the optimization problem can be made linear again by means of simple manipulations of the expression containing absolute values.

Let $g \colon \mathbb{R}^n \to \mathbb{R}$ be a linear function of the decision variables $\boldsymbol{x}$. First, consider the case in which an absolute value of the form $|g(\boldsymbol{x})|$ appears in a constraint. In this case, we can introduce an auxiliary variable $y \in \mathbb{R}$ and replace all the occurrences of $|g(\boldsymbol{x})|$ with y, provided that we add the following two additional constraints to the optimization problem:

$$\begin{cases} y \geq g(\boldsymbol{x}) \\ y \geq -g(\boldsymbol{x}). \end{cases}$$

Note that, even if it is not formulated explicitly, the variable y will always be nonnegative.

As an example, consider the following constraint $2|x_1 + 5x_2 - x_3| - 3|x_2 + 7x_4| \leq 10$. Applying the technique outlined above twice, once for each absolute value (hence introducing two auxiliary variables, y and z), we can obtain the following set of equivalent linear constraints:

$$\begin{cases} 2y - 3z \leq 10 \\ y \geq x_1 + 5x_2 - x_3 \\ y \geq -(x_1 + 5x_2 - x_3) \\ z \geq x_2 + 7x_4 \\ z \geq -(x_2 + 7x_4). \end{cases}$$

Consider now the case of an optimization problem in which an absolute value of the form $|f(x)|$ appears in the objective function, for some linear function $f: \mathbb{R}^n \to \mathbb{R}$ of the decision variables x. This problem can be made linear only if:

- it is a minimization problem and the coefficient in front of $|f(x)|$ is positive; or
- it is a maximization problem and the coefficient in front of $|f(x)|$ is negative.

If we are in one of these two cases, for each absolute value of the form $|f(x)|$ appearing in the objective we can do the following: (i) introduce two new nonnegative variables $y^+, y^- \geq 0$, (ii) replace each occurrence of $|f(x)|$ with $y^+ + y^-$, and (iii) add the linear constraint $f(x) = y^+ - y^-$. It is easy to show that for each pair of new auxiliary variables y^+, y^-, in any solution of the modified problem, either y^+ or y^- is zero.

As an example, consider a minimization problem with $|x_1 - 2x_4| + 3x_2$ as the objective function. Using the expression $f(x) = x_1 - 2x_4$, we can obtain the following equivalent LO problem:

$$\min \quad y^+ + y^- + 3x_2$$
$$\text{s.t.} \quad \ldots$$
$$y^+ - y^- = x_1 - 2x_4$$
$$y^+, y^- \geq 0.$$

If the absolute value of a linear function appears in the objective but does not fit into either of the two cases described above, this approach does not work. Nonetheless, the optimization problem can still be made linear but requires the use of integer variables, which we will introduce in Chapter 3.

Example 2.5 (Least absolute deviation regression) Linear regression is a supervised machine-learning technique that dates back to at least the nineteenth century. It remains a cornerstone of modern data analysis, generating a linear model that predicts the values of a dependent variable based on one or more independent variables. This notebook introduces an alternative approach to traditional linear regression, employing linear optimization to optimize based on the least absolute deviation (LAD) metric.

Unlike standard techniques that aim to minimize the sum of squared errors, this LAD-based method focuses on minimizing the sum of absolute differences between observed and estimated values. This corresponds to considering the L_1 norm of the errors, which is known for its robustness against outliers. The methodology presented here follows closely the survey paper [NW82].

Suppose that we have a finite dataset consisting of n points $\{(X^{(i)}, y^{(i)})\}_{i=1,\ldots,n}$ with $X^{(i)} \in \mathbb{R}^k$ and $y^{(i)} \in \mathbb{R}$. A linear regression model assumes that the relationship between the vector of k regressors X and the dependent variable y is linear. This relationship is modeled through an error or deviation term e_i, which quantifies how much each of the data points diverges from the model prediction and is defined as follows:

$$e_i := y^{(i)} - \boldsymbol{m}^\top \boldsymbol{X}^{(i)} - b = y^{(i)} - \sum_{j=1}^{k} X_j^{(i)} m_j - b, \qquad (2.5)$$

for some real numbers $m_1, \ldots, m_k$ and b.

LAD is a possible statistical optimality criterion for such a linear regression. Like the well-known least-squares technique, it attempts to find a vector of linear coefficients $\boldsymbol{m} = (m_1, \ldots, m_k)$ and intercept b so that the model closely approximates the given set of data. The method minimizes the sum of absolute errors, that is, $\sum_{i=1}^{n} |e_i|$.

The LAD regression can thus be formulated as an optimization problem with the intercept b, the coefficients m_i, and the errors e_i as the decision variables, namely

$$\min \quad \sum_{i=1}^{n} |e_i| \qquad (2.6)$$

$$\text{s.t.} \quad e_i = y^{(i)} - \boldsymbol{m}^\top \boldsymbol{X}^{(i)} - b, \qquad \forall\, i = 1, \ldots, n$$

$$e_i \in \mathbb{R}, \qquad \forall\, i = 1, \ldots, n$$

$$\boldsymbol{m} \in \mathbb{R}^k,$$

$$b \in \mathbb{R}.$$

Since it is a minimization problem and the absolute values appear in the objective function, we can use the absolute value trick to transform it into a linear problem. More specifically, introducing for every term e_i two new variables $e_i^-, e_i^+ \geq 0$, we can rewrite (2.6) as

$$\min \quad \sum_{i=1}^{n} (e_i^+ + e_i^-) \qquad (2.7)$$

$$\text{s.t.} \quad e_i^+ - e_i^- = y^{(i)} - \boldsymbol{m}^\top \boldsymbol{X}^{(i)} - b, \qquad \forall\, i = 1, \ldots, n$$

$$e_i^+, e_i^- \geq 0, \qquad \forall\, i = 1, \ldots, n$$

$$\boldsymbol{m} \in \mathbb{R}^k,$$

$$b \in \mathbb{R}.$$

This LO problem can be implemented in Pyomo as follows, with example visualization in Figure 2.2.

```
m = pyo.ConcreteModel("LAD regression")

# get dimensions of data, assuming the data points have already been
↪  imported as (X,y)
n, k = X.shape

# create index sets (note use of Python style zero based indexing)
model.I = pyo.RangeSet(0, n - 1)
model.J = pyo.RangeSet(0, k - 1)

# create variables
model.ep = pyo.Var(model.I, domain=pyo.NonNegativeReals)
```

```
model.em = pyo.Var(model.I, domain=pyo.NonNegativeReals)
model.m = pyo.Var(model.J)
model.b = pyo.Var()

# constraints
@model.Constraint(model.I)
def residuals(m, i):
    return m.ep[i] - m.em[i] == y[i] - sum(X[i][j] * m.m[j] for j in m.J) -
    ↪   m.b


# objective
@model.Objective(sense=pyo.minimize)
def sum_of_abs_errors(m):
    return sum(m.ep[i] + m.em[i] for i in m.I)


SOLVER.solve(model)
```

In the above code, we introduced two new Pyomo components. First, we used Pyomo's "RangeSet" component to define a set that consists of a range of numerical values. It is a convenient way to specify indices for decision variables, constraints, and other model components. Note that, unlike Python's native range function, RangeSet is inclusive of both the start and end values.

Second, we used a different method, which is supported by Pyomo, that takes advantage of Python decorators to define both the objective functions and the constraints. Utilizing decorators makes the Pyomo models easier to read and maintain and will be our default convention for the code snippets in this book.

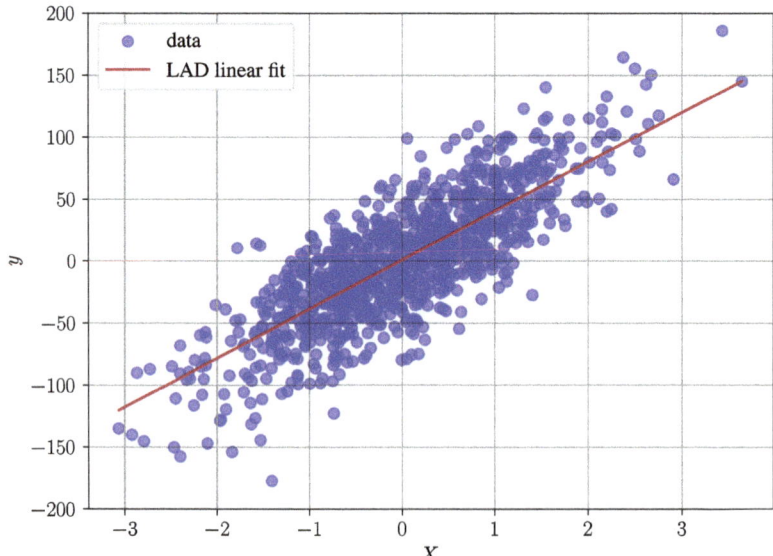

Figure 2.2 The linear approximation resulting from the LAD regression model for $n = 1000$ synthetically generated data points with a single feature ($k = 1$).

Example 2.6 (Portfolio optimization – Minimizing the mean absolute deviation)
Portfolio optimization and modern portfolio theory have a long and important history
in finance and investment. The principal idea is to find a combination of investments in
financial securities that achieves an optimal trade-off between financial risk and return. In
this example, we consider the approach proposed in [KY91], where one tries to minimize
the mean absolute deviation (MAD) in portfolio return as a measure of financial risk. We
build such a model into a large-scale linear optimization problem directly using historical
stock price data, such as the ones displayed in Figure 2.3.

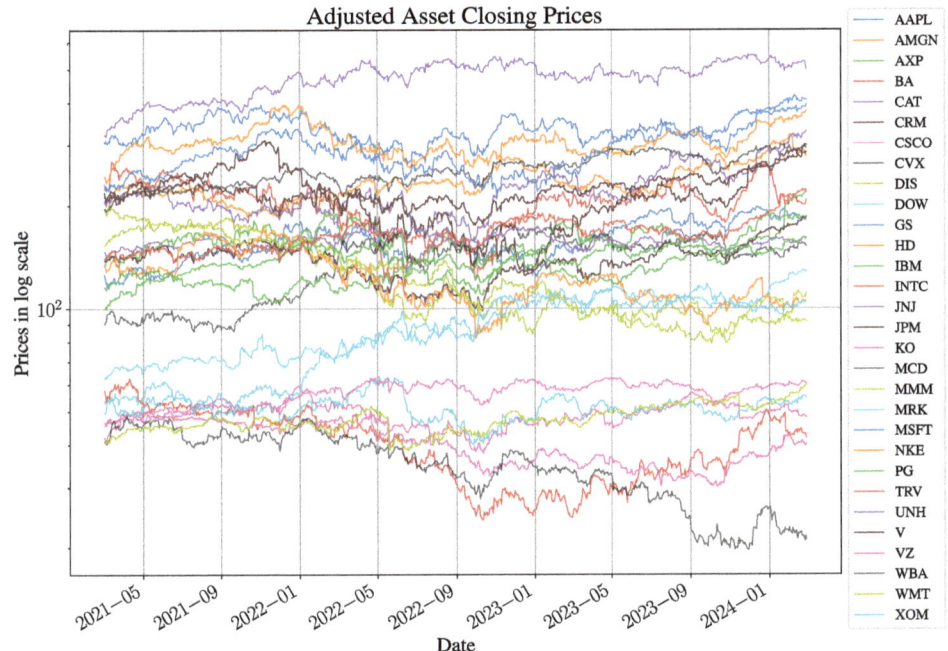

Figure 2.3 Historical prices (in logarithmic scale) over a period of three years for
a selection of stocks.

The portfolio optimization problem is to find an allocation of investment capital to
minimize the portfolio measure of risk subject to constraints on required return and any
other constraints that an investor wishes to impose. Assume that we can make investment
decisions on every trading day t over a fixed time horizon ranging from $t = 1, \dots, T$ and
that there is a set of J assets in which we can choose to invest.

If we want to have a guaranteed minimum portfolio return R, but at the same time
minimize risk, we could choose to have the MAD in portfolio return as the objective
function. More specifically, we can consider the MAD of portfolio return of J assets with
weights w_j for asset j over a period of T intervals given by

$$\text{MAD}(\boldsymbol{w}) = \frac{1}{T} \sum_{t=1}^{T} \left| \sum_{j=1}^{J} w_j (r_{t,j} - \bar{r}_j) \right|,$$

where $r_{t,j}$ is the return on asset j at time t, $\bar{r}_j$ the mean return for asset j, and w_j the fraction of the total portfolio that is invested in asset j. Note that due to the use of absolute values, the MAD for the portfolio is not the weighted sum of the MAD's Δ_j for individual assets, which for every $j = 1, \ldots, J$ is defined as

$$\Delta_j = \frac{1}{T} \sum_{t=1}^{T} \left| r_{t,j} - \bar{r}_j \right|.$$

The MAD portfolio optimization problem is

$$\min \quad \frac{1}{T} \sum_{t=1}^{T} \left| \sum_{j=1}^{J} w_j (r_{t,j} - \bar{r}_j) \right|$$

$$\text{s.t.} \quad \sum_{j=1}^{J} w_j \bar{r}_j \geq R$$

$$\sum_{j=1}^{J} w_j = 1$$

$$w_j \geq 0, \qquad\qquad\qquad \forall j \in J$$

$$w_j \leq w_j^{ub}, \qquad\qquad\qquad \forall j \in J.$$

The lower bound $w_j \geq 0$ is a "no short sales" constraint. The upper bound $w_j \leq w_j^{ub}$ enforces a required level of diversification in the portfolio. Defining two sets of auxiliary variables $u_t \geq 0$ and $v_t \geq 0$ for every $t = 1, \ldots, T$ leads to a reformulation of the problem as an LO problem:

$$\min \quad \frac{1}{T} \sum_{t=1}^{T} (u_t + v_t)$$

$$\text{s.t.} \quad u_t - v_t = \sum_{j=1}^{J} w_j (r_{t,j} - \bar{r}_j), \qquad\qquad \forall t \in 1, \ldots, T$$

$$\sum_{j=1}^{J} w_j \bar{r}_j \geq R$$

$$\sum_{j=1}^{J} w_j = 1$$

$$w_j \geq 0, \qquad\qquad\qquad \forall j \in J$$

$$w_j \leq w_j^{ub}, \qquad\qquad\qquad \forall j \in J$$

$$u_t, v_t \geq 0, \qquad\qquad\qquad \forall t = 1, \ldots, T.$$

We can implement this in Pyomo as follows, where we wrap the entire code into a convenient function:

```python
def mad_portfolio(assets):
    daily_returns = assets.diff()[1:] / assets.shift(1)[1:]
    mean_return = daily_returns.mean()

    m = pyo.ConcreteModel("MAD portfolio optimization")

    m.R = pyo.Param(mutable=True, default=0)
    m.w_lb = pyo.Param(mutable=True, default=0)
    m.w_ub = pyo.Param(mutable=True, default=1.0)

    m.ASSETS = pyo.Set(initialize=assets.columns)
    m.TIME = pyo.RangeSet(len(daily_returns.index))

    m.w = pyo.Var(m.ASSETS)
    m.u = pyo.Var(m.TIME, domain=pyo.NonNegativeReals)
    m.v = pyo.Var(m.TIME, domain=pyo.NonNegativeReals)

    @m.Objective(sense=pyo.minimize)
    def MAD(m):
        return sum(m.u[t] + m.v[t] for t in m.TIME) / len(m.TIME)

    @m.Constraint(m.TIME)
    def portfolio_returns(m, t):
        date = daily_returns.index[t - 1]
        return m.u[t] - m.v[t] == sum(
            m.w[j] * (daily_returns.loc[date, j] - mean_return[j]) for j in
            ↪    m.ASSETS
        )

    @m.Constraint()
    def sum_of_weights(m):
        return sum(m.w[j] for j in m.ASSETS) == 1

    @m.Constraint()
    def mean_portfolio_return(m):
        return sum(m.w[j] * mean_return[j] for j in m.ASSETS) >= m.R

    @m.Constraint(m.ASSETS)
    def no_short(m, j):
        return m.w[j] >= m.w_lb

    @m.Constraint(m.ASSETS)
    def diversify(m, j):
        return m.w[j] <= m.w_ub

    return m
```

```
# assuming historical assets prices are imported as a DataFrame named
↪  assets
m = mad_portfolio(assets)
m.w_lb = 0
m.w_ub = 0.2
m.R = 0.001
SOLVER.solve(m)
```

In the code above, we introduced two new Pyomo components. First, we used Pyomo's "Param" component, which serves as a container for the parameters of the optimization problem. These parameters are immutable within the scope of the model and serve as constants that the solver uses, but do not change. They can be scalar values, arrays, or even indexed collections, and can be used in objective functions and constraints to provide numerical coefficients or bounds.

We also used the "Set" component, which can be used in Pyomo to define sets that serve as the basis for indexing variables, parameters, constraints, and other components of an optimization model. Sets are key for structuring complex models, allowing you to represent multi-dimensional variables and constraints efficiently.

The optimal solution of the portfolio problem above and its properties are illustrated in Figure 2.4.

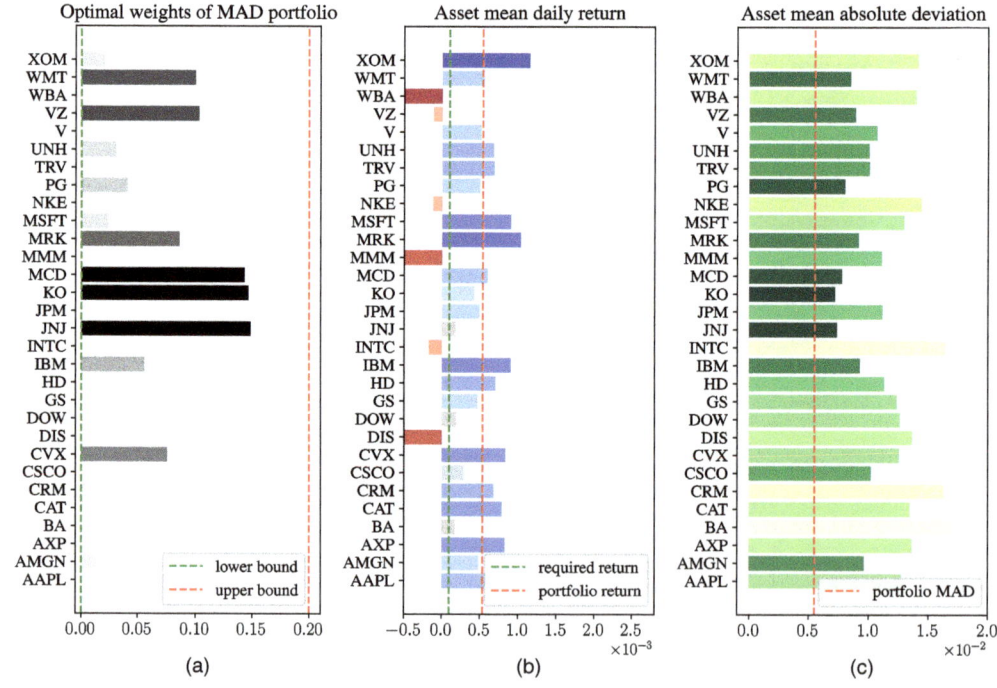

Figure 2.4 The optimal weights found by the MAD portfolio problem for the collection of considered assets (a) with their corresponding mean daily return $\bar{r}_j$ (b) and mean absolute deviations Δ_j (c).

2.2.2 Minmax Objective

Another class of seemingly complicated objective functions that can be easily rewritten as an LO problem are those stated as maxima over several linear functions. Given a finite set of indices $\mathcal{K}$ and a collection of vectors $\{c_k\}_{k \in \mathcal{K}}$, the minmax problem is given by

$$\min \max_{k \in \mathcal{K}} \ c_k^\top x. \tag{2.8}$$

General expressions such as (2.8) can be linearized by introducing an auxiliary variable z and setting

$$\begin{aligned}
\min \quad & z \\
\text{s.t.} \quad & c_k^\top x \leq z, \qquad \forall k \in \mathcal{K}.
\end{aligned}$$

This trick works because if *all* quantities corresponding to different indices $k \in \mathcal{K}$ are below the auxiliary variable z, then we are guaranteed that their maximum is also below z and vice versa. Note that the absolute value function can be rewritten $|x_i| = \max\{x_i, -x_i\}$, therefore the linearization of the optimization problem involving absolute values in the objective functions is a special case of this.

Example 2.7 (Building microchips – Maximizing lowest possible profit) In the same way we can minimize a maximum as we did in (2.8), we can also maximize the minimum. Let us consider the same BIM microchip production problem as in Example 2.1, but suppose that there is uncertainty regarding the selling prices of the microchips. Instead of just nominal prices \$12 and \$9, BIM estimates that there are three scenarios for prices $P = \{(\$12,\$9), (\$11,\$10), (\$8,\$11)\}$. The optimization problem for a production plan that achieves the maximum among the lowest possible profits can be formulated using the trick mentioned above and can be implemented in Pyomo as follows:

```python
def BIM_maxmin(costs):
    m = pyo.ConcreteModel("BIM production planning with maxmin objective")

    m.x1 = pyo.Var(domain=pyo.NonNegativeReals)
    m.x2 = pyo.Var(domain=pyo.NonNegativeReals)
    m.z = pyo.Var()

    m.profit = pyo.Objective(sense=pyo.maximize, expr=m.z)

    m.maxmin = pyo.ConstraintList()
    for c1, c2 in costs:
        m.maxmin.add(expr=m.z <= c1 * m.x1 + c2 * m.x2)

    m.silicon = pyo.Constraint(expr=m.x1 <= 1000)
    m.germanium = pyo.Constraint(expr=m.x2 <= 1500)
    m.plastic = pyo.Constraint(expr=m.x1 + m.x2 <= 1750)
    m.copper = pyo.Constraint(expr=4 * m.x1 + 2 * m.x2 <= 4800)

    return m


BIM = BIM_maxmin([[12, 9], [11, 10], [8, 11]])
SOLVER.solve(BIM)
```

In Pyomo, "ConstraintList" serves as a flexible container for storing a list of constraint expressions within an optimization model. Unlike a regular Constraint component, which usually involves predefined indexing, ConstraintList allows us to dynamically add constraints, making it particularly useful when the number of constraints is not known a priori or varies based on conditional logic. This dynamic nature makes it easier to build and modify models programmatically.

2.2.3 Fractional Objective

In some problems, one might be interested in minimizing a ratio of one quantity to the other, where both depend on the decision variables. Although terms such as

$$\frac{x_1}{2x_2 + 5}$$

are clearly not linear expressions, it is also possible to minimize them using a linear problem under certain conditions. To demonstrate this, take a finite index set $\mathcal{I}$ and consider the optimization problem with a fractional objective function of the form

$$\min \quad \frac{\boldsymbol{c}^\top \boldsymbol{x} + \alpha}{\boldsymbol{d}^\top \boldsymbol{x} + \beta}$$
$$\text{s.t.} \quad A\boldsymbol{x} \leq \boldsymbol{b}$$
$$\boldsymbol{x} \geq 0,$$

where the term $\boldsymbol{d}^\top \boldsymbol{x} + \beta$ is either strictly positive or strictly negative over the entire feasible set of x.

Setting first $t = (\boldsymbol{d}^\top \boldsymbol{x} + \beta)^{-1}$ and then $y_i = x_i t$ for every index i, we obtain the following equivalent linear optimization problem:

$$\min \quad \boldsymbol{c}^\top \boldsymbol{y} + \alpha t$$
$$\text{s.t.} \quad A\boldsymbol{y} \leq t\boldsymbol{b}$$
$$\boldsymbol{d}^\top \boldsymbol{y} + \beta t = 1$$
$$t \geq 0$$
$$\boldsymbol{y} \geq 0.$$

Note that the inequality for t should, in fact, be strict, that is, $t > 0$, but in view of the assumption above for $\boldsymbol{d}^\top \boldsymbol{x} + \beta$, having relaxed the constraint does not change the optimal solution.

Example 2.8 (Building microchips – Different profit formulations) Recall the BIM production model introduced in Example 2.1. Assume now that the pair $(12,9)$ reflects the sales price (revenues) in $ and not the profits made per unit produced. Then we need to account for the production costs. Suppose that the production costs of the chips (x_1, x_2) are equal to a fixed cost of 100 (independent of the number of units produced) plus $7/6x_1$ plus $5/6x_2$. It is reasonable to maximize the difference between revenues and costs. This approach yields the following linear model:

$$\max \quad \left(12 - \frac{7}{6}\right) x_1 + \left(9 - \frac{5}{6}\right) x_2 - 100$$

$$
\begin{aligned}
\text{s.t.} \quad & x_1 \leq 1000 && \text{(silicon)} \\
& x_2 \leq 1500 && \text{(germanium)} \\
& x_1 + x_2 \leq 1750 && \text{(plastic)} \\
& 4x_1 + 2x_2 \leq 4800 && \text{(copper)} \\
& x_1, x_2 \geq 0.
\end{aligned}
$$

```python
def BIM_with_revenues_minus_costs():
    m = pyo.ConcreteModel("BIM production planning with revenues minus
    ↪ costs")

    m.x1 = pyo.Var(domain=pyo.NonNegativeReals)
    m.x2 = pyo.Var(domain=pyo.NonNegativeReals)

    m.revenue = pyo.Expression(expr=12 * m.x1 + 9 * m.x2)
    m.variable_cost = pyo.Expression(expr=7 / 6 * m.x1 + 5 / 6 * m.x2)
    m.fixed_cost = 100

    m.profit = pyo.Objective(
        sense=pyo.maximize, expr=m.revenue - m.variable_cost - m.fixed_cost
    )

    m.silicon = pyo.Constraint(expr=m.x1 <= 1000)
    m.germanium = pyo.Constraint(expr=m.x2 <= 1500)
    m.plastic = pyo.Constraint(expr=m.x1 + m.x2 <= 1750)
    m.copper = pyo.Constraint(expr=4 * m.x1 + 2 * m.x2 <= 4800)

    return m


BIM_linear = BIM_with_revenues_minus_costs()
SOLVER.solve(BIM_linear)
```

This first model has the same optimal solution as the original BIM model, namely $(650, 1100)$ with a revenue of $17,700$ and a cost of 1775.

Alternatively, we may aim to optimize the efficiency of the plan, expressed as the ratio between the revenues and the costs:

$$\max \quad \frac{12x_1 + 9x_2}{\frac{7}{6}x_1 + \frac{5}{6}x_2 + 100}$$

$$
\begin{aligned}
\text{s.t.} \quad & x_1 \leq 1000 && \text{(silicon)} \\
& x_2 \leq 1500 && \text{(germanium)} \\
& x_1 + x_2 \leq 1750 && \text{(plastic)} \\
& 4x_1 + 2x_2 \leq 4800 && \text{(copper)} \\
& x_1, x_2 \geq 0.
\end{aligned}
$$

To solve this second version, we need to deal with the fraction appearing in the objective function by introducing an auxiliary variable $t \geq 0$. More specifically, we reformulate the model as follows:

$$\begin{aligned}
\max \quad & 12y_1 + 9y_2 \\
\text{s.t.} \quad & y_1 \leq 1000 \cdot t & \text{(silicon)} \\
& y_2 \leq 1500 \cdot t & \text{(germanium)} \\
& y_1 + y_2 \leq 1750 \cdot t & \text{(plastic)} \\
& 4y_1 + 2y_2 \leq 4800 \cdot t & \text{(copper)} \\
& \frac{7}{6}y_1 + \frac{5}{6}y_2 + 100 \cdot t = 1 & \text{(fraction)} \\
& y_1, y_2, t \geq 0.
\end{aligned}$$

Despite the change of variables, we can always recover the solution as $(x_1, x_2) = (y_1/t, y_2/t)$.

```python
def BIM_with_revenues_over_costs():
    m = pyo.ConcreteModel("BIM production planning with revenues over
    ↪ costs")

    m.y1 = pyo.Var(domain=pyo.NonNegativeReals)
    m.y2 = pyo.Var(domain=pyo.NonNegativeReals)
    m.t = pyo.Var(domain=pyo.NonNegativeReals)

    m.revenue = pyo.Expression(expr=12 * m.y1 + 9 * m.y2)
    m.variable_cost = pyo.Expression(expr=7 / 6 * m.y1 + 5 / 6 * m.y2)
    m.fixed_cost = 100

    m.profit = pyo.Objective(sense=pyo.maximize, expr=m.revenue)

    m.silicon = pyo.Constraint(expr=m.y1 <= 1000 * m.t)
    m.germanium = pyo.Constraint(expr=m.y2 <= 1500 * m.t)
    m.plastic = pyo.Constraint(expr=m.y1 + m.y2 <= 1750 * m.t)
    m.copper = pyo.Constraint(expr=4 * m.y1 + 2 * m.y2 <= 4800 * m.t)
    m.frac = pyo.Constraint(expr=m.variable_cost + m.fixed_cost * m.t == 1)

    return m


BIM_fractional = BIM_with_revenues_over_costs()
SOLVER.solve(BIM_fractional)
```

The second model has an optimal solution $(250, 1500)$ with a revenue of $16,500$ and a cost of 1641.667.

The efficiency, measured as the ratio of revenue over costs for the optimal solution, is different for the two models. For the first model, the efficiency is equal to $\frac{17,700}{1775} = 9.972$, which is strictly smaller than that of the second model, that is, $\frac{16,500}{1641.667} = 10.051$.

2.3 Linear Duality

In Example 2.1 and all the further variations of it, the solver was certain that the solution it provided us with was an optimal solution. How could that certainty be achieved? We will now develop a general methodology for constructing such **optimality certificates**. First, we will demonstrate how to manually guess/create such a certificate for a small problem.

Example 2.9 (Building microchips – Optimality certificate) We consider again the BIM production problem presented in Example 2.1 and show how manipulating its constraints we can derive an optimality certificate of the solution we already found. Consider first the constraints $x_1 \leq 1000$ and $x_2 \leq 1500$. If we multiply the first constraint by 12 and the second one by 9, we obtain the system of inequalities

$$12x_1 \leq 12{,}000$$
$$9x_2 \leq 13{,}500.$$

Adding the two inequalities side-wise yields

$$12x_1 + 9x_2 \leq 25{,}500.$$

Note that the left-hand side is the same as the objective function. Also, because we obtained this inequality by adding nonnegative multiples of inequalities that hold for any feasible (x_1, x_2), this inequality must hold for any feasible (x_1, x_2) as well. Therefore, it is not possible to find a solution that achieves an objective function value greater than 25,500. This gives us our first upper limit on the optimal value. Is there a solution that can reach this value? You can verify that this is not possible.

Now, let us look at other constraints. Summing side by side the copper constraint and the germanium constraint we get $4x_1 + 3x_2 \leq 6300$, which is equivalent to

$$12x_1 + 9x_2 \leq 18{,}900.$$

Again, the left-hand side takes the same form as the objective function. At the same time, the right-hand side is lower than 25,500. Therefore, we get a tighter upper bound than the previous one. This explains why it was not possible to find a solution that attains the previous bound.

But can the new bound be attained by some solution or can we lower this upper bound further? The latter is true – by taking six times the plastic constraint and adding $3/2$ times the copper constraint one gets

$$12x_1 + 9x_2 \leq 17{,}700.$$

In this case, we can find a feasible solution x_1, x_2 that achieves the value 17,700 prescribed by the upper bound, namely $(x_1, x_2) = (650, 1100)$. At the same time, because this is an upper bound, we know that no feasible solution can achieve a higher objective value. Therefore, the solution found is optimal and the inequality $12x_1 + 9x_2 \leq 17{,}700$ serves as a proof of its optimality.

In the above example, we implicitly constructed a **dual problem** for our original problem, which we will refer to as **primal problem**. As we shall see, the dual problem is itself an LO problem that aims to determine the best bound on the objective. Let us formally derive the dual problem for the example above.

Example 2.10 (Building microchips – Dual problem) Consider again the microchip production problem:

$$\begin{aligned}
\max \quad & 12x_1 + 9x_2 \\
\text{s.t.} \quad & x_1 \leq 1000 \\
& x_2 \leq 1500 \\
& x_1 + x_2 \leq 1750 \\
& 4x_1 + 2x_2 \leq 4800 \\
& x_1, x_2 \geq 0.
\end{aligned}$$

As illustrated in Example 2.9, we can construct bounds for the value of the objective function by multiplying the constraints by nonnegative numbers and adding them to each other so that the left-hand side looks like the objective function, while the right-hand side is the corresponding bound.

Let $\lambda_1, \lambda_2, \lambda_3, \lambda_4$ be nonnegative numbers. We multiply each of these variables by one of the four constraints of the original problem and sum all of them side by side to obtain the inequality

$$(\lambda_1 + \lambda_3 + 4\lambda_4)x_1 + (\lambda_2 + \lambda_3 + 2\lambda_4)x_2 \leq 1000\lambda_1 + 1500\lambda_2 + 1750\lambda_3 + 4800\lambda_4.$$

It is clear that if $\lambda_1, \lambda_2, \lambda_3, \lambda_4 \geq 0$ satisfy

$$\lambda_1 + \lambda_3 + 4\lambda_4 \geq 12,$$
$$\lambda_2 + \lambda_3 + 2\lambda_4 \geq 9,$$

then it holds that

$$\begin{aligned}
12x_1 + 9x_2 \leq \ & (\lambda_1 + \lambda_3 + 4\lambda_4)x_1 + (\lambda_2 + \lambda_3 + 2\lambda_4)x_2 \qquad\qquad (2.9) \\
\leq \ & 1000\lambda_1 + 1500\lambda_2 + 1750\lambda_3 + 4800\lambda_4,
\end{aligned}$$

where the first inequality follows from the fact that $x_1, x_2 \geq 0$, and the most right-hand expression becomes an upper bound on the optimal value of the objective. To make this upper bound as tight as possible, we can search for the numbers $\lambda_1, \lambda_2, \lambda_3, \lambda_4 \geq 0$ that *minimize* the expression $1000\lambda_1 + 1500\lambda_2 + 1750\lambda_3 + 4800\lambda_4$. By doing so, we have formulated the following LO problem, which we name the **dual problem**:

$$\begin{aligned}
\min \quad & 1000\lambda_1 + 1500\lambda_2 + 1750\lambda_3 + 4800\lambda_4 \\
\text{s.t.} \quad & \lambda_1 + \lambda_3 + 4\lambda_4 \geq 12 \\
& \lambda_2 + \lambda_3 + 2\lambda_4 \geq 9 \\
& \lambda_1, \lambda_2, \lambda_3, \lambda_4 \geq 0.
\end{aligned}$$

It is easy to solve and find the optimal solution $(\lambda_1, \lambda_2, \lambda_3, \lambda_4) = (0, 0, 6, 1.5)$, for which the objective function takes the value $17,700$. In view of (2.9), such a value is (the tightest) upper bound for the original problem. Here, we present the Pyomo code for this example.

```
model = pyo.ConcreteModel("BIM production planning dual problem")

# Decision variables and their domain
model.y1 = pyo.Var(domain=pyo.NonNegativeReals)
model.y2 = pyo.Var(domain=pyo.NonNegativeReals)
model.y3 = pyo.Var(domain=pyo.NonNegativeReals)
model.y4 = pyo.Var(domain=pyo.NonNegativeReals)

# Objective function
model.obj = pyo.Objective(
    sense=pyo.minimize,
    expr=1000 * model.y1 + 1500 * model.y2 + 1750 * model.y3 + 4800 *
    ↪   model.y4,
)

# Constraints
model.x1 = pyo.Constraint(expr=model.y1 + model.y3 + 4 * model.y4 >= 12)
model.x2 = pyo.Constraint(expr=model.y2 + model.y3 + 2 * model.y4 >= 9)

SOLVER.solve(m)
```

In the above example, we constructed a dual for a specific maximization problem where all variables were nonnegative and all constraints were inequalities of the $\leq$ form.

For a general minimization LO problem in the canonical form (2.4), the procedure is analogous and the dual can be formulated as follows:

Primal problem:
$$\min \quad c^\top x \qquad (2.10)$$ $$\text{s.t.} \quad Ax \geq b$$ $$x \geq 0.$$

Dual problem:
$$\max \quad b^\top \lambda \qquad (2.11)$$ $$\text{s.t.} \quad A^\top \lambda \leq c$$ $$\lambda \geq 0.$$

For a general minimization–maximization problem with diverse constraints/variables, we can provide the following procedure to construct the dual problem:

1. Rewrite the primal LO problem such that all terms with variables are on the left-hand side of each constraint and all constants on the right-hand side.
2. For each primal constraint $A_i x \geq b_i$ (excluding domain restrictions) add a dual variable λ_i and set its coefficient in the objective function of the dual equal to the constant on the right-hand side of the constraint, that is, b_i. If the primal problem has m constraints (i.e., the matrix A is $m \times n$), then the resulting vector λ of dual variables is m-dimensional.
3. The objective switches between primal and dual from minimization to maximization and vice versa.

Table 2.2 *Duality in linear optimization.*

(a) Case of primal minimization			(b) Case of primal maximization		
		Dual constraint			Dual constraint
Primal variable	$x_j \geq 0$ $x_j \leq 0$ $x_j \in \mathbb{R}$	$(A^\top \lambda)_j \leq c_j$ $(A^\top \lambda)_j \geq c_j$ $(A^\top \lambda)_j = c_j$	Primal variable	$x_j \geq 0$ $x_j \leq 0$ $x_j \in \mathbb{R}$	$(A^\top \lambda)_j \geq c_j$ $(A^\top \lambda)_j \leq c_j$ $(A^\top \lambda)_j = c_j$
		Dual variable			Dual variable
Primal constraint	$(Ax)_j \leq b_j$ $(Ax)_j \geq b_j$ $(Ax)_j = b_j$	$\lambda_j \leq 0$ $\lambda_j \geq 0$ $\lambda_j \in \mathbb{R}$	Primal constraint	$(Ax)_j \leq b_j$ $(Ax)_j \geq b_j$ $(Ax)_j = b_j$	$\lambda_j \geq 0$ $\lambda_j \leq 0$ $\lambda_j \in \mathbb{R}$

4. Add dual constraints and domain restrictions depending on the primal domain restrictions and primal constraints, respectively, as in Tables 2.2a and 2.2b. In particular, if the primal problem is n-dimensional and has m primal constraints, the dual problem will be m-dimensional and have n dual constraints.

These procedural rules can be derived in two ways, either by constructing "manually" the dual problem using bounds as in Example 2.10, or, more simply, by using the Lagrangian relaxation technique that will be introduced later in Chapter 5.

Let us now derive the dual of an LO problem using these rules. Consider an LO problem with three variables x_1, x_2, x_3 (on the left) which display all types of constraints, and derive its dual (on the right):

$$\begin{array}{lll}
\min & x_1 - 2x_2 + 4x_3 & (2.12) \\
\text{s.t.} & -x_1 + 3x_2 = 3 \\
& 2x_1 - x_2 + 3x_3 \geq -5 \\
& x_3 \leq 7 \\
& x_1 + x_2 - x_3 \leq 11 \\
& x_1 \geq 0 \\
& x_2 \leq 0 \\
& x_3 \in \mathbb{R}.
\end{array}$$

$$\begin{array}{lll}
\max & 3\lambda_1 - 5\lambda_2 + 7\lambda_3 + 11\lambda_4 & (2.13) \\
\text{s.t.} & -\lambda_1 + 2\lambda_2 + \lambda_4 \leq 1 \\
& 3\lambda_1 - \lambda_2 + \lambda_4 \geq -2 \\
& 3\lambda_2 + \lambda_3 - \lambda_4 = 4 \\
& \lambda_1 \in \mathbb{R} \\
& \lambda_2 \geq 0 \\
& \lambda_3 \leq 0 \\
& \lambda_4 \leq 0.
\end{array}$$

Since the primal problem is a minimization problem, we need to look at Tables 2.2a.

Note in particular that the dual has four variables (because the primal had four linear constraints) and three linear constraints (because the primal had three variables).

Finally, we remark that if one takes a dual problem, treats it as if it was a primal problem, and derives its dual, then the original primal problem is reobtained.

We come now to the central part of this section – how can all this duality be used to build optimality certificates for the solutions? Mathematically, what makes the optimality certificate search work is the fact that the primal and dual problems are related to each other via theorems known as the weak and strong duality. The first theorem states formally that we can use the dual problem to obtain bounds on the optimal value of the primal problem.

Theorem 2.1 (Weak duality theorem – Linear case) *For all feasible solutions x of the primal problem and all feasible solutions λ of the dual problem, the following inequality holds:*

$$c^\top x \leq b^\top \lambda.$$

The proof is very easy and uses the inequality constraint of the dual problem first and then that of the primal problem to get the following chain of inequalities $c^\top x \leq \lambda^\top Ax \leq \lambda^\top b = b^\top \lambda$, proving the desired result.

As mentioned earlier, weak duality is useful to certify that a candidate solution pair (x, λ) is in fact optimal. Suppose that x and λ are feasible solutions to primal and dual, respectively. Then, if the equality $c^\top x = b^\top \lambda$ holds, weak duality implies that x and λ are optimal solutions to the primal and dual, respectively.

An optimization problem is said to be **unbounded** if the objective function may be improved indefinitely without violating the constraints and is said to be **infeasible** if the feasible set is empty, that is, there is no feasible point that satisfies all constraints. Using these definitions, we can state three immediate consequences of the theorem above:

1. if the primal problem is unbounded, then the dual is infeasible;
2. if the dual problem is unbounded, then the primal is infeasible;
3. if a primal solution x attains the same value of a dual solution λ, that is, $c^\top x = b^\top \lambda$, then they are *both* optimal.

The following result shows under which conditions **strong duality** holds, that is, under which conditions the primal and dual problems have exactly the same optimal value.

Theorem 2.2 (Strong duality – Linear case) *If a linear optimization problem has an optimal solution, so does its dual, and the respective optimal values are equal, that is, $c^\top x = b^\top \lambda$. Therefore, strong duality holds for an LO problem if either of the primal or dual problems is feasible and bounded.*

Having strong duality guaranteed at the cost of hardly any assumptions is one of the huge benefits of working with linear problems. As we shall see in Chapter 5, we do not always have this privilege for general optimization problems, where, only weak duality can be proven "with no assumptions added."

Strong duality allows us to derive an important relation between primal and dual optimal solutions, the so-called **complementary slackness** conditions. If the optimal solution of the primal problem exists and is known, these conditions can be used to determine the unique solution to the dual problem.

Proposition 2.1 (Complementary slackness – Linear case) *Let x be a feasible solution of the primal problem and λ a feasible solution of the dual problem. They are both optimal for the respective problems if and only if*

$$\begin{cases} \lambda^\top (b - Ax) = 0 \\ (A^\top \lambda - c)^\top x = 0. \end{cases} \tag{2.14}$$

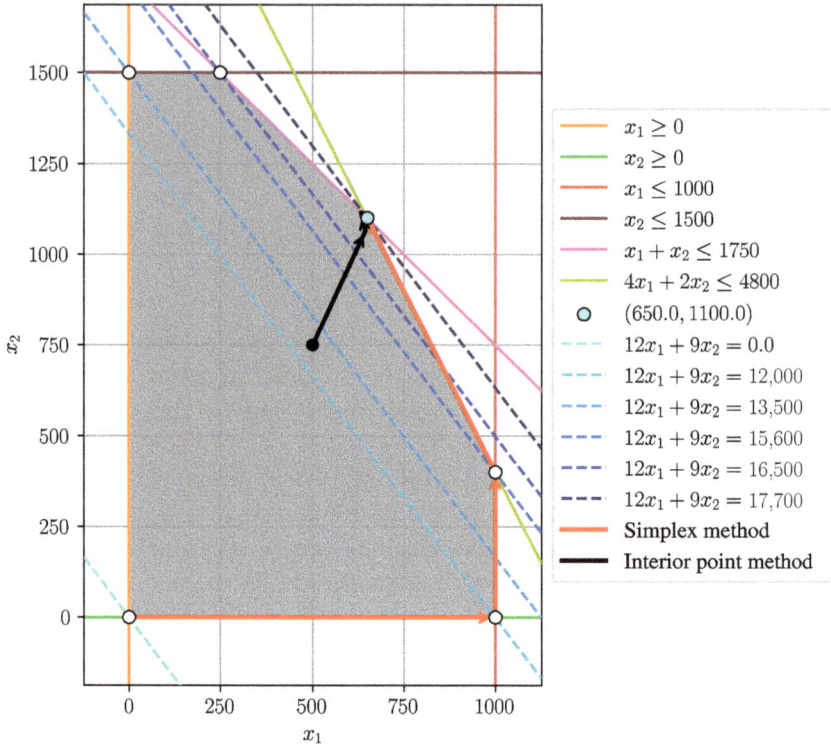

Figure 2.5 The trajectories of two LO solution methods.

2.4 Solution Methods

We finish this chapter by giving a high-level view of the two basic algorithms used to solve general LO problems. Although we do not go into the details, understanding how they work is helpful in understanding the solver output, which we will show in the last part of this chapter.

These two most used methods to solve linear optimization problems are called **simplex method** and **interior point method**, for more details, see [BJS10]. First, we graphically illustrate the difference between them on the example of the feasible set of the problem of Example 2.1 in Figure 2.5. We see that the simplex algorithm operates by selecting subsequent vertices (and only vertices) of the feasible set. The interior-point method, as the name suggests, explores the solution space moving between points in the interior of the feasible set (not on the boundary), converging to the same optimal solution as the one obtained by a simplex method.

How do the two different algorithms arrive at the same optimal solution?

2.4.1 Simplex Method

As you can see in Figure 2.5, the simplex method visits only the vertices of the feasible set, that is, the points at which m linearly independent constraints (including the nonnegativity constraints on x) hold with equality. Indeed, it can be shown that, for any LO problem, an optimal solution, if it exists, can be found among such points.

How can one search the list of such points only and do it efficiently? To enumerate all possible sets of such constraints efficiently, we rewrite the LO problem into the form

$$\begin{aligned} \max \quad & c^\top x \\ \text{s.t.} \quad & Ax + x' = b \\ & x, x' \geq 0, \end{aligned}$$

where the inequalities have been rewritten as equalities using a vector $x' \in \mathbb{R}^m$ of **slack variables**, which, intuitively, take any (nonnegative) value equal to the difference between the original left- and right-hand sides. By merging x and x' into a single vector, the problem is equivalent to

$$\begin{aligned} \max \quad & (c')^\top x \\ \text{s.t.} \quad & A'x = b \\ & x \geq 0, \end{aligned}$$

where

$$c' := \begin{bmatrix} c \\ 0 \end{bmatrix}, \quad x := \begin{bmatrix} x \\ x' \end{bmatrix}, \quad \text{and} \quad A' := [A \ I],$$

with I being the $m \times m$ identity matrix. In this notation, a **vertex** corresponds to a **basic solution**, which corresponds to a selection of m linearly independent columns of matrix A with indices belonging to a set $\mathcal{B}$. If we denote $\mathcal{N} = \{1, \ldots, n+m\} \setminus \mathcal{B}$ then the basic solution corresponding to $\mathcal{B}$ is one obtained by solving the linear system

$$\begin{cases} A_{\mathcal{B}} x_{\mathcal{B}} = b \\ x_{\mathcal{N}} = 0. \end{cases}$$

The simplex method cycles through different bases $\mathcal{B}_1, \mathcal{B}_2, \ldots$ until it finds the optimal solution. It uses rules to decide which columns enter and leave the basis, so that each step leads to a basic solution with a better value of the objective. An effective implementation strives to minimize the number of computations necessary to invert the matrix B. In the worst-case scenario, it may still require visiting all the points of the feasible set.

The simplex method was the first algorithm developed for general LO problems. While it is possible for this method to traverse all vertices of the feasible set in highly challenging problems, it exhibits excellent performance on practical problems, even when they are exceedingly large.

2.4.2 Interior Point Methods

Interior point methods, as the name suggests, attack the LO problem by searching only points that are "strictly inside" the feasible set. They iteratively solve auxiliary optimization problems

$$\min_{x \in \mathbb{R}^n} c^\top x + \mu \left(\sum_{i=1}^{m} \log \left(b_i - a_i^\top x \right) + \sum_{j=1}^{n} \log(x_j) \right). \tag{2.15}$$

In this auxiliary problem, the constraints have been moved to the objective function as an additional **barrier term** parametrized by $\mu > 0$.

For the barrier term to make sense, all entries in x need to be positive, and all constraints need to be strictly satisfied: $a_i^\top x < b_i$. This explains why a particular point needs to lie inside the feasible set. Furthermore, the smaller the parameter μ, the "less important" the barrier term for the actual value of the objective function, and the larger the role of the original objective function. In this way, for a decreasing μ the role of the barrier term becomes smaller and smaller, while still ensuring that the solution meets the problem constraints. In this way, with decreasing μ, the optimal solution of the auxiliary problem can approach the optimal solution.

In their actual implementation, interior point methods sequentially solve a sequence of problems (2.15) for $\mu_1 > \mu_2 > \cdots > 0$, at each step using the optimal solution for the previous μ_{k-1} as the initial point for the problem with μ_k.

Interior point methods, although they look more complicated than the simplex method, can be proven to converge to the optimal solution efficiently. Mathematically, this means that the number of iterations needed can be upper-bounded by a polynomial whose argument is the problem size (number of variables and constraints).

2.5 A Complete Example: BIM Yearly Production Plan

This example is a continuation of the BIM chip production problem illustrated in Example 2.1. Recall that BIM produces logic and memory chips using four types of raw materials. BIM needs to carefully manage the acquisition and inventory of these raw materials based on the forecasted demand for microchips. Data analysis led to the following prediction of monthly demands given in Table 2.3.

Table 2.3 *Monthly demand forecasts for BIM chips.*

chip type	Jan	Feb	Mar	Apr	May	Jun	Jul	Aug	Sep	Oct	Nov	Dec
logic chip	88	125	260	217	238	286	248	238	265	293	259	244
memory chip	47	62	81	65	95	118	86	89	82	82	84	66

Table 2.4 stores the inventory of each material at the beginning of the year and the desired inventory left at the end of the year.

Table 2.4 *Initial and end-state inventory levels for BIM raw materials.*

	copper	silicon	germanium	plastic
initial inventory	480	1000	1500	1750
end of the year minimum	200	500	500	1000

Each raw material can be acquired each month, but the unit prices of each material type vary month by month as described in Table 2.5.

The inventory is limited by a capacity of a total of 9000 units per month, regardless of the type of material of the products in stock. Inventory holding costs are $0.05 per unit per

Table 2.5 *Monthly unit prices of the raw materials in the BIM example.*

material	Jan	Feb	Mar	Apr	May	Jun	Jul	Aug	Sep	Oct	Nov	Dec
copper	1	1	1	2	2	3	3	2	2	1	1	2
silicon	4	3	3	3	5	5	6	5	4	3	3	5
germanium	5	5	5	3	3	3	3	2	3	4	5	6
plastic	0.1	0.1	0.1	0.1	0.1	0.1	0.1	0.1	0.1	0.1	0.1	0.1

month, regardless of the type of material. Due to budget constraints, BIM cannot spend more than 5000 per month on new material acquisition.

BIM aims at minimizing the acquisition and holding costs of the materials while meeting the required quantities for production. The production is made to order, meaning that no inventory of chips is kept. Let us model material acquisition planning and solve it optimally based on the forecasted chip demand.

Let us now build the optimization model. Define the set of raw materials $P = \{\text{copper}, \text{silicon}, \text{germanium}, \text{plastic}\}$ and T the set of 12 months of the year. Let:

- $x_{pt} \geq 0$ be the variable describing the amount of raw material $p \in P$ acquired in month $t \in T$;
- $s_{pt} \geq 0$ be the variable describing the amount of raw material $p \in P$ left in stock at the end of month $t \in T$. Note that these values are uniquely determined by the x variables, but we keep these additional variables to simplify the modeling.

The objective function of our optimal acquisition and production problem is to minimize the total cost. If π_{pt} is the unit price of product $p \in P$ in month $t \in T$ and h_{pt} the unit holding costs we can express the total cost as

$$\sum_{p \in P} \sum_{t \in T} \pi_{pt} x_{pt} + \sum_{p \in P} \sum_{t \in T} h_{pt} s_{pt}.$$

Let us now focus on the constraints. If $\beta \geq 0$ denotes the monthly acquisition budget, the budget constraint can be expressed as

$$\sum_{p \in P} \pi_{pt} x_{pt} \leq \beta \quad \forall t \in T.$$

Further, we constrain the inventory to be always the storage capacity $\ell \geq 0$ using

$$\sum_{p \in P} s_{pt} \leq \ell \quad \forall t \in T.$$

Next, we add another constraint to fix the value of the variables s_{pt} by balancing the acquired amounts with the previous inventory and the demand δ_{pt}, which for each month is implied by the total demand for the chips of both types. Here, $s_{p,t-1}$ is defined as the initial stock when t is the first period, that is, January:

$$x_{pt} + s_{p,t-1} = \delta_{pt} + s_{pt}, \quad \forall p \in P, t \in T.$$

Finally, we capture the required minimum inventory levels in December with the constraint

$$s_{p\text{Dec}} \geq \Omega_p, \quad \forall p \in P,$$

where $(\Omega_p)_{p \in P}$ is the vector with the desired end inventories.

Overall, we obtain the following optimization problem:

$$\min \sum_{p \in P} \sum_{t \in T} \pi_{pt} x_{pt} + \sum_{p \in P} \sum_{t \in T} h_{pt} s_{pt}$$

$$\sum_{p \in P} \pi_{pt} x_{pt} \leq \beta, \qquad\qquad\qquad \forall t \in T$$

$$\sum_{p \in P} s_{pt} \leq \ell, \qquad\qquad\qquad\qquad \forall t \in T$$

$$x_{pt} + s_{p,t-1} = \delta_{pt} + s_{pt}, \qquad\qquad \forall p \in P, t \in T$$

$$s_{p\text{Dec}} \geq \Omega_p, \qquad\qquad\qquad\qquad \forall p \in P$$

$$x_{pt}, s_{pt} \geq 0, \qquad\qquad\qquad\qquad \forall p \in P, t \in T.$$

The following is a Pyomo implementation:

```python
demand_data = """chip,Jan,Feb,Mar,Apr,May,Jun,Jul,Aug,Sep,Oct,Nov,Dec
Logic,88,125,260,217,238,286,248,238,265,293,259,244
Memory,47,62,81,65,95,118,86,89,82,82,84,66"""
demand_chips = pd.read_csv(StringIO(demand_data), index_col="chip")
display(demand_chips)


price = pd.read_csv(StringIO(price_data), index_col="product")
display(price)


price_data = """product,Jan,Feb,Mar,Apr,May,Jun,Jul,Aug,Sep,Oct,Nov,Dec
copper,1,1,1,2,2,3,3,2,2,1,1,2
silicon,4,3,3,3,5,5,6,5,4,3,3,5
germanium,5,5,5,3,3,3,3,2,3,4,5,6
plastic,0.1,0.1,0.1,0.1,0.1,0.1,0.1,0.1,0.1,0.1,0.1,0.1"""
price = pd.read_csv(StringIO(price_data), index_col="product")


# Define a simple dataframe describing the material usage for each microchip
#    type
use = dict()
use["Logic"] = {"silicon": 1, "plastic": 1, "copper": 4}
use["Memory"] = {"germanium": 1, "plastic": 1, "copper": 2}
use = pd.DataFrame.from_dict(use).fillna(0).astype(int)


# Calculate how much of each raw material we need each month using a matrix
#    multiplication
demand = use.dot(demand_chips)


def BIMProductAcquisitionAndInventory(
    demand, acquisition_price, existing, desired, stock_limit, month_budget
):
    m = pyo.ConcreteModel("BIM product acquisition and inventory")
```

```python
periods = demand.columns
products = demand.index
first = periods[0]
prev = {j: i for i, j in zip(periods, periods[1:])}
last = periods[-1]

m.T = pyo.Set(initialize=periods)
m.P = pyo.Set(initialize=products)
m.PT = m.P * m.T   # to avoid internal set bloat

m.x = pyo.Var(m.PT, domain=pyo.NonNegativeReals)
m.s = pyo.Var(m.PT, domain=pyo.NonNegativeReals)

@m.Param(m.PT)
def pi(m, p, t):
    return acquisition_price.loc[p][t]

@m.Param(m.PT)
def h(m, p, t):
    return 0.05   # the holding cost

@m.Param(m.PT)
def delta(m, p, t):
    return demand.loc[p, t]

@m.Expression()
def acquisition_cost(m):
    return pyo.quicksum(m.pi[p, t] * m.x[p, t] for p in m.P for t in m.T)

@m.Expression()
def inventory_cost(m):
    return pyo.quicksum(m.h[p, t] * m.s[p, t] for p in m.P for t in m.T)

@m.Objective(sense=pyo.minimize)
def total_cost(m):
    return m.acquisition_cost + m.inventory_cost

@m.Constraint(m.PT)
def balance(m, p, t):
    if t == first:
        return existing[p] + m.x[p, t] == m.delta[p, t] + m.s[p, t]
    return m.x[p, t] + m.s[p, prev[t]] == m.delta[p, t] + m.s[p, t]

@m.Constraint(m.P)
def finish(m, p):
    return m.s[p, last] >= desired[p]
```

```
        @m.Constraint(m.T)
        def inventory(m, t):
            return pyo.quicksum(m.s[p, t] for p in m.P) <= stock_limit

        @m.Constraint(m.T)
        def budget(m, t):
            return pyo.quicksum(m.pi[p, t] * m.x[p, t] for p in m.P) <=
            ↪   month_budget

        return m


budget = 5000
m = BIMProductAcquisitionAndInventory(
    demand,
    price,
    {"silicon": 1000, "germanium": 1500, "plastic": 1750, "copper": 4800},
    {"silicon": 500, "germanium": 500, "plastic": 1000, "copper": 2000},
    9000,
    budget,
)
SOLVER.solve(m)
```

Tables 2.6 and 2.7 report the optimal solution in terms of material acquisition and stock levels for a budget of 5000, see also Figure 2.6.

Table 2.6 *The optimal material acquisitions over the entire year.*

material	Jan	Feb	Mar	Apr	May	Jun	Jul	Aug	Sep	Oct	Nov	Dec
silicon	0	0	0	965	0	0	0	0	0	1078.1	217.9	0
plastic	0	0	0	0	0	0	266	327	347	375	343	1310
copper	0	0	3548	0	0	0	0	0	962	1336	4312	0
germanium	0	0	0	0	0	0	0	0	0	0	0	0

Table 2.7 *The optimal stock levels over the entire year.*

material	Jan	Feb	Mar	Apr	May	Jun	Jul	Aug	Sep	Oct	Nov	Dec
silicon	912	787	527	1275	1037	751	503	265	0	785.1	744	500
plastic	1615	1428	1087	805	472	68	0	0	0	0	0	1000
copper	4354	3730	6076	5078	3936	2556	1392	262	0	0	3108	2000
germanium	1453	1391	1310	1245	1150	1032	946	857	775	693	609	543

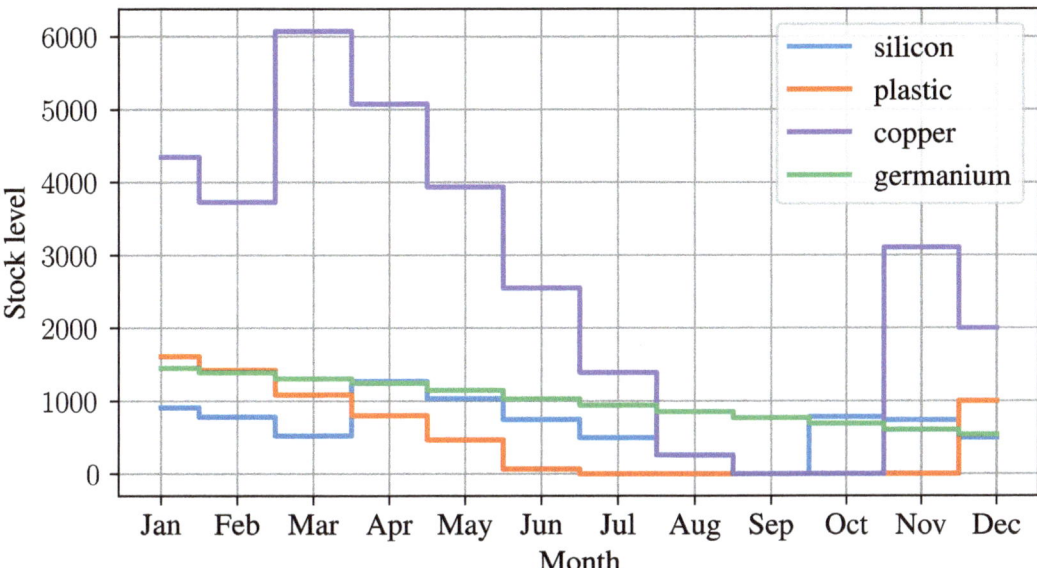

Figure 2.6 Visualization of the optimal stock levels over the entire year.

Exercises

Exercise 2.1 The manager of a smartphone store wants to sell off her stock of 8 phones, 4 hands-free kits, and 19 prepaid cards, due to the arrival of new models. Thanks to a market study, she knows that she can propose an offer with a phone and two prepaid cards and that this offer will generate a profit of $7. Similarly, she can prepare a box with a phone, a hands-free kit, and three prepaid cards, which yields a profit of $9. She is sure to be able to sell any quantity of these two offers within the availability of the stock.

(a) Give an LO model that determines the amount of each offer that the manager should prepare to maximize her net profit. What is the optimal solution?
(b) A sales representative of a supermarket chain proposes to buy the stock in the store (the products, not the offers). The store manager wants to know the minimum unit prices that she should negotiate for each product (phone, hands-free kits, and prepaid cards) to obtain at least the same per-package profit as she could make with her own offer.
(c) What is the relationship between the two models in the previous two parts of the exercise?
(d) How would you advise the store manager to take an offer from the supermarket to buy the entire stock for:

• $65
• $62
• $60?

Exercise 2.2 Furnco makes tables and chairs using wood and labor. Wood costs $1 per board foot, with 40,000 board feet available. Making an unfinished table or chair takes 2 hours of labor; finishing them needs 3 and 2 more hours, respectively, with 6000 labor hours available. Unfinished/finished tables sell for $70/$140, chairs for $60/$110. Formulate an LO problem to maximize profit from the manufacture of tables and chairs.

Exercise 2.3 BigFarm has two farms that grow wheat and corn. Due to the different soil conditions, there are differences in the yields and costs of growing crops on the two farms. Yields and costs are shown as follows.

Each farm has 100 acres available for cultivation. To satisfy the demand, 11,000 bushels of wheat and 7000 bushels of corn must be grown. Determine a planting plan that minimizes the cost of meeting these demands.

	Farm 1	Farm 2
Corn yield/acre (in bushels)	500	650
Cost/acre of corn (in $)	100	120
Wheat yield/acre (in bushels)	400	350
Cost/acre of wheat (in $)	90	80

Exercise 2.4 Gromicho Electronics produces three products. Each product must be processed on each of three types of machines. When a machine is in use, it must be operated by a worker. The time (in hours) required to process each product on each machine and the profit associated with each product are shown in the following table. Currently, five type 1 machines, three type 2 machines, and four type 3 machines are available. The company has 10 workers available and must determine how many workers to assign to each machine. The plant is open 40 hours per week and each worker works 35 hours per week. Formulate an LO model that will enable Gromicho Electronics to assign workers to machines in a way that maximizes weekly profits. (Note: A worker need not spend the entire working week operating a single machine.)

	Product 1	Product 2	Product 3
M1	2	3	4
M2	3	5	6
Machine 3	4	7	9
Profit ($)	6	8	10

Exercise 2.5 You are a financial analyst tasked with optimizing a portfolio of financial assets. The portfolio includes various stocks, bonds, and other investments. Your goal is to allocate a certain amount of funds to each asset in the portfolio to maximize the expected return while maintaining a specified level of risk. The portfolio consists of the following assets:

Stock	Mean annual return (%)	Annual standard deviation (%)
A	12	18
B	8	15
C	5	6

You have a total investment budget of $1,000,000, and you want to allocate these funds to the assets to maximize the expected return while ensuring that the standard deviation of the portfolio does not exceed 12%. Formulate this as an LO problem. You can assume that the riskiness of different assets is combined linearly.

Exercise 2.6 A company manufactures two products, Product A and Product B, using two types of machines, M1 and M2. Each unit of Product A requires 2 hours on M1 and 1 hour on M2. Each unit of Product B requires 1 hour on M1 and 3 hours on M2. The company has 40 hours of M1 time available and 30 hours of M2 time available each day. The profit per unit for Product A is $100, and for Product B is $150. Determine the number of units of each product to produce to maximize profit while respecting the machine's time constraints.

Exercise 2.7 Consider a company that pursues a project consisting of two activities that can be worked on simultaneously. The project ends when both activities are completed. The company aims to complete the project as soon as possible.

According to current estimates, activity 1 will be completed in 200 days, while activity 2 requires 100 days. To speed up the process, the company has a budget of $500,000 to reduce the completion times. The company estimates that:

- for each $10,000 spent on activity 1, its completion time is reduced by 3 days;
- for each $10,000 spent on activity 2, its completion is reduced by 2 days.

What is the budget allocation that allows the company to complete the project as soon as possible? Formulate this problem as an optimization problem.

Exercise 2.8 A truck driver must deliver packages to houses 1, 2, and 3, in order. The normal travel times are a_1, a_2, and a_3 time units respectively, with deadlines d_i at each house i. The driver can adjust speed, taking $a_i x_i$ time units for $x_i \in \mathbb{R}^+$ from one house to the next. If $x_i > 1$, the driver is slower, and if $0 < x_i < 1$, the driver is faster, with constraints $x_1 + x_2 + x_3 \geq 2.5$ and $x_1, x_2, x_3 \geq 0.5$. The objective is to minimize the total delivery delay, defined as the positive difference between the arrival time and d_i at each house. It is assumed the driver proceeds to the next house immediately upon arrival.

(a) Formulate the above problem as an LO model.
(b) Assume the driver may arrive before d_1 at the first house, to be on time for the second delivery. Adjust the previous problem so that the driver is penalized not only when he/she is too late, but also (proportionally) when he/she is too early.

Exercise 2.9 Consider the following problem:

$$\begin{aligned} \min \quad & |x_1| + |x_2| \\ \text{s.t.} \quad & 2x_1 + x_2 \geq 3 \\ & x_1 \geq -1 \\ & x_2 \geq 2. \end{aligned}$$

(a) Represent it graphically and solve.
(b) Reformulate the problem as an LO problem.

Exercise 2.10 Consider the following optimization problem:

$$\begin{aligned} \max \quad & \frac{5x_1 + 6x_2}{2x_2 + 7} \\ \text{s.t.} \quad & 2x_1 + 3x_2 \leq 6 \\ & 2x_1 + x_2 \leq 3 \\ & x_1, x_2 \geq 0. \end{aligned}$$

Formulate it as an LO problem and solve it.

3

Mixed-Integer Linear Optimization

3.1 Formulation

The particular feature of linear optimization problems is that as long as the decision variables satisfy all the constraints, they can take any value. However, there are many situations in which it makes sense to restrict the solution space in a way that cannot be expressed using linear (in)equality constraints. For example, some numbers might need to be integers, such as the number of people to be assigned to a task. Another situation is when certain constraints need to hold only if another constraint holds. For example, the amount of power generated by a power plant must not be less than a certain minimum threshold only if that generator is turned on. Neither of these two examples can be expressed using only linear constraints, as we have seen up to this point. In these cases, it is often still possible to formulate the problem as an LO problem, although some additional restrictions may be needed on certain variables, requiring them to take integer values only. We will refer to this type of LO problem in which some variables are constrained to be integers as **mixed-integer linear optimization (MILO)** problems.

Consider the following slightly modified version of the chip production problem introduced in Example 2.1.

Example 3.1 (Building microchips – Adjusting for wasted material) The company BIM realizes that a 1% fraction of copper is always wasted when producing both types of microchip, more specifically 1% of the required amount. This means that it takes 4.04 g of copper to produce a logic chip and 2.02 g of copper to produce a memory chip. If we rewrite the LO problem in Example 2.1 and modify accordingly the coefficients in the relevant constraints, we obtain the following problem:

$$\begin{array}{lll}
\max & 12x_1 + 9x_2 & \\
\text{s.t.} & x_1 \leq 1000 & \text{(silicon)} \\
& x_2 \leq 1500 & \text{(germanium)} \\
& x_1 + x_2 \leq 1750 & \text{(plastic)} \\
& 4.04x_1 + 2.02x_2 \leq 4800 & \text{(copper with waste)} \\
& x_1, x_2 \geq 0. &
\end{array}$$

If we solve it, we obtain an optimal solution different from the original, namely $(x_1, x_2) \approx (626.238, 1123.762)$ and an optimal value of approximately $17,628.713$. The new optimal

---

solution is not an integer, but, in fact, there is no constraint that requires x_1 and x_2 to be integers.

   In terms of production, we want to manufacture an integer number of microchips, but it is not clear how to implement the fractional optimal solution $(x_1, x_2) \approx (626.238, 1123.762)$. Rounding down to $(x_1, x_2) = (626, 1123)$ will intuitively yield a feasible solution, but this could lead to a minor loss of profit and/or an inefficient use of the available material. Rounding up to $(x_1, x_2) = (627, 1124)$ could possibly lead to an infeasible solution for which the available material is not enough. We could, of course, examine all the potential integer solutions by hand. However, if the problem had a more intricate structure or a larger number of decision variables, this would be much more difficult and may not lead to the true optimal solution.

   A safer approach is to explicitly require the two decision variables to be nonnegative integers, thus transforming the original LO problem into the following MILO:

$$\begin{aligned}
\max \quad & 12x_1 + 9x_2 \\
\text{s.t.} \quad & x_1 \le 1000 & \text{(silicon)} \\
& x_2 \le 1500 & \text{(germanium)} \\
& x_1 + x_2 \le 1750 & \text{(plastic)} \\
& 4.04x_1 + 2.02x_2 \le 4800 & \text{(copper with waste)} \\
& x_1, x_2 \in \mathbb{N}_+.
\end{aligned}$$

The optimal solution of this new MILO problem is $(x_1, x_2) = (626, 1124)$ with a profit of 17,628. Note that for this specific problem, both the naive rounding strategies outlined above would not have yielded the true optimal solution. The Python code to obtain the optimal solution is given below.

```
m = pyo.ConcreteModel("BIM production planning with perturbed
↪    coefficients")

m.x1 = pyo.Var(domain=pyo.NonNegativeIntegers)
m.x2 = pyo.Var(domain=pyo.NonNegativeIntegers)

m.profit = pyo.Objective(expr=12 * m.x1 + 9 * m.x2, sense=pyo.maximize)

m.silicon = pyo.Constraint(expr=m.x1 <= 1000)
m.germanium = pyo.Constraint(expr=m.x2 <= 1500)
m.plastic = pyo.Constraint(expr=m.x1 + m.x2 <= 1750)
m.copper = pyo.Constraint(expr=4.04 * m.x1 + 2.02 * m.x2 <= 4800)

SOLVER.solve(m)
```

   Just as in the case of "ordinary" LO problems introduced in Chapter 2, it is much more concise to formulate MILO problems using matrices and vectors:

$$\begin{aligned}
\min \quad & \boldsymbol{c}^\top \boldsymbol{x} \\
\text{s.t.} \quad & A\boldsymbol{x} \ge \boldsymbol{b} \\
& x_i \in \mathbb{Z}, \quad i \in \mathcal{I},
\end{aligned}$$

where $\mathcal{I} \subseteq \{1, \ldots, n\}$ is the subset of indices identifying the variables that take integer values. The remaining variables are tacitly assumed to be real variables, that is, $x_i \in \mathbb{R}$ for $i \notin \mathcal{I}$. Of course, if the decision variables are required to be nonnegative, we could use the set $\mathbb{N}$ instead of $\mathbb{Z}$.

Example 3.2 (Building microchips – Matrix formulation) To formulate the problem Example 3.1 in matrix notation, we set

$$c = \begin{pmatrix} -12 \\ -9 \end{pmatrix}, \qquad A = - \begin{bmatrix} 1 & 0 \\ 0 & 1 \\ 1 & 1 \\ 4.04 & 2.02 \end{bmatrix}, \qquad \text{and} \qquad b = - \begin{pmatrix} 1000 \\ 1500 \\ 1750 \\ 4800 \end{pmatrix}.$$

to obtain the problem:

$$\begin{aligned} \min \quad & c^\top x \\ \text{s.t.} \quad & Ax \geq b \\ & x_i \in \mathbb{N}, \quad i \in \mathcal{I} = \{1, 2\}. \end{aligned}$$

A special case of integer variables are **binary variables**, which can take only values in $\mathbb{B} = \{0, 1\}$. MILO naturally applies to situations in which we need to deal with integer numbers, as when scheduling people, as the following extensive example illustrates.

Example 3.3 (Workforce shift scheduling) Consider the following model to schedule weekly shifts in a store inspired by [Loz18].

The scheduling problem can be described as follows. A new store has been opened and plans to be open 24 hours a day, 7 days a week. Each day there are three 8-hour shifts. The morning shift is from 6:00 to 14:00, the evening shift is from 14:00 to 22:00, and the night shift is from 22:00 to 6:00 the next day. During the night, there is only one worker, while during the day there are two, except on Sunday when there is only one for each shift. Each worker will not exceed a maximum of 40 hours per week and will have to rest for 12 hours between two shifts. As for the weekly rest days, an employee who rests one Sunday will also prefer to do the same on Saturday. In principle, there are 10 employees available, which is clearly too many. The fewer workers are needed, the more resources are available to other stores.

Assuming that we have N available workers, the problem requires the assignment of these available workers to a predetermined set of shifts with specific staffing requirements. Let $R_{d,s}$ describe the minimum number of workers required for the day-shift pair (d, s).

There are three shifts per day, seven days per week. These observations suggest the need for three ordered sets:

- W with N elements representing workers.
- D labeling the days of the week.
- S labeling the shifts each day.

It is convenient to add the following additional sets to improve the readability of the model:

- T is an ordered set of (day, shift) pairs describing all of the available shifts during the week.
- B is an order set of all overlapping 24-hour periods in the week. An element of the set contains the period (day, shift) in the corresponding period. This set will be used to limit worker assignments to no more than one for each 24-hour period.
- E is the set of all pairs (day, shift) on a weekend. This set will be used to implement worker preferences in the weekend schedule.

To recap, the sets that we will be using are defined as follows:

$$W = \{w_1, w_2, \ldots, w_1\} \text{ set of all workers}$$
$$D = \{\text{Mon}, \text{Tues}, \ldots, \text{Sun}\} \text{ days of the week}$$
$$S = \{\text{morning}, \text{evening}, \text{night}\} \text{ 8-hour daily shifts}$$
$$T = D \times S \text{ ordered set of all (day, shift) pairs}$$
$$B \subset T \times T \times T \text{ all 24 blocks of consecutive slots}$$
$$E \subset T \text{ subset of slots corresponding to weekends.}$$

The decision variables are

$$a_{w,d,s} = \begin{cases} 1 & \text{if worker } w \text{ is assigned to day, shift pair } (d,s) \in T \\ 0 & \text{otherwise} \end{cases}$$

$$e_w = \begin{cases} 1 & \text{if worker } w \text{ is assigned to a weekend day, shift pair } (d,s) \in E \\ 0 & \text{otherwise} \end{cases}$$

$$n_w = \begin{cases} 1 & \text{if worker } w \text{ is needed during the week} \\ 0 & \text{otherwise.} \end{cases}$$

Note, in particular, that we have only binary decision variables.

With such variables, the objective is to minimize the overall number of workers needed to fill the shift and work requirements while also trying to meet worker preferences regarding weekend shift assignments. This is achieved, for example, by minimizing a weighted sum of the number of workers needed to meet all shift requirements and the number of workers assigned to weekend shifts. The resulting objective function is

$$\min \quad \sum_{w \in W} n_w + \gamma \sum_{w \in W} e_w,$$

where the weight γ is a fixed positive parameter that determines the relative importance of these two measures in a desirable shift schedule.

Let us now formulate the constraints using the variables we have created:

- For each (day, shift) pair we need to have enough workers to meet staffing requirements:

$$\sum_{w \in W} a_{w,d,s} \geq R_{d,s}, \quad \forall (d,s) \in T.$$

- No worker can be assigned more than 40 hours per week, which means that the total number of day-shifts assigned times 8 hours cannot be greater than 40:

$$8 \sum_{d,s \in \mathrm{T}} a_{w,d,s} \leq 40, \quad \forall w \in \mathrm{W}.$$

- No worker can be assigned more than one shift in each 24-hour period, to enforce which we need to loop over all 24-hour periods of three consecutive work shifts and make sure that for each worker the total number of assignments is less than or equal to one:

$$a_{w,d_1,s_1} + a_{w,d_2,s_2} + a_{w,d_3,s_3} \leq 1, \quad \forall w \in \mathrm{W}, \forall \left((d_1,s_1),(d_2,s_2),(d_3,s_3)\right) \in \mathrm{B}.$$

- The variable n_w needs to correctly be equal to 1 if a given worker is assigned to any of the day-shifts during the week, and 0 otherwise. Keeping in mind that 21 is the total number of day-shifts during the week, we can implement this relationship as follows:

$$\sum_{d,s \in \mathrm{T}} a_{w,d,s} \leq 21 \cdot n_w, \quad \forall w \in \mathrm{W}. \tag{3.1}$$

Indeed, whenever any of the variables $a_{w,d,s}$ is equal to 1, n_w has to become equal to 1, while at the same time, the number 21 is big enough for the right-hand side not to be a restriction (beyond what the other constraints do) on the total number of shifts assigned to worker w in this constraint. At the same time, because in our objective function we minimize, the variable n_w will naturally become equal to 0 in an optimal solution where the left-hand side would be equal to 0, so the constraint correctly enforces tracking the number of employed workers during the week.

- The variable e_w must be equal to 1 when worker w is assigned to any of the weekend day-shifts. We can formulate a constraint that will enforce this relationship correctly in the context of our problem as follows:

$$\sum_{d,s \in \mathrm{E}} a_{w,d,s} \leq 6 \cdot e_w, \quad \forall w \in \mathrm{W}. \tag{3.2}$$

Indeed, variable e_w is forced to be equal to 1 if any day-shift on the weekend is assigned to worker w, and since there are six day-shifts on the weekend if that happens, this constraint does not impose extra limitations (beyond what the other constraints do) on the number of shifts assigned to w on the weekend. At the same time, since our objective is a minimization one, e_w will become equal to 0 in an optimal solution where w does not have any weekend day-shifts.

The resulting MILO problem then is

$$\min \quad \sum_{w \in \mathrm{W}} n_w + \gamma \sum_{w \in \mathrm{W}} e_w$$

$$\text{s.t.} \quad \sum_{w \in \mathrm{W}} a_{w,d,s} \geq \mathrm{R}_{d,s}, \qquad \forall (d,s) \in \mathrm{T}$$

$$8 \sum_{d,s \in \mathrm{T}} a_{w,d,s} \leq 40, \qquad \forall w \in \mathrm{W}$$

$$a_{w,d_1,s_1} + a_{w,d_2,s_2} + a_{w,d_3,s_3} \leq 1, \qquad \forall w, \forall ((d_1,s_1),(d_2,s_2),(d_3,s_3)) \in \mathbf{B}$$

$$\sum_{d,s \in \mathrm{T}} a_{w,d,s} \leq 21 \cdot n_w, \qquad \forall w \in \mathbf{W}$$

$$\sum_{d,s \in \mathrm{E}} a_{w,d,s} \leq 6 \cdot e_w, \qquad \forall w \in \mathbf{W}$$

$$a_{w,d,s} \in \mathbb{B}, \qquad \forall d \in \mathbf{D}, \forall s \in \mathbf{S}, \forall w \in \mathbf{W}$$

$$n_w, e_w \in \mathbb{B}, \qquad \forall w \in \mathbf{W}.$$

The Pyomo implementation of this model is as follows:

```python
def shift_schedule(N=10, hours=40):
    m = pyo.ConcreteModel("Workforce Shift Scheduling")

    # ordered set of avaiable workers
    m.W = pyo.Set(initialize=[f"W{i:02d}" for i in range(1, N + 1)])

    # ordered sets of days and shifts
    m.D = pyo.Set(initialize=["Mon", "Tue", "Wed", "Thu", "Fri", "Sat",
    ↪    "Sun"])
    m.S = pyo.Set(initialize=["morning", "evening", "night"])

    # ordered set of day, shift time slots
    m.T = pyo.Set(initialize=m.D * m.S)

    # ordered set of 24-hour time blocks
    m.B = pyo.Set(
        initialize=[
            [m.T.at(i), m.T.at(i + 1), m.T.at(i + 2)] for i in range(1,
            ↪    len(m.T) - 1)
        ]
    )

    # ordered set of weekend shifts
    m.E = pyo.Set(initialize=m.T, filter=lambda m, day, shift: day in
    ↪    ["Sat", "Sun"])

    # parameter of worker requirements
    @m.Param(m.T)
    def R(m, day, shift):
        if shift in ["night"] or day in ["Sun"]:
            return 1
        return 2

    # max hours per week per worker
    m.H = pyo.Param(mutable=True, default=hours)
```

```python
# decision variable: a[worker, day, shift] = 1 assigns worker to a time
↪    slot
m.a = pyo.Var(m.W, m.T, domain=pyo.Binary)

# decision variables: e[worker] = 1 worker is assigned weekend shift
m.e = pyo.Var(m.W, domain=pyo.Binary)

# decision variable: n[worker] = 1
m.n = pyo.Var(m.W, domain=pyo.Binary)

# assign a sufficient number of workers for each time slot
@m.Constraint(m.T)
def required_workers(m, day, shift):
    return m.R[day, shift] == sum(m.a[worker, day, shift] for worker in
    ↪    m.W)

# workers limited to forty hours per week assuming 8 hours per shift
@m.Constraint(m.W)
def forty_hour_limit(m, worker):
    return 8 * sum(m.a[worker, day, shift] for day, shift in m.T) <=
    ↪    m.H

# workers are assigned no more than one time slot per 24 time block
@m.Constraint(m.W, m.B)
def required_rest(m, worker, d1, s1, d2, s2, d3, s3):
    return m.a[worker, d1, s1] + m.a[worker, d2, s2] + m.a[worker, d3,
    ↪    s3] <= 1

# determine if a worker is assigned to any shift
@m.Constraint(m.W)
def is_needed(m, worker):
    return (
        sum(m.a[worker, day, shift] for day, shift in m.T) <= len(m.T) *
        ↪    m.n[worker]
    )

# determine if a worker is assigned to a weekend shift
@m.Constraint(m.W)
def is__weekend(m, worker):
    return sum(m.a[worker, day, shift] for day, shift in m.E) <= 6 *
    ↪    m.e[worker]

# minimize a blended objective of needed workers and needed weekend
↪    workers
@m.Objective(sense=pyo.minimize)
def minimize_workers(m):
    return sum(
        i * m.n[worker] + 0.1 * i * m.e[worker] for i, worker in
        ↪    enumerate(m.W)
```

```
        )

    return m


m = shift_schedule(10, 40)
SOLVER.solve(m)
```

Figure 3.1 is a visual representation of the optimal shift schedule obtained for a specific instance of the problem. Although $N = 10$ workers were available, only seven of them are necessary to cover the given staffing requirements and shifts.

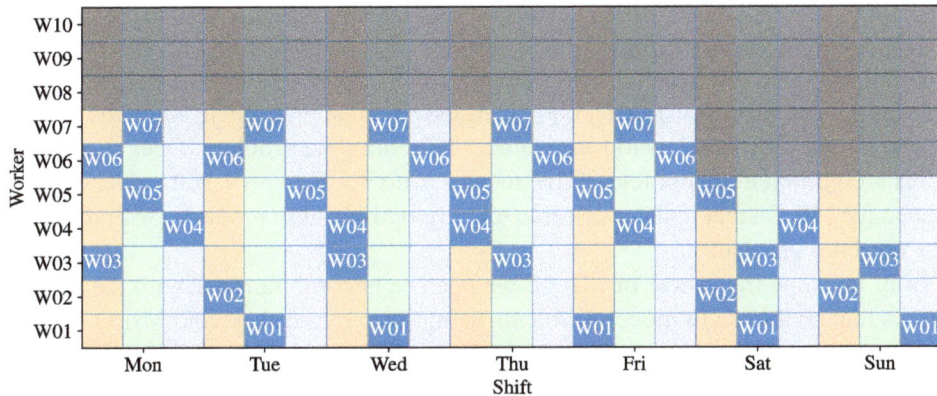

Figure 3.1 The optimal shift schedule obtained for a specific instance of the problem.

In the previous chapter, we claimed that every optimization problem that can be formulated as an LO problem is "easy" to solve. Does that mean that, in contrast, every MILO problem is easy to solve? Not necessarily – due to the significantly greater modeling capacities of MILO, it can be indeed used to model problems that are fundamentally "difficult," that is, for which no efficient solution procedure is known, even if tools other than MILO are allowed. We illustrate here a classic problem, the knapsack problem, which can be used to represent situations in which resources need to be allocated optimally as, for instance, in staff scheduling.

Example 3.4 (Resource allocation – Knapsack problem) A traveler can only bring a single knapsack with a weight limit and must fill it with the most valuable items. Given a finite set of n items, where each item i has a weight w_i and a value v_i, we want to select the subset of items to put in the knapsack so that the total weight is less than or equal to a given limit W and the total value is as large as possible. It can be formulated as a MILO problem as follows:

$$\max \quad \sum_{i=1}^{n} v_i x_i$$

$$\text{s.t.} \quad \sum_{i=1}^{n} w_i x_i \leq W$$

$$x_i \in \mathbb{B}, \quad i = 1, \ldots, n.$$

The **knapsack problem** is one of the most fundamental combinatorial optimization problems whose variants often arise in resource allocation where the decision makers have to choose a subset of nondivisible tasks/resources under a fixed time/budget constraint, respectively. General versions of this problem are routinely solved in online computational clusters, for example. This problem is known to be **NP-complete**, which implies that it is generally accepted that for such a problem there is no algorithm that in the worst-case instance would not need to check all 2^n solutions.

As is clear from the last example, our eagerness to model a problem we come across as a MILO may sometimes lead us to formulate an NP-complete problem. Does that mean that MILO is an inefficient technology? The answer is no because powerful solvers have been developed for MILOs that make it possible to solve optimization problems with thousands of integer variables efficiently.

To fully leverage the power of MILO, we need to become familiar with the techniques and tricks that enable us to model problems using integer variables and MILO constraints. Similarly to the previous chapter, we first present the new MILO modeling opportunities in Section 3.2 and then the key solution methods in Section 3.3. At the same time, we want the reader to be aware that in certain situations there exist other modeling alternatives that are better than "forcing" a given constraint into a MILO form through complicated tricks. One of these alternatives is called **disjunctive programming**, which is supported by Pyomo. Later in this chapter, we will illustrate how in specific suitable examples disjunctive programming can be equivalently used in place of MILO constraints.

3.2 Modeling Techniques

3.2.1 Variables Taking a Set of Discontinuous Values

MILO can be used to model variables taking discontinuous values. Consider, for example, the case in which for a decision variable x we must have either $x = 0$ or $l \leq x \leq u$. By introducing a binary variable $y \in \mathbb{B}$, with the interpretation that

$$y = \begin{cases} 0, & \text{if } x = 0, \\ 1, & \text{if } l \leq x \leq u, \end{cases} \tag{3.3}$$

we can model the discontinuous variable x using the following linear constraints:

$$x \leq uy$$
$$x \geq ly$$
$$y \in \mathbb{B}.$$

With this system of constraints, the relationship (3.3) is enforced.

3.2.2 Variable Enforcing a Given Constraint or Not

In many optimization problems, we want to ensure that a certain constraint $a^\top x \leq b$ holds *only* under specific conditions, which we can capture with a yes–no decision variable, say a binary variable $y \in \mathbb{B}$. The so-called **big-M method** gives a way to construct such an "optional" constraint by writing the original constraint as

$$a^\top x \leq b + M(1 - y),$$

where $M > 0$ is a large number. In this constraint, if the binary variable y takes the value 1 (corresponding to a "yes"), then the right-hand side is equal to b and we recover the original constraints, which then "needs to hold." Otherwise, if $y = 0$ (corresponding to a "no"), then the right-hand side becomes so large due to the $M(1 - y) = M$ term that effectively the constraint does not impose any restriction on x.

The value M can typically be easily guessed or derived from the properties of the problem: It should be large enough to make the trick work, that is, greater than $a^\top x - b$ for all reasonable x, but it should not be too large because this will deteriorate the solver performance.

In Example 3.3, we already implicitly used the big-M method in two of the constraints, compare (3.1) and (3.2), for which we individually chose the smallest possible value of M. We will now see this trick applied to several specific situations.

3.2.3 Cost Function With a Fixed Component

In production planning problems, the cost of producing a given good often consists of a fixed component (e.g., machine setup costs) and a component that scales with the production size. Such a cost function $f(x)$ with a per-unit cost c and a fixed component k given by

$$f(x) = \begin{cases} 0, & \text{if } x = 0 \\ k + cx, & \text{if } x > 0 \end{cases}$$

can be modeled by adding a binary variable $y \in \mathbb{B}$ as follows:

$$f(x,y) = ky + cx$$
$$x \leq My$$
$$y \in \mathbb{B},$$

where $M > 0$ is a large positive constant that should be an upper bound on x.

3.2.4 Either–Or Constraints

The either–or constraint requires that at least one of the following constraints holds:

$$a_1^\top x \leq b_1 \qquad \text{or} \qquad a_2^\top x \leq b_2.$$

Also in this case we can use the big-M method. Adding a new binary variable $y \in \mathbb{B}$ and two large positive constants $M_1, M_2 > 0$, the equivalent linear constraints are

$$a_1^\top x \leq b_1 + M_1 y$$
$$a_2^\top x \leq b_2 + M_2(1 - y)$$
$$y \in \mathbb{B}.$$

3.2.5 If–Then Constraints

The if–then condition requires that if one condition, say

$$A: \qquad a_1^\top x \leq b_1,$$

has to hold, then *another* condition, say

$$B: \qquad a_2^\top x \leq b_2$$

must hold. Such a relationship between linear constraints can still be encoded in a MILO as follows.

The core idea is to recognize that the implication $A \Rightarrow B$ is logically equivalent to $\overline{A} \vee B$. Using this fact, the if–then condition can be equivalently rewritten as the following either–or constraint:

$$a_1^\top x > b_1 \qquad \text{or} \qquad a_2^\top x \leq b_2.$$

In turn, introducing two large constants $M_1, M_2 > 0$ and a binary variable y, this either–or constraint is equivalent to

$$a_1^\top x > b_1 - M_1 y$$
$$a_2^\top x \leq b_2 + M_2(1 - y)$$
$$y \in \mathbb{B}.$$

It is important to be careful because, in MILOs, strict constraints of the form $a_1^\top x > b_1 - M_1 y$ cannot be enforced as such, and are always implemented as weak inequalities $a_1^\top x \geq b_1 - M_1 y$, which in most contexts is fine.

Example 3.5 (Logical constraints in machine scheduling problems) In machine scheduling problems, such as the one presented later in Example 3.8, there are often logical constraints of the following form: If task A is started before task B, then also task C must be started before task B. Such a situation can arise, for example, when two groups of tasks require a different machine setup that nobody wants to change too often. This can be expressed using an if–then constraint

$$\text{finish}_A \leq \text{start}_B \qquad \Rightarrow \qquad \text{finish}_C \leq \text{start}_B,$$

which can be reformulated as a MILO constraint as

$$\text{finish}_A > \text{start}_B - M_1 y$$
$$\text{finish}_C \leq \text{start}_B + M_2(1 - y).$$

In the next example, we illustrate the utilization of if–then logic in the context of an optimization problem in which we want to identify the most effective recharging strategy for an electric vehicle.

Example 3.6 (Recharging strategy for electric vehicle) Given the current location x, battery charge c, and planning horizon D, a driver needs to plan ahead when to rest and when to charge his electric vehicle (EV). Data are provided for the location and charging rate available at each charging station. The distances to the charging stations are measured relative to an arbitrary location. The objective is to drive from location x to location $x + D$ in as little time as possible, subject to the following constraints:

- The maximum charge is $c^{\mathrm{max}} = 150\,\mathrm{kW}$.
- To accommodate unforeseen events, the state of charge should never drop below 20% of the maximum battery capacity, so $c^{\mathrm{min}} = 30\,\mathrm{kW}$.
- For comfort, no more than $r^{\mathrm{max}} = 3$ hours should pass between stops, and a rest stop should last at least $t_{\mathrm{rest}} = 20$ minutes.
- Any stop includes $t_{\mathrm{lost}} = 10$ minutes of "lost time."

We make several simplifying assumptions that can be relaxed at a later time:

- Travel is carried out at a constant speed $v = 100\,\mathrm{km}$ per hour and a constant discharge rate $R = 0.24\,\mathrm{kW/km}$.
- Batteries are recharged at a constant rate determined by the charging station.
- Only consider stopping at the recharging stations and always recharging while resting.

The problem statement identifies four state variables:

- c the current battery charge;
- r the elapsed time since the last rest stop;
- t elapsed time since the start of the trip;
- x the current location.

The charging stations are located in positions d_i for $i \in I$ and have charging rates per time unit C_i. The arrival time at charging station i is given by

$$c_i^{\mathrm{arr}} = c_{i-1}^{\mathrm{dep}} - R(d_i - d_{i-1})$$

$$r_i^{\mathrm{arr}} = r_{i-1}^{\mathrm{dep}} + \frac{d_i - d_{i-1}}{v}$$

$$t_i^{\mathrm{arr}} = t_{i-1}^{\mathrm{dep}} + \frac{d_i - d_{i-1}}{v},$$

where the script t_{i-1}^{dep} refers to departure from the prior location. At each charging location, there is a decision to make about whether to stop, rest, and recharge. If the decision is positive, then

$$c_i^{\mathrm{dep}} \leq c^{\mathrm{max}}$$

$$r_i^{\mathrm{dep}} = 0$$

$$t_i^{\text{dep}} \geq t_i^{\text{arr}} + t_{\text{lost}} + \frac{c_i^{\text{dep}} - c_i^{\text{arr}}}{C_i}$$

$$t_i^{\text{dep}} \geq t_i^{\text{arr}} + t_{\text{rest}},$$

which accounts for the battery charge, the lost time, and the time required for battery charging, and allows for a minimum rest time. On the other hand, if a decision is made to skip the charging and rest opportunity, then we have:

$$c_i^{\text{dep}} = c_i^{\text{arr}}$$
$$r_i^{\text{dep}} = r_i^{\text{arr}}$$
$$t_i^{\text{dep}} = t_i^{\text{arr}}.$$

The latter sets of constraints have an exclusive–or relationship. That is, either one or the other of the constraint sets holds, but not both.

$$
\begin{aligned}
\min \quad & t_{n+1}^{\text{arr}} \\
\text{s.t.} \quad & r_i^{\text{arr}} \leq r^{\text{max}}, & \forall i \in I \\
& c_i^{\text{arr}} \geq c^{\text{min}}, & \forall i \in I \\
& c_i^{\text{arr}} = c_{i-1}^{\text{dep}} - R(d_i - d_{i-1}), & \forall i \in I \\
& r_i^{\text{arr}} = r_{i-1}^{\text{dep}} + \frac{d_i - d_{i-1}}{v}, & \forall i \in I \\
& t_i^{\text{arr}} = t_{i-1}^{\text{dep}} + \frac{d_i - d_{i-1}}{v}, & \forall i \in I \\
& \begin{bmatrix} c_i^{\text{dep}} & \leq & c^{\text{max}} \\ r_i^{\text{dep}} & = & 0 \\ t_i^{\text{dep}} & \geq & t_i^{\text{arr}} + t_{\text{lost}} + \frac{c_i^{\text{dep}} - c_i^{\text{arr}}}{C_i} \\ t_i^{\text{dep}} & \geq & t_i^{\text{arr}} + t_{\text{rest}} \end{bmatrix} \vee \begin{bmatrix} c_i^{\text{dep}} = c_i^{\text{arr}} \\ r_i^{\text{dep}} = r_i^{\text{arr}} \\ t_i^{\text{dep}} = t_i^{\text{arr}} \end{bmatrix}, & \forall i \in I \\
& c_i^{\text{dep}}, c_i^{\text{arr}}, r_i^{\text{dep}}, r_i^{\text{arr}}, t_i^{\text{dep}}, t_i^{\text{arr}} \geq 0, & \forall i \in I.
\end{aligned}
$$

A Pyomo implementation of this model is as follows:

```
n_charging_stations = 20

# randomly distribute charging stations along a fixed route
np.random.seed(2023)
d = np.round(np.cumsum(np.random.triangular(20, 150, 223,
↪    n_charging_stations)), 1)

# randomly assign changing changing rates to the charging stations
c = np.random.choice([50, 100, 150, 250], n_charging_stations, p=[0.2, 0.4,
↪    0.3, 0.1])

# assign names to the charging stations
s = [f"S{i:02d}" for i in range(n_charging_stations)]
```

```python
stations = pd.DataFrame([s, d, c]).T
stations.columns = ["name", "location", "kw"]


x = 0   # current location (km)
D = 2000   # planning horizon
c_max = 150   # max charge limits (kWh)
c_min = 0.2 * c_max   # min charge limits (kWh)
c = c_max   # battery level at x=0
v = 100.0   # velocity km/hr
R = 0.24   # discharge rate kWh/km
t_lost = 10 / 60   # lost time
t_rest = 20 / 60   # min stop time
r_max = 3   # max time between stops


def ev_plan(stations, x, D):
    m.D = D
    # find stations between x and x + D
    on_route = stations[(stations["location"] >= x) & (stations["location"]
    ↪   <= x + m.D)]

    m = pyo.ConcreteModel("Recharging EV strategy")

    m.n = pyo.Param(default=len(on_route))

    # locations and road segments between location x and x + D
    m.STATIONS = pyo.RangeSet(1, m.n)
    m.LOCATIONS = pyo.RangeSet(0, m.n + 1)
    m.SEGMENTS = pyo.RangeSet(1, m.n + 1)

    # distance traveled
    m.x = pyo.Var(m.LOCATIONS, domain=pyo.NonNegativeReals, bounds=(0,
    ↪   10000))

    # arrival and departure charge at each charging station
    m.c_arr = pyo.Var(m.LOCATIONS, domain=pyo.NonNegativeReals,
    ↪   bounds=(c_min, c_max))
    m.c_dep = pyo.Var(m.LOCATIONS, domain=pyo.NonNegativeReals,
    ↪   bounds=(c_min, c_max))

    # arrival and departure times from each charging station
    m.t_arr = pyo.Var(m.LOCATIONS, domain=pyo.NonNegativeReals, bounds=(0,
    ↪   100))
    m.t_dep = pyo.Var(m.LOCATIONS, domain=pyo.NonNegativeReals, bounds=(0,
    ↪   100))

    # arrival and departure rest from each charging station
    m.r_arr = pyo.Var(m.LOCATIONS, domain=pyo.NonNegativeReals, bounds=(0,
    ↪   r_max))
```

```python
m.r_dep = pyo.Var(m.LOCATIONS, domain=pyo.NonNegativeReals, bounds=(0,
↪    r_max))

# initial conditions
m.x[0].fix(x)
m.t_dep[0].fix(0.0)
m.r_dep[0].fix(0.0)
m.c_dep[0].fix(c)

@m.Param(m.STATIONS)
def C(m, i):
    return on_route.loc[i - 1, "kW"]

@m.Param(m.LOCATIONS)
def location(m, i):
    if i == 0:
        return x
    elif i == m.n + 1:
        return x + m.D
    else:
        return on_route.loc[i - 1, "location"]

@m.Param(m.SEGMENTS)
def dist(m, i):
    return m.location[i] - m.location[i - 1]

@m.Objective(sense=pyo.minimize)
def min_time(m):
    return m.t_arr[m.n + 1]

@m.Constraint(m.SEGMENTS)
def drive_time(m, i):
    return m.t_arr[i] == m.t_dep[i - 1] + m.dist[i] / v

@m.Constraint(m.SEGMENTS)
def rest_time(m, i):
    return m.r_arr[i] == m.r_dep[i - 1] + m.dist[i] / v

@m.Constraint(m.SEGMENTS)
def drive_distance(m, i):
    return m.x[i] == m.x[i - 1] + m.dist[i]

@m.Constraint(m.SEGMENTS)
def discharge(m, i):
    return m.c_arr[i] == m.c_dep[i - 1] - R * m.dist[i]

@m.Disjunction(m.STATIONS, xor=True)
def recharge(m, i):
```

```
        # list of constraints thtat apply if there is no stop at station i
        disjunct_1 = [
            m.c_dep[i] == m.c_arr[i],
            m.t_dep[i] == m.t_arr[i],
            m.r_dep[i] == m.r_arr[i],
        ]

        # list of constraints that apply if there is a stop at station i
        disjunct_2 = [
            m.t_dep[i] == t_lost + m.t_arr[i] + (m.c_dep[i] - m.c_arr[i]) /
            ↪    m.C[i],
            m.c_dep[i] >= m.c_arr[i] + t_rest,
            m.r_dep[i] == 0,
        ]

        # return a list disjuncts
        return [disjunct_1, disjunct_2]

    return m

m = ev_plan(stations, 0, 2000)
pyo.TransformationFactory("gdp.bigm").apply_to(m)
SOLVER.solve(m)
```

Figure 3.2(a) summarizes the charging stations' locations and charging capacities for the specific instance implemented above. The optimal charging strategy is reported in Table 3.1 and visualized in Figure 3.2(b).

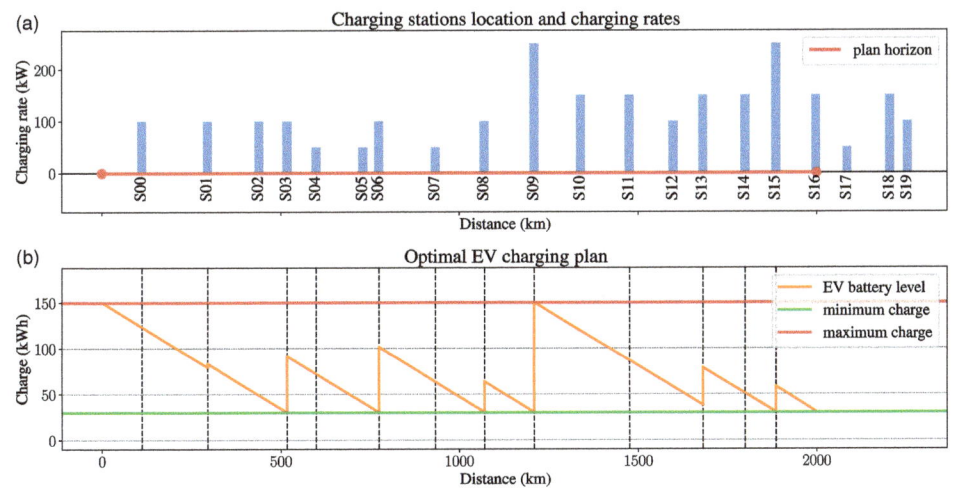

Figure 3.2 Visualization of the optimal EV changing strategy.

Table 3.1 *Details of the optimal EV charging strategy.*

	location	t^{arr}	t^{dep}	c^{arr}	c^{dep}	t^{stop}
0	0.00	–	0.00	–	150.00	–
1	112.20	1.12	1.12	123.07	123.07	0.00
2	294.90	2.95	3.16	79.22	83.38	0.21
3	439.50	4.60	4.60	48.67	48.67	0.00
4	517.30	5.38	6.16	30.00	91.51	0.78
5	598.30	6.97	6.97	72.07	72.07	0.00
6	729.50	8.28	8.28	40.58	40.58	0.00
7	773.60	8.73	9.61	30.00	101.30	0.88
8	933.00	11.20	11.20	63.05	63.05	0.00
9	1070.70	12.58	13.08	30.00	63.58	0.50
10	1210.60	14.48	15.12	30.00	150.00	0.65
11	1340.30	16.42	16.42	118.87	118.87	0.00
12	1475.40	17.77	17.94	86.45	86.78	0.17
13	1597.40	19.16	19.16	57.50	57.50	0.00
14	1680.50	19.99	20.43	37.56	78.72	0.44
15	1798.10	21.61	21.61	50.50	50.50	0.00
16	1883.50	22.46	22.74	30.00	57.96	0.28
17	1998.00	23.89	23.89	30.48	30.48	0.00
18	2000.00	23.91	–	30.00	–	–

3.2.6 Products of Variables

If the optimization problem contains the product of two variables, in a few special cases, it still is possible to "linearize" it at the cost of adding a new variable and a set of additional constraints.

The product $x_1 x_2$ of two binary variables $x_1, x_2 \in \mathbb{B}$ can be replaced by a new variable y and the following additional constraints:

$$y \leq x_1$$
$$y \leq x_2$$
$$y \geq x_1 + x_2 - 1$$
$$y \in \mathbb{B}.$$

Similarly, the product $x_1 x_2$ with $x_1 \in \mathbb{B}$ and $0 \leq l \leq x_2 \leq u$ can be replaced by a new variable y and the following additional constraints:

$$y \leq u x_1$$
$$y \geq l x_1$$
$$y \leq x_2 - l(1 - x_1)$$
$$y \geq x_2 - u(1 - x_1)$$
$$y \in \mathbb{R}$$
$$x_1 \in \mathbb{B}$$
$$x_2 \geq 0.$$

3.2.7 Disjunctive Modeling

Having seen several examples of transforming logical relationships into MILO constraints, it should now be clear that although useful, they are somewhat artificial. Indeed, (i) they require determining the value of the constant M and (ii) it might not be straightforward to immediately recognize from the final problem formulation whether they correspond to an "either–or" or "if–then" constraint.

The advantage of formulating constraints in this way is the versatility, since a great variety of generic MILO solvers can be used. Sometimes, however, it may be beneficial to "remember" the logical structure when formulating the model. This is the core idea of disjunctive programming, which is a class of optimization problems that include disjunctive ("or") constraints.

The benefit of using a disjunctive programming formulation lies in its ability to preserve and leverage the intrinsic logical framework of the problem at hand, thereby mitigating its combinatorial complexity. This advantage is most effectively elucidated through the subsequent example.

Example 3.7 (Multi-product factory optimization) A small production facility produces two products, X and Y. With current technology α, the facility is subject to the following conditions and constraints:

- Product X requires 1 hour of labor A, 2 hours of labor B, and $100 of raw material. Product X sells for $27 per unit. Daily demand is limited to 40 units.
- Product Y requires 1 hour of labor A, 1 hour of labor B, and $90 of raw material. Product Y sells for $210 per unit with unlimited demand.
- There are 50 hours per day of labor A available at a cost of $50/hour.
- There are 100 hours per day of labor B available at a cost of $40/hour.

Based on the available data, we deduce that the net profit per unit for X and Y amounts to $40 and $30, respectively.

Consider the following production problem:

$$
\begin{aligned}
\max \quad & 40x + 30y \\
\text{s.t.} \quad & x \leq 40 && \text{(demand)} \\
& x + y \leq 80 && \text{(labor A)} \\
& 2x + y \leq 100 && \text{(labor B)} \\
& x, y \geq 0,
\end{aligned}
$$

and its Pyomo implementation:

```
m = pyo.ConcreteModel("Multi-Product Factory")

m.production_x = pyo.Var(domain=pyo.NonNegativeReals)
m.production_y = pyo.Var(domain=pyo.NonNegativeReals)
```

```
@m.Objective(sense=pyo.maximize)
def maximize_profit(m):
    return 40 * m.production_x + 30 * m.production_y


@m.Constraint()
def demand(m):
    return m.production_x <= 40


@m.Constraint()
def laborA(m):
    return m.production_x + m.production_y <= 80


@m.Constraint()
def laborB(m):
    return 2 * m.production_x + m.production_y <= 100


SOLVER.solve(m)
```

The optimal solution is $(x,y) = (20,60)$, which results in a profit of $2600.

Labor B is a relatively high cost for the production of product X. Suppose a new technology β has been developed with the potential to cut costs by reducing the time required to finish product X to 1.5 hours, but requires more highly skilled labor with a unit cost of $60 per hour.

The net profit for a unit of product X with technology α is equal to $270 - 100 - 50 - 2 \cdot 40 = \$40$, while with technology β is equal to $270 - 100 - 50 - 1.5 \cdot 40 = \$60$.

We need to assess whether the new technology is beneficial, that is, whether adopting it would lead to higher profits. The decision here is whether to use technology α or β.

In this situation we have an "either–or" structure for both the objective and the labor B constraint:

$$\underbrace{p = 40x + 30y, \ 2x + y \le 100}_{\alpha \text{ technology}} \quad \text{or} \quad \underbrace{p = 60x + 30y, \ 1.5x + y \le 100}_{\beta \text{ technology}}.$$

There are several commonly used techniques for embedding disjunctions in mixed-integer linear optimization problems.

The first approach using the "big-M" technique introduces a single binary decision variable z associated with choosing technology α ($z = 0$) or technology β ($z = 1$). Using MILO, we can formulate this problem as follows:

$$
\begin{aligned}
\max \quad & p \\
\text{s.t.} \quad & x \le 40 & \text{(demand)} \\
& x + y \le 80 & \text{(labor A)} \\
& p \le 40x + 30y + Mz & \text{(profit with technology } \alpha)
\end{aligned}
$$

$$p \leq 60x + 30y + M(1-z) \qquad \text{(profit with technology } \beta)$$
$$2x + y \leq 100 + Mz \qquad \text{(labor B with technology } \alpha)$$
$$1.5x + y \leq 100 + M(1-z) \qquad \text{(labor B with technology } \beta)$$
$$x, y \geq 0$$
$$z \in \mathbb{B}$$
$$p \in \mathbb{R},$$

where the variable $z \in \mathbb{B}$ "activates" the constraints related to the old or new technology, respectively, and M is a large enough constant. It can be implemented in Pyomo as follows:

```python
m = pyo.ConcreteModel("Multi-Product Factory - MILO formulation")

m.profit = pyo.Var(domain=pyo.NonNegativeReals)
m.production_x = pyo.Var(domain=pyo.NonNegativeReals)
m.production_y = pyo.Var(domain=pyo.NonNegativeReals)

m.z = pyo.Var(domain=pyo.Binary)
M = 10000


@m.Objective(sense=pyo.maximize)
def maximize_profit(m):
    return m.profit


@m.Constraint()
def profit_constr_1(m):
    return m.profit <= 40 * m.production_x + 30 * m.production_y + M * m.z


@m.Constraint()
def profit_constr_2(m):
    return m.profit <= 60 * m.production_x + 30 * m.production_y + M * (1 -
    ↪   m.z)


@m.Constraint()
def demand(m):
    return m.production_x <= 40


@m.Constraint()
def laborA(m):
    return m.production_x + m.production_y <= 80
```

```
@m.Constraint()
def laborB_1(m):
    return 2 * m.production_x + m.production_y <= 100 + M * m.z


@m.Constraint()
def laborB_2(m):
    return 1.5 * m.production_x + m.production_y <= 100 + M * (1 - m.z)


SOLVER.solve(m)
```

Alternatively, we can formulate our problem using a disjunction, preserving the logical structure, as follows:

$$
\begin{aligned}
\max \quad & p \\
\text{s.t.} \quad & x \le 40 & \text{(demand)} \\
& x + y \le 80 & \text{(labor A)} \\
& \begin{bmatrix} p = 40x + 30y \\ 2x + y \le 100 \end{bmatrix} \veebar \begin{bmatrix} p = 60x + 30y \\ 1.5x + y \le 100 \end{bmatrix} \\
& x, y \ge 0.
\end{aligned}
$$

This formulation, should the software be capable of handling it, has the benefit that the solver can intelligently partition the problem's solution into various subcases, based on the given disjunction. Pyomo natively supports disjunctions, as illustrated in the following implementation.

```
m = pyo.ConcreteModel("Multi-Product Factory - Disjunctive Programming")

m.profit = pyo.Var(bounds=(-1000, 10000))
m.x = pyo.Var(domain=pyo.NonNegativeReals, bounds=(0, 1000))
m.y = pyo.Var(domain=pyo.NonNegativeReals, bounds=(0, 1000))


@m.Objective(sense=pyo.maximize)
def maximize_profit(m):
    return m.profit


@m.Constraint()
def demand(m):
    return m.x <= 40


@m.Constraint()
def laborA(m):
    return m.x + m.y <= 80
```

```
# Define a disjunction using Pyomo's Disjunction component
# The 'xor=True' indicates that only one of the disjuncts must be true
@m.Disjunction(xor=True)
def technologies(m):
    # The function returns a list of two disjuncts
    # each containing a profit and a constraint
    return [
        [m.profit == 40 * m.x + 30 * m.y, 2 * m.x + m.y <= 100],
        [m.profit == 60 * m.x + 30 * m.y, 1.5 * m.x + m.y <= 100],
    ]


# Transform the Generalized Disjunctive Programming (GDP) model using
# the big-M method into a MILO problem and solve it
pyo.TransformationFactory("gdp.bigm").apply_to(m)
SOLVER.solve(m)
```

The new optimal solution is $(x,y) = (40,40)$, which results in a profit of \$3600.

Disjunctive programming is particularly powerful when there are multiple logical constraints that are related to each other, that is, where there are many logical constraints that interact/might restrict the options for other logical constraints. The following example is a good illustration.

Example 3.8 (Machine scheduling) Consider the problem of scheduling a set of jobs on a single machine given the release time, duration, and due time for each job. Our goal is to find a sequence of the jobs on the machine that meets the due dates. If no such schedule exists, then the objective is to find the "least bad" schedule that minimizes a designated performance metric. The problem data, summarized in Table 3.2, are taken from [GPS02, chapter 5].

Table 3.2 *Summary of the time constraints of all jobs.*

j	release$_j$	duration$_j$	due$_j$
A	2	5	10
B	5	6	21
C	4	8	15
D	0	4	10
E	0	2	5
F	8	3	15
G	9	2	22

There exist many well-known empirical rules for scheduling jobs on a single machine, such as first in, first out (FIFO) and earliest due date (EDD). Figure 3.3 shows the scheduling resulting from implementing the FIFO rule using a Gantt chart. The time window in which each job should be completed is shown in gray, while the actual schedule for the job is either green or red, depending on whether it is within that window or not.

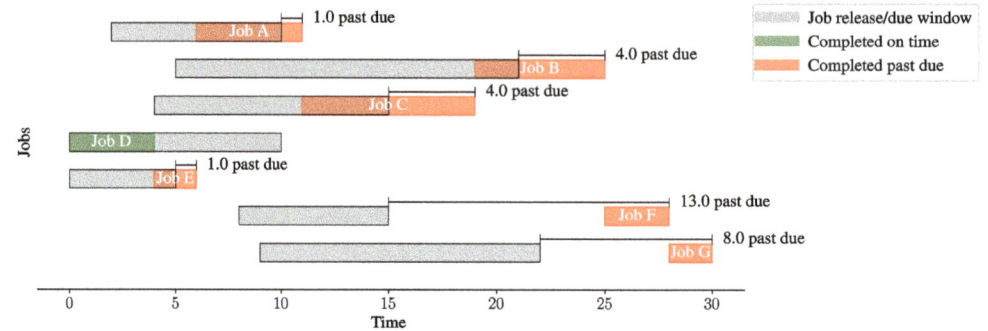

Figure 3.3 The schedule obtained using the FIFO policy for the instance described in Table 3.2. The resulting total past due time is 31.

Following a specific rule, we are not guaranteed to obtain an optimal schedule. This motivates us to formulate a proper optimization problem. Let us introduce the following three decision variables for every job j, with the essential one being the time at which the job starts processing:

- $\text{start}_j \geq 0$ is the time when job j starts;
- $\text{finish}_j \geq 0$ is the time when job j finishes;
- $\text{past}_j \geq 0$ is how long job j is past due.

Depending on application and circumstances, one could have many different choices for the objective function. Minimizing the number of past due jobs, minimizing the maximum past due time, or the total amount of time past due would all be appropriate objectives.

We choose here to minimize the total time past due, that is,

$$\min \sum_j \text{past}_j.$$

The constraints describe the relationships among the decision variables. For example, a job cannot start until it is released for processing:

$$\text{start}_j \geq \text{release}_j.$$

Once started, the processing continues until the job is finished. The completion time is compared with the due time, and the result is stored in the decision variable past_j. These decision variables are needed to handle cases where it might not be possible to complete all jobs by their due time:

$$\text{finish}_j = \text{start}_j + \text{duration}_j$$
$$\text{past}_j \geq \text{finish}_j - \text{due}_j$$
$$\text{past}_j \geq 0.$$

The final set of constraints requires that no pair of jobs operate on the same machine simultaneously. For this purpose, we consider each unique pair (i, j) where the constraint $i < j$ is imposed to avoid considering the same pair twice. Then for any unique pair i and j, either i finishes before j starts or j finishes before i starts. This is expressed as the family of disjunctions:

$$\left[\text{finish}_i \leq \text{start}_j\right] \underline{\vee} \left[\text{finish}_j \leq \text{start}_i\right], \qquad \forall i < j.$$

Such constraints can be enforced using the extra binary variable $z_{ij} \in \mathbb{B}$ as

$$\text{finish}_i \leq \text{start}_j + M z_{ij}$$
$$\text{finish}_j \leq \text{start}_i + M(1 - z_{ij}).$$

However, all these disjunctions are highly related to each other because if we have that

$$\text{finish}_i \leq \text{start}_j, \quad \text{finish}_j \leq \text{start}_k,$$

then automatically we must also have

$$\text{finish}_i \leq \text{start}_k,$$

which disjunctive programming is able to utilize very efficiently, as opposed to standard MILO reformulations blind to this structure.

The disjunctive programming version of the model can be implemented as follows:

```python
# The problem data as a nested Python dictionary of jobs.
# Each job is labeled by a key and is associated to a dictionary with:
# - the time at which the job is released to the for machine processing,
# - the expected duration of the job, and
# - the due date.

jobs = pd.DataFrame(
    {
        "A": {"release": 2, "duration": 5, "due": 10},
        "B": {"release": 5, "duration": 6, "due": 21},
        "C": {"release": 4, "duration": 8, "due": 15},
        "D": {"release": 0, "duration": 4, "due": 10},
        "E": {"release": 0, "duration": 2, "due": 5},
        "F": {"release": 8, "duration": 3, "due": 15},
        "G": {"release": 9, "duration": 2, "due": 22},
    }
).T


def machine_schedule(jobs):
    m = pyo.ConcreteModel("Job machine scheduling")

    # Create a set of jobs using the dataframe index and a
    # set of orderd pairs of distinct jobs (i,j) with i < j
    m.JOBS = pyo.Set(initialize=jobs.index)
```

```python
    m.PAIRS = pyo.Set(initialize=m.JOBS * m.JOBS, filter=lambda m, i, j: i <
    ↪  j)

    # We set an upper bound on the time horizon to 100
    m.maxtime = pyo.Param(initialize=100)
    m.start = pyo.Var(m.JOBS, domain=pyo.NonNegativeReals, bounds=(0,
    ↪  m.maxtime))
    m.finish = pyo.Var(m.JOBS, domain=pyo.NonNegativeReals, bounds=(0,
    ↪  m.maxtime))
    m.past = pyo.Var(m.JOBS, domain=pyo.NonNegativeReals, bounds=(0,
    ↪  m.maxtime))

    @m.Constraint(m.JOBS)
    def job_release(m, job):
        return m.start[job] >= jobs.loc[job, "release"]

    @m.Constraint(m.JOBS)
    def job_duration(m, job):
        return m.finish[job] == m.start[job] + jobs.loc[job, "duration"]

    @m.Constraint(m.JOBS)
    def past_due_constraint(m, job):
        return m.past[job] >= m.finish[job] - jobs.loc[job, "due"]

    @m.Disjunction(m.PAIRS, xor=True)
    def machine_deconflict(m, job_a, job_b):
        return [m.finish[job_a] <= m.start[job_b], m.finish[job_b] <=
        ↪  m.start[job_a]]

    @m.Objective(sense=pyo.minimize)
    def minimize_past(m):
        return sum(m.past[job] for job in m.JOBS)

    return m


m = machine_schedule(jobs)
pyo.TransformationFactory("gdp.bigm").apply_to(m)
SOLVER.solve(m)

optimalschedule = pd.DataFrame(
    {
        "start": m.start.extract_values(),
        "finish": m.finish.extract_values(),
        "past": m.past.extract_values(),
    }
)
```

The solution obtained by solving the optimization problem outperforms that derived from the FIFO policy; compare Figure 3.3. Nonetheless, heuristic techniques become essential when dealing with large scheduling problems.

3.3 Solution Methods

MILO problems are much more difficult to solve than LO problems. At the same time, the requirements for a solution method remain the same: to find the optimal solution quickly, along with a certificate of optimality. How can one go about this? A straightforward yet brute-force approach is to inspect each potential value of the integer variables and solve an LO problem for the remaining continuous variables (if any).

Example 3.9 (Building microchips – Brute-force solution) If we were to enumerate all integer solutions in the feasible set for Example 3.1, we would end up with $1,150,751$ (x_1, x_2) pairs. For each of them, we can compute the corresponding value of the objective function and identify the optimum. For such a small two-dimensional problem, such a brute-force search quickly returns the optimal solution, but this approach rapidly becomes impractical when the number of decision variables and constraints grows large.

However, for problems with many dimensions and/or many feasible solutions, such an approach would be too time-consuming. We will now present the two most well-known and popular methods for solving MILO problems. Both use, as their key building block, LO-solving algorithms applied to the **linear relaxation** of the given MILO problem, which is the problem obtained "relaxing" the integrality constraints on x, that is, replacing it with $x \geq 0$ instead.

Since the feasible set of a MILO problem is a subset of the feasible set of its linear relaxation, it should be clear that in the case of minimization, the optimal value of the linear relaxation is less than or equal to the optimal value of MILO. To exclude such "good" solutions that violate the integrality constraints, both methods skillfully add extra constraints when needed.

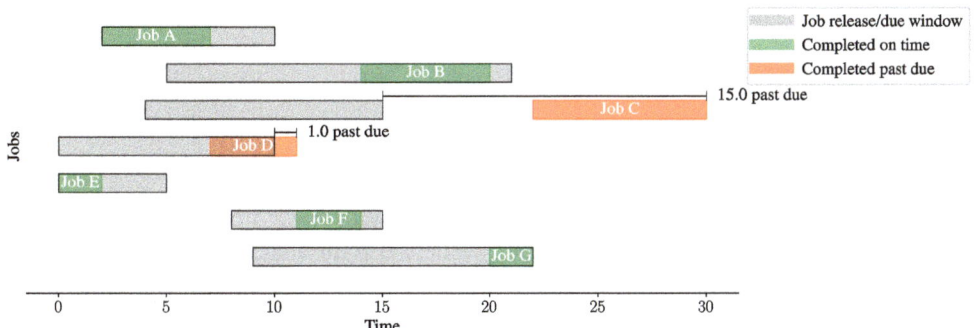

Figure 3.4 The optimal schedule for the instance described in Table 3.2. The resulting total past due time is 16.

Branch and Bound

The most common solution method is called **branch and bound**. Here, first, the linear relaxation of the MILO is solved. Clearly, if the optimal solution of the relaxed problem is integer in the variables that were originally required to be integer, an optimal solution to the original problem has been found – this situation, although rare, occurs in problems with special structure, some examples of which we discuss in Chapter 4.

The optimal value of the linear relaxation provides a lower bound to the optimal value of the MILO. At this moment, if we can, we use any heuristic to construct a primal feasible solution to the (unrelaxed) MILO problem to obtain an upper bound ("some" solution is always at least as good as an "optimal" solution). Again, if this solution attains the lower bound, it is an optimal solution to the MILO problem.

If the two bounds differ, we proceed with the *branching* step, which consists of selecting an integer variable x_j that has a fractional value $\bar{x}_j$ in the optimal solution of the LO problem and creating two new subproblems, one with a constraint $\bar{x}_j \leq \lfloor \bar{x}_j \rfloor$, and the other with $x_j \geq \lceil \bar{x}_j \rceil$.

Both subproblems have a stricter feasible region than the original one, and the optimum to the original MILO must lie in one of the branches. For both subproblems, we can obtain bounds just as in the first step and improve (tighten) the global lower and upper bounds. These steps are repeated until we have found the optimal MILO solution.

The advantage of the branch-and-bound method over an enumeration of all possible solutions is the use of the lower/upper bounds. These allow us to "prune" certain branches from consideration, and thus many of the possible solutions do not need to be checked. There are three reasons to prune: by infeasibility, by (local) optimality, and by bound.

First, if any subproblem is infeasible, it should be discarded. Second, if the lower and upper bounds for a subproblem are equal, we have determined an optimal integer solution for that subproblem and we do not further branch this subproblem, since we cannot improve any further (hence the terminology "local optimal"). Third, the bounds for a subproblem tell us what is the best possible objective value for all integer solutions in the corresponding feasible region. If this best possible objective value is not better than the best feasible solution found so far anywhere else in the tree, we can discard this subproblem again.

An example of the branch-and-bound process is given in Figures 3.6 and 3.7 for the following LO problem:

$$\begin{aligned}
\max \quad & 2x_1 + 3x_2 & (3.4)\\
\text{s.t.} \quad & 2x_1 + x_2 \leq 10 \\
& 3x_1 + 6x_2 \leq 40 \\
& x_1, x_2 \geq 0 \\
& x_1, x_2 \in \mathbb{N}.
\end{aligned}$$

The performance of the branch-and-bound method can greatly improve by having a tight problem formulation, that is, one where the gap between the feasible region of the original problem and its linear relaxation is small. Such a tight formulation typically leads to better bounds and, therefore, more pruning of subproblems, speeding up computation. The extreme case is to use the convex hull for the formulation. The convex hull of a set S is the smallest convex set containing S or, equivalently, the set of all convex combinations of its points; see Figure 3.5.

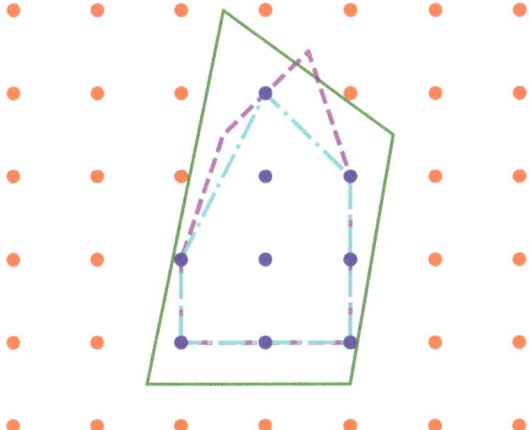

Figure 3.5 Three possible feasible regions (in green, purple, and blue) for the linear relaxation of a MILO problem with nine feasible points (in blue). The blue region is the convex hull of these points. If the problem is formulated so that the feasible region is the blue one, the linear relaxation would be exact and find the optimal solution of the original MILO.

Figure 3.5 illustrates three distinct **polytopes**, each corresponding to a different set of inequalities. These polytopes correspond to three separate MILO problems. Remarkably, these three MILO problems share an identical optimal solution, since the feasible regions, defined as the integer points enclosed within these polytopes, perfectly coincide. On the other hand, the three corresponding linear relaxations are different.

Given a MILO problem, there always exists a polytope corresponding to the convex hull of the feasible region (i.e., the blue one in Figure 3.5). If the model happens to be expressed in this form, there is no need to enforce integrality constraints on the variables for the optimal solution to have integral values. This property is referred to as being "naturally integral." This property underscores the significance of the specific formulation of MILO problems, where certain representations can lead to more efficient solutions without the need for additional integrality constraints. In Chapter 4 we show that some important optimization models are naturally integral, a feature that makes them very special.

However, finding such a polytope if the model is expressed in a different way (say, with feasible green or purple feasible regions) is not easy. Finding a tight formulation for a specific problem is an art that requires experience with mathematical modeling. Some "rules of thumb" here are to (i) have as few as possible big-M constraints with the large constants M, (ii) avoid having superfluous decision variables, and (iii) avoid formulating as inequality a relation that should hold with equality for the optimal solution.

In theory, each MILO has an equivalent LO formulation, for example, the convex hull of the integer feasible region. Solving the linear relaxation with the convex-hull formulation would immediately lead to the optimal integer solution. However, this formulation is either difficult to find or requires excessively many variables and/or constraints. Nevertheless, if we have the convex-hull formulation, we can simply use a linear optimization method to solve our MILO.

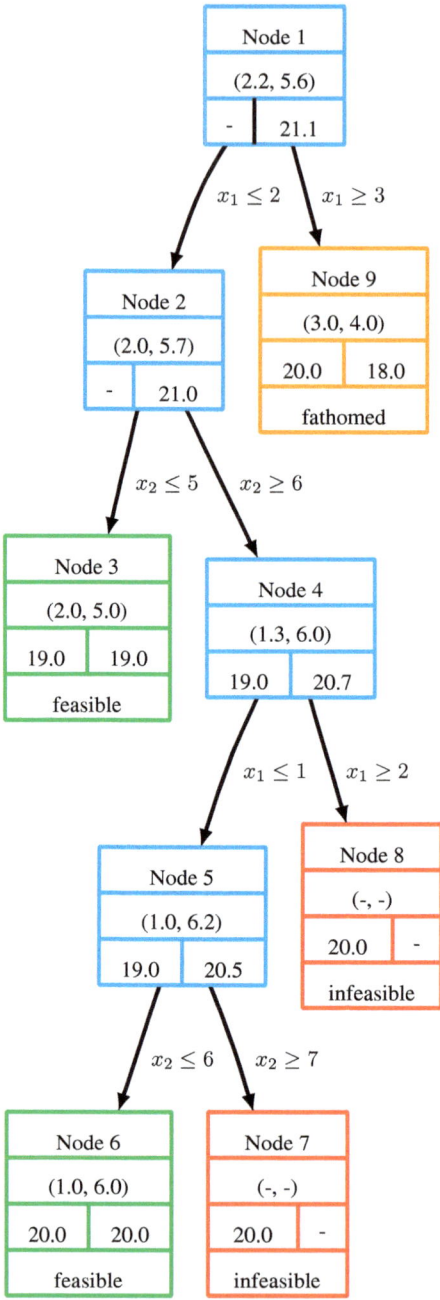

Figure 3.6 A branch-and-bound tree solving Problem (3.4) with a summary of each node's values and the cuts added at each branch.

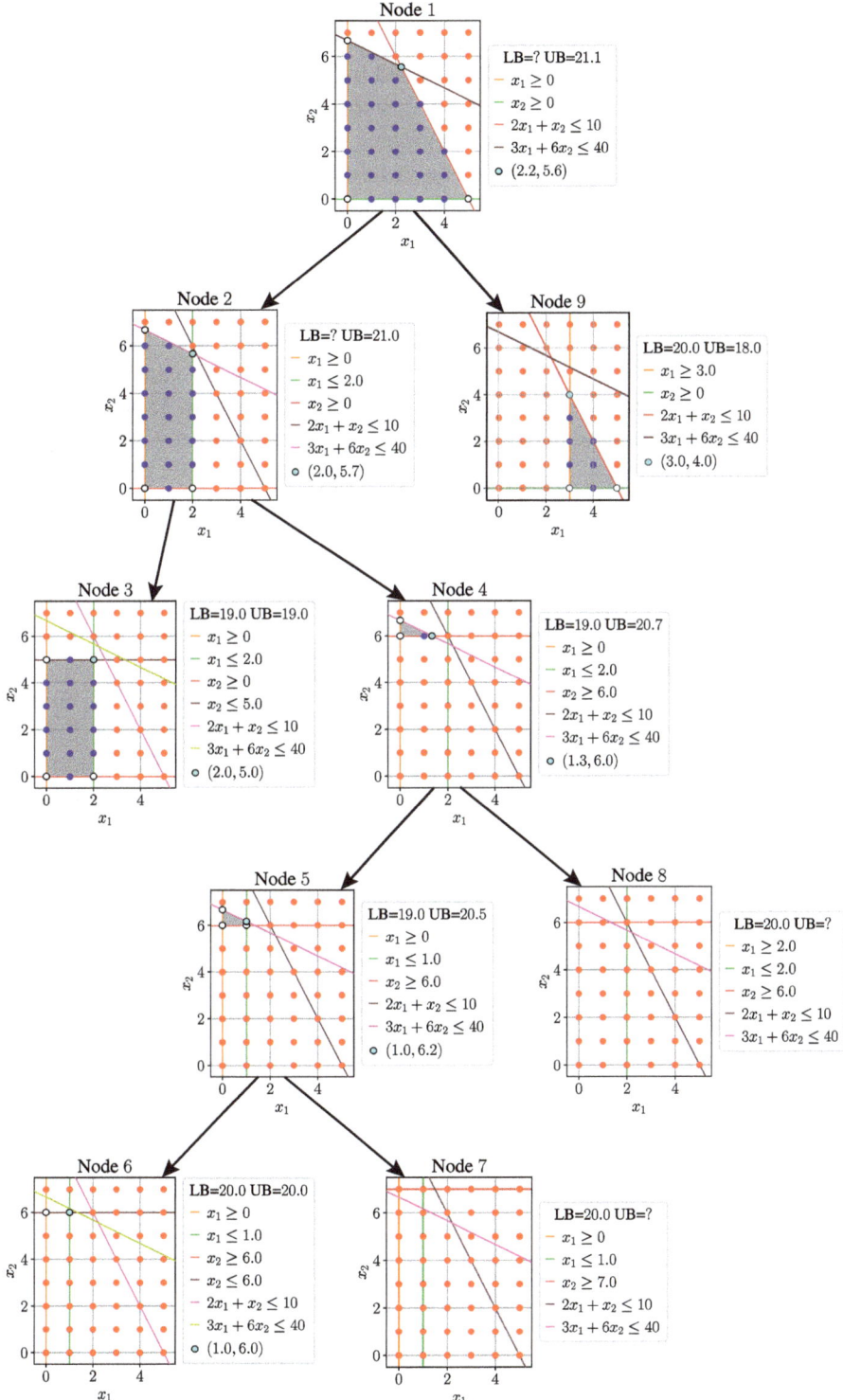

Figure 3.7 Details on which problem is solved at each node of the branch-and-bound tree for Problem (3.4) including their graphic resolutions.

Cutting Planes

Another solution method is to directly address the gap between the original problem and its linear relaxation. An inequality $a^\top x \le b$ is a **valid inequality** for a MILO if $a^\top x \le b$ holds for all x in its (integer) feasible region. In other words, valid inequalities cut off noninteger points from the linear relaxation's feasible region.

However, to use this method, one needs to know how to generate such inequalities efficiently. A systematic way to generate valid inequalities is to use **Chvátal–Gomory cuts**. Consider the constraints $Ax \le b$ with $x \in \mathbb{N}^n$, where $A \in \mathbb{R}^{m \times n}$. The expression $\lambda^\top A x \le \lambda^\top b$ with $\lambda \in \mathbb{R}^m_{\ge 0}$ is a valid inequality. Since x is integer and nonnegative, $\lfloor \lambda^\top A \rfloor x \le \lambda^\top b$ is valid as well. Now realize that the right-hand side can be rounded as well: $\lfloor \lambda^\top A \rfloor x \le \lfloor \lambda^\top b \rfloor$, which is a Chvátal–Gomory cut.

In certain cases, iteratively determining useful valid inequalities can be done efficiently, which opens the way to a cutting-plane method to solve such problems. This method iteratively solves the linear relaxation and adds new suitable valid inequalities whenever a fractional solution is found. However, a "pure" cutting plane method typically only works for very specific problems and will not converge fast enough in general.

Performance of MILO Algorithms

Most efficient approaches to general MILO problems combine branch and bound with valid inequalities, that is, they simultaneously analyze the branch-and-bound tree and add valid inequalities to the resulting subproblems.

The performance of a particular algorithm is often intricately tied to the initial model formulation. Crafting a suitable problem formulation is a blend of art and science, requiring a combination of experience and intuition. This process also involves familiarity with the intricacies of the specific solver, as the computational efficiency ultimately hinges on the intricacies of the solver's inner workings.

For instance, it might be intuitive to assume that a good problem formulation should entail a minimal number of decision variables and/or constraints. However, the following example illustrates that the intuitive belief that an optimal problem formulation involves the fewest possible constraints can sometimes be misleading.

Example 3.10 (Facility location – Fewer constraints are not always good for MILO)
Consider the problem of a supplier facing the task of fulfilling specific customer demands with minimal costs while simultaneously deciding how many facilities to build and where. Given a set I of customers and a set J of possible locations, let c_j be the cost of building facility j and h_{ij} be the cost incurred to satisfy the demands of customer i at facility j. Introduce two sets of binary variables,

$$x_j := \begin{cases} 1 & \text{if facility } j \text{ is built} \\ 0 & \text{otherwise,} \end{cases}$$

and

$$y_{ij} := \begin{cases} 1 & \text{if customer } i \text{ is served at facility } j \\ 0 & \text{otherwise.} \end{cases}$$

The resulting MILO is

$$\min \quad \sum_{j \in J} c_j x_j + \sum_{i \in I} \sum_{j \in J} h_{ij} y_{ij}$$

$$\text{s.t.} \quad \sum_{j \in J} y_{ij} = 1, \qquad \forall i \in I \qquad \text{(every customer is served)}$$

$$y_{ij} \le x_j, \qquad \forall i \in I, \forall j \in J \qquad \text{(facility built before use)}$$

$$x_j \in \mathbb{B}, \qquad \forall i \in I$$

$$y_{ij} \in \mathbb{B}, \qquad \forall i \in I, \forall j \in J.$$

Let $n = |J|$ be the number of possible facility locations and $m = |I|$ the number of customers. Note that, since $\sum_j y_{ij} = 1$ for every $i \in I$, we can replace the $n \times m$ constraints $y_{ij} \le x_j$ by only n constraints, namely

$$\sum_i y_{ij} \le m x_j, \qquad \forall j \in J.$$

This approach leads to a more concise mathematical formulation of the model. However, it may not be a good idea if we want to solve the problem using its linear relaxation. Indeed, by reducing the number of constraints, we inadvertently made the feasible region of the relaxation larger and less tight around the feasible integer points. This fact becomes clearly evident in the increased run-time required to solve the optimization problem when working with these weaker constraints, see Figure 3.8.

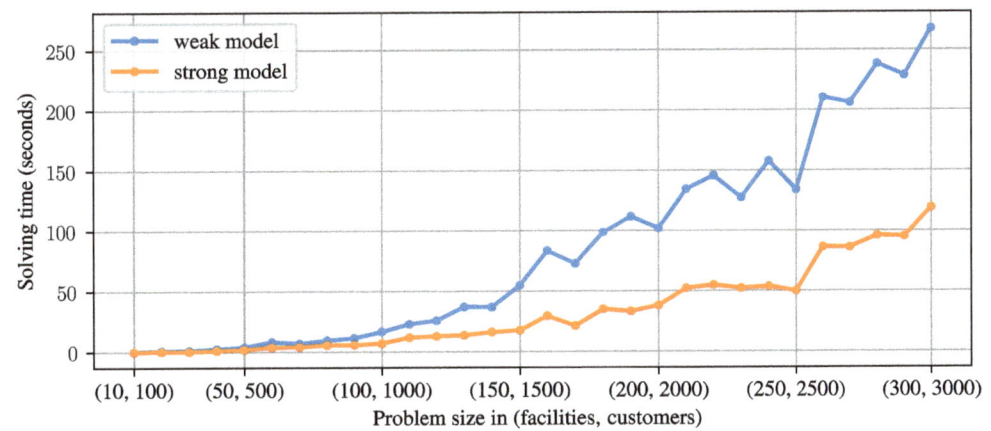

Figure 3.8 Comparison of the run-times of the commercial solver Gurobi for the two different problem formulations for identical instances of increasing size. We disable cuts to get a more transparent comparison of the solving time.

3.4 A Complete Example: BIM Production Revisited

We again consider the BIM raw material planning problem from Section 2.5 but with more sophisticated pricing and acquisition protocols. There are now three suppliers, each of which can deliver the following materials:

- A: silicon, germanium, and plastic;
- B: copper;
- C: all of the above.

For the suppliers, the following conditions apply: copper should be acquired in multiples of 100 g since it is delivered in sheets of that weight.

Unitary products such as silicon, germanium, and plastic may be acquired in any real nonnegative number, but the price is charged for each "started" batch of 100 units, regardless of whether the full batch was used or not. This means that 30 units of silicon with 10 units of germanium and 50 units of plastic cost as much as 1 unit of silicon but half as much as 30 units of silicon with 30 units of germanium and 50 units of plastic.

Furthermore, supplier C sells all products and offers a discount if two products are purchased together: 100 g of copper and a batch of 100 units of a unitary product cost together only 7. This set price is only applied to pairs, which means that 100 g of copper and two batches cost 13. The summary of the prices in $ is given in the following table:

Supplier	Copper per sheet of 100 g	Batches	Together
A	–	5	–
B	3	–	–
C	4	6	7

Next, for stocked products, the inventory costs are summarized in the following table:

Copper per 10 g	Silicon per unit	Germanium per unit	Plastic per unit
0.1	0.02	0.02	0.02

The holding price of copper is per 10 g, and the copper stocked is rounded up to multiples of 10 g, which means that 12 g cost as much as 20 g. The capacity limitations of the warehouse allow for a maximum of 10 kg of copper in stock at any time, but there are no practical limitations to the number of units of unitary products in stock. Production is made-to-order, which means that no inventory of chips is kept.

Recall the inventory of each material at the beginning of the year and the desired inventory left at the end of the year:

	Copper	Silicon	Germanium	Plastic
initial inventory	480	1000	1500	1750
end-of-the-year minimum	200	500	500	1000

The goal is to build an optimization model using these data and solve it to minimize the acquisition and holding costs of the products while meeting the required quantities for production.

As usual, we need to start modeling by naming our decisions and identifying the relevant index sets. In Section 2.5 we already had a set T time periods (i.e., the months) and a set P

for the products. Since we want to keep track of how much we buy for which material and from which supplier, we need an additional index set S for the suppliers.

In the complete example in Section 2.5 all decision variables were continuous, but now we also need to take care of several integer decision variables:

- number of batches of the unitary products;
- number of copper sheets;
- number of "volume units" in stock, since the inventory of copper pays per $10\,\mathrm{g}$;
- number of copper sheet-batch of unitary product pairs acquired in the same month from supplier C, yielding discount.

Given the above, we can reuse the variables s_{pt} from the complete example in Section 2.5 to account for the total number of grams/units of each product. However, with respect to purchasing variables, we now need to keep track of unitary products and suppliers. For this reason, we introduce integer variables y_{ts} for the number of copper sheets purchased from supplier s, and continuous variables x_{pts} for the number of units of unitary product p purchased from supplier s. To translate these variables into "amounts of a product purchased regardless of the supplier," we introduce another variable u_{pt}.

Further, we introduce the variables that we use to account for the "tricky" integer quantities in our model. To account for the number of batches of unitary products at supplier s, we introduce integer variables b_{ts}, and, similarly, we introduce an integer variable r_t for the number of batches of $10\,\mathrm{g}$ of copper in the inventory. Buying a copper sheet together with a batch of unitary products from supplier C is advantageous, as we get a discount of $\beta = 4 + 6 - 7$. For this reason, we introduce an extra integer variable p_t to count the number of such pairs, so that βp_t describes the amount to the total "discount" acquired in month t. We need to include additional constraints to make sure that these three variables "take the values they should" in relation to the purchasing and inventory variables.

Taking all the above into account, we shall use the following decision variables:

- y_{ts}: number of sheets of copper purchased in month t from supplier s;
- x_{pts}: number of units of unitary product p purchased in month t from supplier s;
- u_{pt}: total number of units of product p purchased in month t (from all suppliers combined);
- b_{ts}: number of batches of 100 unitary products purchased in month t from supplier s;
- p_t: number of batch–copper sheet pairs purchased in month t from supplier C;
- r_t: number of 10-unit amounts of copper (rounded up) in month t.

We begin by formulating the objective function using \$ cents as unit. If we denote by α_s the per-batch cost of unitary products at supplier s, then the total material acquisition cost is

$$\sum_{t \in T} \left(\sum_{s \in S} \pi_s b_{ts} + \alpha_s y_{ts} \right) - \beta p_t,$$

where we count the acquisition costs for supplier C as the total cost of the sheets of copper plus the batches of unitary materials minus the discount due to purchasing the pairs:

$$\pi_C b_{t,C} + \alpha_C y_{t,C} - \beta p_t.$$

The total inventory cost is

$$\sum_{t \in T} \left(0.1 r_t + \sum_{p \in P \setminus \{\text{copper}\}} h_p s_{pt} \right).$$

Now, we need to construct a system of constraints to ensure that the above costs are computed correctly.

To transform the amounts of unitary materials purchased into batches of 100, we construct the following constraints:

$$\sum_{p \in P \setminus \{\text{copper}\}} x_{pts} \leq 100 b_{ts}.$$

Because the inventory costs for copper are computed by rounding up to $10\,\text{g}$ amounts, we account for this via the following constraint:

$$s_{\text{copper}, t} \leq 10 r_t.$$

In the end, we can account for the number of sheets of copper–unitary material batch pairs purchased at supplier C, using the following constraints:

$$p_t \leq b_{t,\text{C}}, \quad \text{and} \quad p_t \leq y_{t,\text{C}}.$$

Choosing the maximum possible amount p_t that satisfies such constraints will naturally lead to maximizing the number of pairs and, thus, the total discount.

The remaining constraints such as the inventory balance constraints, total inventory capacity, etc., are straightforward, which gives us the following complete model formulation:

$$
\min \; \sum_{t \in T} \left(\left(\sum_{s \in S} \pi_s b_{ts} + \alpha_s y_{ts} \right) - \beta p_t \right) + \sum_{t \in T} \left(\gamma_{\text{copper}} r_t + \sum_{p \in P \setminus \{\text{copper}\}} \gamma_p s_{pt} \right)
$$

$$
\text{s.t.} \quad \sum_{p \in P \setminus \{\text{copper}\}} x_{pts} \leq 100 \cdot b_{ts}, \qquad\qquad \forall\, t \in T, \forall\, s \in S
$$

$$
s_{\text{copper}, t} \leq 10 \cdot r_t, \qquad\qquad \forall\, t \in T
$$

$$
s_{\text{copper}, t} \leq 10{,}000, \qquad\qquad \forall\, t \in T
$$

$$
u_{\text{copper}, t} = 100 \sum_{s \in S} y_{ts}, \qquad\qquad \forall\, t \in T
$$

$$
p_t \leq b_{t,\text{C}}, \quad p_t \leq y_{t,\text{C}}, \qquad\qquad \forall\, t \in T
$$

$$
u_{pt} = \sum_{s \in S} x_{pts}, \qquad\qquad \forall\, t \in T, \forall\, P \setminus \{\text{copper}\}
$$

$$
s_{p, t-1} + u_{p,t} = \delta_{pt} + s_{pt}, \qquad\qquad \forall\, t \in T, \forall\, p \in P
$$

$$
s_{p\text{Dec}} \geq \Omega_p, \qquad\qquad \forall\, p \in P
$$

$$
b_{ts}, r_t \in \mathbb{Z}_+, \qquad\qquad \forall\, t \in T, \forall\, s \in S,
$$

where we assume that $s_{p,t-1}$ represents the initial inventory state. Note that because r_t, b_{ts} appear with a nonnegative coefficient and p_t with a positive coefficient in the objective functions, at the optimal solution these variables will automatically take their minimal (r_t, b_{ts}) and maximal (p_t) values allowed by the corresponding constraints.

This optimization problem has a very special structure, that is, for every month $t \in T$ two types of decisions are made: acquisition and inventory. Therefore, we can see the model as composed of $|T| = 12$ pairs of smaller interconnected models. Pyomo includes the concept of *block*, which is ideal for modeling such types of structures. In our implementation below, we use a Pyomo `block` component to have the model decomposed into two submodels, [A]cquisition and [I]nventory.

```python
def BIMproduction(
    demand,
    existing,
    desired,
    stock_limit,
    supplying_copper,
    supplying_batches,
    price_copper_sheet,
    price_batch,
    discounted_price,
    batch_size,
    copper_sheet_mass,
    copper_bucket_size,
    unitary_products,
    unitary_holding_costs,
):
    m = pyo.ConcreteModel(
        "BIM product acquisition and inventory with sophisticated prices"
    )

    periods = demand.columns
    products = demand.index
    first = periods[0]
    prev = {j: i for i, j in zip(periods, periods[1:])}
    last = periods[-1]

    m.T = pyo.Set(initialize=periods)
    m.P = pyo.Set(initialize=products)

    m.PT = m.P * m.T  # to avoid internal set bloat

    @m.Block(m.T)
    def A(b):
        b.x = pyo.Var(supplying_batches, products,
        ↪   domain=pyo.NonNegativeReals)
        b.b = pyo.Var(supplying_batches, domain=pyo.NonNegativeIntegers)
        b.y = pyo.Var(supplying_copper, domain=pyo.NonNegativeIntegers)
        b.p = pyo.Var(domain=pyo.NonNegativeIntegers)

        @b.Constraint(supplying_batches)
        def in_batches(b, s):
            return pyo.quicksum(b.x[s, p] for p in products) <= batch_size *
            ↪   b.b[s]
```

```python
        @b.Constraint()
        def pairs_in_batches(b):
            return b.p <= b.b["C"]

        @b.Constraint()
        def pairs_in_sheets(b):
            return b.p <= b.y["C"]

        @b.Expression(products)
        def u(b, p):
            if p == "copper":
                return copper_sheet_mass * pyo.quicksum(
                    b.y[s] for s in supplying_copper
                )
            return pyo.quicksum(b.x[s, p] for s in supplying_batches)

        @b.Expression()
        def cost(b):
            discount = price_batch["C"] + price_copper_sheet["C"] -
            ↪ discounted_price
            return (
                pyo.quicksum(price_copper_sheet[s] * b.y[s] for s in
                ↪ supplying_copper)
                + pyo.quicksum(price_batch[s] * b.b[s] for s in
                ↪ supplying_batches)
                - discount * b.p
            )

@m.Block(m.T)
def I(b):
    b.s = pyo.Var(products, domain=pyo.NonNegativeReals)
    b.r = pyo.Var(domain=pyo.NonNegativeIntegers)

    @b.Constraint()
    def copper_in_buckets(b):
        return b.s["copper"] <= copper_bucket_size * b.r

    @b.Constraint()
    def capacity(b):
        return b.s["copper"] <= stock_limit

    @b.Expression()
    def cost(b):
        return unitary_holding_costs["copper"] * b.r + pyo.quicksum(
            unitary_holding_costs[p] * b.s[p] for p in unitary_products
        )

@m.Param(m.PT)
def delta(m, t, p):
```

```
        return demand.loc[t, p]

    @m.Expression()
    def acquisition_cost(m):
        return pyo.quicksum(m.A[t].cost for t in m.T)

    @m.Expression()
    def inventory_cost(m):
        return pyo.quicksum(m.I[t].cost for t in m.T)

    @m.Objective(sense=pyo.minimize)
    def total_cost(m):
        return m.acquisition_cost + m.inventory_cost

    @m.Constraint(m.PT)
    def balance(m, p, t):
        if t == first:
            return existing[p] + m.A[t].u[p] == m.delta[p, t] + m.I[t].s[p]
        else:
            return m.A[t].u[p] + m.I[prev[t]].s[p] == m.delta[p, t] +
            ↪   m.I[t].s[p]

    @m.Constraint(m.P)
    def finish(m, p):
        return m.I[last].s[p] >= desired[p]

    return m
```

We can then pass the parameters and data to this Pyomo model and solve it, as illustrated in the following code snippet.

```
from io import StringIO

demand_data = """
chip, Jan, Feb, Mar, Apr, May, Jun, Jul, Aug, Sep, Oct, Nov, Dec
logic, 88, 125, 260, 217, 238, 286, 248, 238, 265, 293, 259, 244
memory, 47, 62, 81, 65, 95, 118, 86, 89, 82, 82, 84, 66
"""
demand_chips = pd.read_csv(StringIO(demand_data), index_col="chip")
use = dict()
use["logic"] = {"silicon": 1, "plastic": 1, "copper": 4}
use["memory"] = {"germanium": 1, "plastic": 1, "copper": 2}
use = pd.DataFrame.from_dict(use).fillna(0).astype(int)
demand = use.dot(demand_chips)

m = BIMproduction(
    demand=demand,
    existing={"silicon": 1000, "germanium": 1500, "plastic": 1750, "copper":
    ↪   4800},
    desired={"silicon": 500, "germanium": 500, "plastic": 1000, "copper":
    ↪   2000},
```

```
        stock_limit=10000,
        supplying_copper=["B", "C"],
        supplying_batches=["A", "C"],
        price_copper_sheet={"B": 300, "C": 400},
        price_batch={"A": 500, "C": 600},
        discounted_price=700,
        batch_size=100,
        copper_sheet_mass=100,
        copper_bucket_size=10,
        unitary_products=["silicon", "germanium", "plastic"],
        unitary_holding_costs={"copper": 10, "silicon": 2, "germanium": 2,
        ↪  "plastic": 2},
    )
    SOLVER.solve(m)
```

The optimal solution has a total cost of $1102.16. The solution prescribes the acquisitions of unitary products and copper sheets as detailed in the Tables 3.3 and 3.4.

The acquisition plans of units and sheets lead to the following batches and these to the pairs of acquisitions to supplier C, yielding a discount, as described in Tables 3.5 and 3.6.

Table 3.3 *Acquisition plan for unitary materials.*

supplier	materials	Jan	Feb	Mar	Apr	May	Jun	Jul	Aug	Sep	Oct	Nov	Dec
	silicon	0	0	0	0	0	0	0	0	0	0	0	0
A	germanium	0	0	0	0	0	0	0	0	0	0	0	0
	plastic	0	0	0	0	0	0	0	0	0	0	0	0
	silicon	0	0	0	0	0	214	249	237	265	349	257	690
C	germanium	0	0	0	0	0	0	0	0	0	0	0	0
	plastic	0	0	0	0	0	15	251	363	335	351	343	1310

Table 3.4 *Acquisition plan for copper sheets materials.*

supplier	Jan	Feb	Mar	Apr	May	Jun	Jul	Aug	Sep	Oct	Nov	Dec
B	0	0	0	0	0	7	7	5	7	6	6	11
C	0	0	0	0	0	3	5	6	6	7	6	20

Table 3.5 *Acquisition plan for batches.*

supplier	Jan	Feb	Mar	Apr	May	Jun	Jul	Aug	Sep	Oct	Nov	Dec
A	0	0	0	0	0	0	0	0	0	0	0	0
C	0	0	0	0	0	3	5	6	6	7	6	20

Table 3.6 *Acquisition plan for copper sheet and batch pairs from supplier C.*

Jan	Feb	Mar	Apr	May	Jun	Jul	Aug	Sep	Oct	Nov	Dec
0	0	0	0	0	3	5	6	6	7	6	20

Table 3.7 *The stock levels at the end of each month resulting from the optimal acquisition and production plan.*

material	Jan	Feb	Mar	Apr	May	Jun	Jul	Aug	Sep	Oct	Nov	Dec
silicon	912	787	527	310	72	0	1	0	0	56	54	500
plastic	1615	1428	1087	805	472	83	0	36	24	0	0	1000
copper	4354	3730	2528	1530	388	8	44	14	90	54	50	2042
germanium	1453	1391	1310	1245	1150	1032	946	857	775	693	609	543

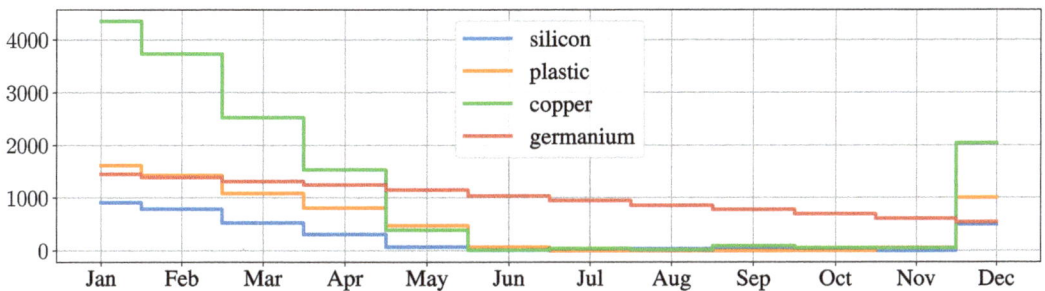

Figure 3.9 The stock levels at the end of each month resulting from the optimal acquisition and production plan.

Table 3.7 conclude the description of the solution with the products kept in stock each month, visualized graphically in Figure 3.9.

Exercises

Exercise 3.1 Recall Example 3.10 and assume that the same supplier has a limited budget for installing the facilities: Only a number p of facilities may be installed. Furthermore, potential customers will only buy from that supplier if the facility is within a given distance threshold of S from the closest facility. A simple example should convince you that serving all potential customers may be infeasible in this case. Therefore, the supplier differentiates the potential customers in terms of importance by assigning each a weight. Since it is not always possible to serve all customers, the objective is to maximize the total weight of the customers served for a given number of installed facilities.

(a) Show how the models from Example 3.10 can be adapted to solve this problem. You should obtain two models: the aggregated one that you have seen in Example 3.10 to be weaker and the deaggregated one that you have seen to be stronger.
(b) The model of the previous question can, in fact, be substantially simplified: Check the well-known paper [CR74] for the so-called **maximum covering** model to see how to express it using single indices. Implement the three models, and analyze their scalability in terms of the instance size on randomly generated problems. What do you observe?
(c) Notice that the models, including the one from [CR74], install exactly p facilities. However, it may happen that maximal attainable coverage is less than 100%, even if installing all facilities. Show how to modify the models to pursue two objectives in strict lexicographic order: First, maximize covering and, next, for the maximum

attainable covering minimize the number of facilities while observing the budget as a maximum.

(d) The **lexicographic optimization** approach of the previous point was a specific way of dealing with the trade-off between costs and benefits. Another common way is to determine the so-called **efficient frontier** or **Pareto frontier** for the highest attainable coverage for each budget. Determine this frontier for an instance of your choice of this problem.

Exercise 3.2 The goal of a **Sudoku** puzzle is to complete a 9×9 grid with digits so that each column, each row, and each of the nine 3×3 subgrids that make up the grid contain all the digits from 1 to 9. Each digit in a row, column, or subgrid must be unique and cannot be repeated. An example of a Sudoku puzzle is shown.

7				2		4	8	
2		6			8			5
5			9					
			1	5				
	2						6	
				6	7			
					6			3
6			5			1		4
	9	3		4				7

(a) Formulate the Sudoku puzzle as an integer optimization model. Formulate the problem as an integer optimization model using the notation stated above. What is the objective function? How can we incorporate the fixed/given elements of a Sudoku into the model?
(b) Your model will give you a solution, if one exists, for each set of given elements. A Sudoku is called *proper* if it has only one feasible solution. How can you establish whether the solution you found is proper?
(c) Solve the Sudoku example above using your MILO formulation and verify that you have obtained a valid solution. Are there multiple feasible solutions?

Exercise 3.3 A company has N projects under consideration. The profit of project i added during the next two years is given by p_i. Project i will require expenditures of c_i during year 1, and d_i during year 2. For year 1, $\$B_1$ are available for projects and $\$B_2$ are available during year 2. It is not possible to undertake a fraction of each project. Moreover, if project 4 is chosen, then project 5 should not be selected. If project 7 is chosen, then project 8 should also be chosen. Finally, project 3 should be selected if both projects 1 and 2 are chosen and may not be chosen when one of the projects 1 or 2 is not chosen. Formulate an integer linear optimization problem to maximize profits.

Exercise 3.4 A manufacturer can produce n different products and sell product i at a profit of $\$p_i$ $(i = 1, \ldots, n)$. Moreover, r_i units of raw material are needed to manufacture 1 unit of product i. A total of R units of raw material are available. If any product i is produced, a setup cost of $\$s_i$ is incurred. The manufacturer can only sell at most M_i units of product i. Formulate an integer optimization problem to maximize profits.

Exercise 3.5 A big city has been divided into d districts. The time (in minutes) it takes an ambulance to travel from district i to district j is given by t_{ij}. The travel time within a district is negligible, so we assume $t_{ii} = 0$. The population of district i is given by p_i. The city has n ambulances and wants to locate them to maximize the number of people who live within α minutes of an ambulance. Since district 1 has a hospital, there should be an ambulance located in district 1. The costs to locate an ambulance in district i are $\$c_i$ per year. The available budget is $\$C$ per year. Formulate an (integer) linear optimization problem to accomplish the goal mentioned above.

Exercise 3.6 There are n cities in a given county. The county must determine where to build fire stations. The county wants to build the minimum number of fire stations needed to ensure that at least one fire station is within 15 minutes (driving time) of each city. The time (in minutes) required to drive between the cities in that county is precomputed and given in a table. Formulate an integer linear optimization model that will define how many fire stations should be built and where they should be located.

Exercise 3.7 In a medium-paced eatery, meal preparation precedes customer arrival due to demand surges and limited cooking capacity. The 8-hour operation is split into 48, 10-minute slots, with an additional hour before opening, adding six more slots.

During each slot, up to C meals can be initiated for cooking, ready for sale at interval $t + 2$, and remain salable until interval $t + 8$, after which they are discarded. At any moment, only C meals can be under preparation, regardless of their start time.

Given the historical average demand d_t for each slot, formulate a MILO problem to determine the number of meals to begin cooking in each interval to meet this demand.

Exercise 3.8 Formulate as a mixed-integer optimization problem and solve the following problem:

$$\frac{abc}{def} = \frac{1}{5},$$

where a, b, c, d, e, and f should take distinct integer values in the set $\{1, 2, \dots, 6\}$. *Hint*: $\log(ab) = \log(a) + \log(b)$ and $2x - 1$ transforms $\mathbb{B}$ into $\{-1, 1\}$.

Exercise 3.9 Assume there are three constraints of the type $\sum_{j \in J} a_{ij} x_j \leq b_i$, $i = 1, 2, 3$, where a_{ij} is a generic real number and x_j is a decision variable for each $j \in J$. We want to construct an objective function that reflects the way the constraints are satisfied. More specifically:

- for each constraint that is satisfied, the objective function increases by 1;
- if at least one constraint is satisfied, then the objective function increases by 1;
- if constraints 1 and 2 are satisfied simultaneously, the objective function gets an extra 1.

Examples:

- *Satisfying all constraints yields* $3 \times 1 + 1 + 1 = 5$.
- *Satisfying only constraint 3 yields* $1 + 1 = 2$.
- *Satisfying constraints 1 and 2 yields* $2 \times 1 + 1 + 1 = 4$.
- *Satisfying no constraint yields* 0.

Formulate a MILO for which the objective function is maximized.

Exercise 3.10 A company produces and sells m different products. For each product $j = 1, \ldots, m$, the demand forecast is equal to d_j. To produce each product, the company uses n types of ingredients. In particular, the amount of ingredient i needed for product j is denoted by a_{ij}. The company purchases ingredients from a supplier that applies the following pricing policy: Buying one unit of ingredient i costs c_i, but, if more than q_i units of ingredient i are purchased, the company gets a volume discount and the cost per unit of ingredient i decreases to c'_i for the entire order. The supply chain department must decide on the amount of ingredients to purchase to satisfy the demand for each product while minimizing the purchasing costs. Formulate a MILO problem that can be used to help the supply chain department.

4

Network Optimization

4.1 Introduction

In the previous chapters, we dealt with general problems by first formulating all necessary constraints and then passing the problem to an LO or MILO solver, but in a way, we have been oblivious to the problem's structure. However, it is often advantageous to analyze this structure, as it can enable us to find better solution methods. In this chapter, we consider a very general class of problems with special structure – the **network** problems. In the following example, we illustrate the key ideas.

Example 4.1 (Dinner seating arrangement) Assume that we are organizing a wedding dinner and our goal is to have guests from different families mingle with each other as much as possible. A possible way to achieve this is to seat people at tables so that no more than a given threshold k_{max} of individuals from the same family sit at the same table. How could we solve a problem like this?

First, we need the problem data – for each family $f \in F$ we need to know the number of its members m_f, and for each table $t \in T$ we need to know its capacity c_t. Using these data and the tools we have learned so far, we can formulate this problem as an LO problem.

If we are not concerned about specific individuals, but only about the number of members in a given family, then we can use the variable x_{ft} to determine the number of individuals in family f who will sit at table t. In the problem formulation, we are not given any objective function since our goal is to find a feasible seating arrangement. For this reason, we can set the objective function to a constant value, say 0, without distinguishing between the various feasible solutions.

The mathematical formulation of this optimization problem is the following:

$$\min \quad 0 \tag{4.1}$$
$$\text{s.t.} \quad \sum_f x_{ft} \le c_t, \qquad \forall t \in T$$
$$\sum_t x_{ft} = m_f, \qquad \forall f \in F$$
$$0 \le x_{ft} \le k_{max}, \qquad \forall f \in F, t \in T.$$

The three constraints ensure that:

• the seating capacity is not exceeded at each table;

- each family is fully seated; and
- the number of members of each family seated at each table does not exceed k_{max}.

Consider a specific instance of this problem with family sizes $m = (6, 8, 2, 9, 13, 1)$, table capacities $c = (8, 8, 10, 4, 9)$, and threshold $k_{max} = 3$. The optimal solution can be found in Table 4.1.

Table 4.1 *A seating plan that satisfies the $k_{max} = 3$ requirement.*

f/t	0	1	2	3	4
0	2	3	1	0	0
1	2	0	3	0	3
2	2	0	0	0	0
3	0	2	3	1	3
4	1	3	3	3	3
5	1	0	0	0	0

A peculiar fact is that, although we did not explicitly require that all variables x_{ft} be integer, the optimal solution turned out to be integer anyway. This is not a coincidence, as it follows from a specific structural property of the problem at hand. This means that, even for larger versions of the same problem, we can solve them with LO instead of MILO solvers to find integer solutions, gaining a huge computational advantage.

Our objective was to make members of different families mingle as much as possible. Is $k_{max} = 3$ the lowest possible number for which a feasible table allocation exists, or can we make the tables even more diverse by bringing this number down? In order to find out, we can make $k = k_{max}$ a decision variable and change the objective function of (4.1) to minimize k, obtaining the following problem:

$$\min \quad k \qquad\qquad\qquad\qquad\qquad\qquad\qquad (4.2)$$
$$\text{s.t.} \quad \sum_f x_{ft} \leq c_t, \qquad\qquad\qquad \forall t \in T$$
$$\sum_t x_{ft} = m_f, \qquad\qquad\qquad \forall f \in F$$
$$0 \leq x_{ft} \leq k, \qquad\qquad\qquad \forall f \in F, t \in T$$
$$k \geq 0.$$

The optimal solution to (4.2) is given in Table 4.2. Unfortunately, this solution is no longer integer.

Mathematically, this is due to the fact that the structure, which previously ensured the integrality of solutions at no extra cost, has been lost as a result of making k a decision variable. To find the solution to this problem, we need to use a MILO solver, which returns the same optimal value $k = 3$ as before.

However, using a MILO solver is not necessarily the best approach for problems like this. Many real-life situations (e.g., assigning people to groups/teams) require solving really large problems similar to (4.1) and there exist algorithms that can take advantage of the special **network structure** of the problem better than LO solvers.

Table 4.2 *Optimal x_{ft} solution for the assignment problem where k was minimized.*

f/t	0	1	2	3	4
0	2.6	2.2	0.0	0.0	1.2
1	1.0	2.6	1.8	0.0	2.6
2	0.0	0.0	2.0	0.0	0.0
3	1.8	0.6	2.6	1.4	2.6
4	2.6	2.6	2.6	2.6	2.6
5	0.0	0.0	1.0	0.0	0.0

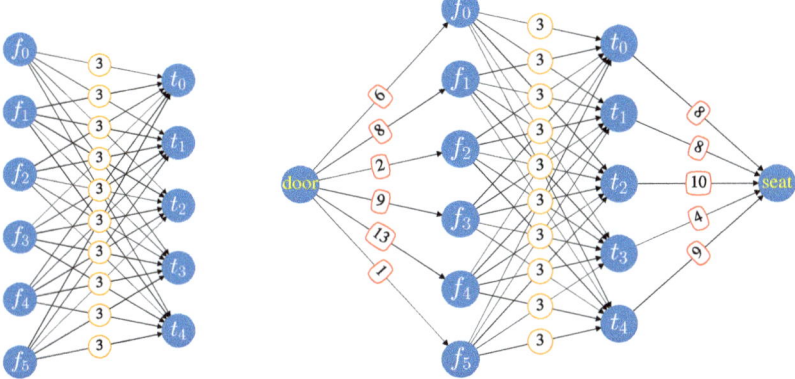

(a) Network reformulation   (b) Max-flow equivalent reformulation

Figure 4.1 Visualizations of the dinner seating allocation problem as a network problem.

To illustrate this, we first visualize the problem using a graph, see Figure 4.1a, where:

- the nodes on the left-hand side represent the families;
- the nodes on the left-hand side stand for the tables;
- each left-to-right arrow stand comes with a number denoting the capacity of arc (f,t), that is at most how many individuals from family f can be assigned to table t (which is equal to $k_{\max}$).

If we think of each family as a "supply of individuals" and each table as a "demand of individuals," then we can rephrase our original task as the problem of sending individuals from families f to tables t so that everyone is assigned to some table, the capacities of the tables are respected, and no table gets more than $k_{\max} = 3$ members of the same family.

By adding two more nodes to Figure 4.1a, we can formulate the problem as a slightly different flow problem where all the problem parameters are captured as arc capacities; see Figure 4.1b. In such a network, consider the problem of sending resources from the **source node** "door" to the **target node** "seat," subject to the restriction that for any node that is neither the start nor the target, the sum of incoming and outgoing flows are equal (**balance constraint**). If there exists a flow that respects the arc capacities and the sum of outgoing flows at the source node is equal to the total number of individuals, that is, $\sum_{f \in F} m_f$, it means that there exists a family-to-table assignment that meets our requirements.

Thus, if we maximize the total flow going through "door" and reaching "seat" and it matches the total number of individuals, the problem is solved. As unimpressive as this sounds, this means that it can be treated as a special case of a famous maximum-flow problem, for which there exist algorithms that are way more efficient than a generic LO solver. One such algorithm is the Bellman–Ford algorithm, implicitly invoked in the following code using the Python package `networkx`.

```python
def model_as_network(members, capacity, kmax):
    families = [f"f{i}" for i in range(len(members))]
    tables = [f"t{j}" for j in range(len(capacity))]

    G = nx.DiGraph()
    G.add_edges_from(["door", f, {"capacity": n}] for f, n in zip(families,
    ↪    members))
    G.add_edges_from([(f, t) for f in families for t in tables],
    ↪    capacity=kmax)
    G.add_edges_from([t, "seat", {"capacity": n}] for t, n in zip(tables,
    ↪    capacity))

    return G


members = [6, 8, 2, 9, 13, 1]
capacity = [8, 8, 10, 4, 9]
G = model_as_network(members, capacity, kmax=3)
flow_value, flow_dict = nx.maximum_flow(G, "door", "seat")
```

In Figure 4.2 we present the average runtime of solving randomly generated instances of the family–table problem of increasing sizes, solved using various MILO solvers and the Bellman–Ford algorithm. Realizing that the optimization problem we are tackling is a particular problem class for which a tailored algorithm is available can result in solving times orders of magnitude faster, particularly for large instances.

In the previous example, we encountered an optimization problem that could be easily modeled as a network problem. In general, a network $G = (V, E)$ is a graph consisting of a set V of **node** and a collection $E \subseteq V \times V$ of **directed arcs** or **undirected edges** (which is the naming convention that we will stick to from now on). A directed arc from node i to node j is denoted by (i, j). Edges typically have costs and/or capacities associated with them. If we decide to use directed arcs, the flows are always nonnegative but could be required to be integer or not (depending on the specific application). Similarly, if we use undirected edges (i, j) equivalent to (j, i), we can set the convention that a positive flow on this edge corresponds to a flow from i to j, and a negative flow corresponds to a flow from j to i. Networks are powerful concepts for modeling a variety of problems, typically related to flows, matchings, or assignments.

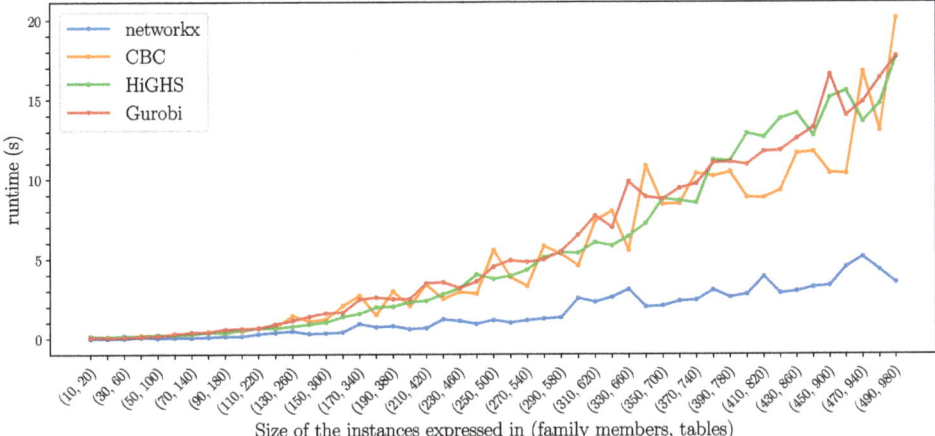

Figure 4.2 The solution time (in seconds) using a network formulation (using `networkx` built-in functions) vs. a standard MILO formulation (using CBC, HiGHS, and Gurobi solvers). Horizontal axis: number of family members and table pairs.

Example 4.2 (Transport logistics – Minimum-cost flow problem) In the context of logistics optimization, a company aims to identify the most cost-effective strategy for transporting goods from its production facilities to its retail locations across an entire continent. This problem can be naturally formulated using a graph. Each node $j \in V$ of this network represents a manufacturing facility, a distribution center, or a retail outlet. Correspondingly, node j is characterized by having a supply $b_j > 0$, a demand $b_j < 0$, or just serving as a transshipment point with $b_j = 0$. Each directed arc $(i,j) \in E$ represents a possible mode of transport (rail, airway, road) between locations i and j, with an associated maximum capacity $u_{ij} \geq 0$ and cost $c_{ij} \geq 0$ per unit of good sent using edge (i,j). Note that multiple edges are possible between the same pair of nodes (i,j), modeling different means of transport available between those locations, each with a specific cost and maximum capacity. We refer to Figure 4.3 for a small instance of one of these graphs.

For every edge $(i,j) \in E$, we introduce a decision variable $x_{ij} \geq 0$ describing the nonnegative amount of goods sent from node i to node j via edge $(i,j) \in E$. These quantities are usually referred to as **flows**. The minimum-cost flow problem aims to find the set of flows $\{x_{ij}\}_{(i,j)\in E}$ with the minimum cost through the network such that the available supply is used to satisfy the demand. It can be formulated as follows:

$$\min \quad \sum_{(i,j)\in E} c_{ij}x_{ij} \tag{4.3a}$$

$$\text{s.t.} \quad \sum_{i\in V \,:\, (i,j)\in E} x_{ij} - \sum_{i\in V \,:\, (j,i)\in E} x_{ji} = b_j, \qquad \forall j \in V \tag{4.3b}$$

$$0 \leq x_{ij} \leq u_{ij}, \qquad \forall (i,j) \in E. \tag{4.3c}$$

Equations (4.3b) are the key constraints, usually referred to as **flow conservation** or **flow-balance** constraints, expressing the fact that the net balance of the goods arriving at node j minus those departing from node j should be exactly equal to the supply/demand b_j set for that node.

The following code snippet implements the minimum-cost flow problem in Pyomo for a small network instance.

```
network_fig4 = {
    "nodes": {
        "a": {"b": 1},
        "b": {"b": 0},
        "c": {"b": 0},
        "d": {"b": 2},
        "e": {"b": -2},
        "f": {"b": 0},
        "g": {"b": -1},
        "h": {"b": 0},
    },
    "edges": {
        ("a", "b"): {"u": 2, "c": 3},
        ("a", "c"): {"u": 2, "c": 1},
        ("a", "d"): {"u": 3, "c": 2},
        ("b", "e"): {"u": 1, "c": 1},
        ("c", "b"): {"u": 2, "c": 1},
        ("c", "e"): {"u": 2, "c": 4},
        ("c", "d"): {"u": 1, "c": 1},
        ("d", "g"): {"u": 3, "c": 2},
        ("d", "f"): {"u": 2, "c": 3},
        ("c", "f"): {"u": 2, "c": 1},
        ("f", "e"): {"u": 2, "c": 4},
        ("g", "f"): {"u": 1, "c": 8},
        ("f", "h"): {"u": 3, "c": 2},
        ("e", "h"): {"u": 2, "c": 2},
        ("g", "h"): {"u": 2, "c": 0},
    },
}


def mincostflow(network):
    model = pyo.ConcreteModel("Minimum cost flow")

    model.x = pyo.Var(network["edges"], domain=pyo.NonNegativeReals)

    @model.Objective(sense=pyo.minimize)
    def objective(m):
        return sum(data["c"] * m.x[e] for e, data in
        ↪   network["edges"].items())

    @model.Expression(network["nodes"])
    def incoming_flow(m, j):
```

```
        return sum(m.x[i, j] for i in network["nodes"] if (i, j) in
        ↪    network["edges"])

    @model.Expression(network["nodes"])
    def outgoing_flow(m, j):
        return sum(m.x[j, i] for i in network["nodes"] if (j, i) in
        ↪    network["edges"])

    @model.Constraint(network["nodes"])
    def flow_conservation(m, j):
        return m.outgoing_flow[j] - m.incoming_flow[j] ==
        ↪    network["nodes"][j]["b"]

    @model.Constraint(network["edges"])
    def flow_upper_bound(m, *e):
        return m.x[e] <= network["edges"][e]["u"]

    return model


model = mincostflow(network)
SOLVER.solve(model)
```

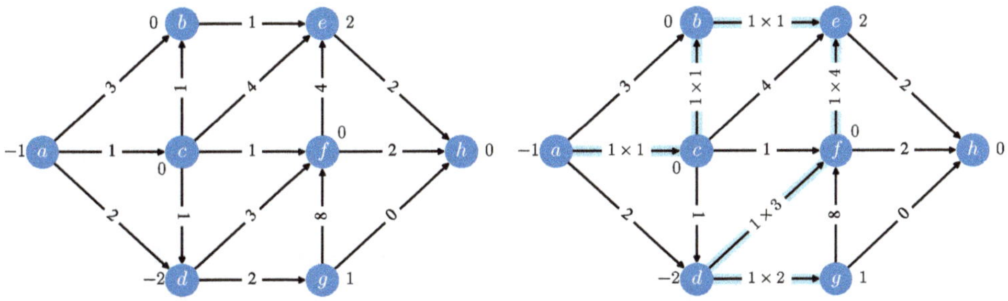

(a) Network instance, with numbers on the arcs denoting the unit cost and numbers next to the nodes denoting the supply (+)/demand (−) of each node.

(b) Minimum-cost flow solution. The arcs with a positive flow are highlighted and the costs are obtained multiplying the flow values by the corresponding unit costs.

Figure 4.3 Example of a min-cost flow problem for a small network instance.

What is special about network problems? First, they enable us to visualize and structure optimization problems and their constraints in a convenient way. A typical network problem consists of the following components:

- a (directed or undirected) graph with:
  - arcs/edges, with upper and lower bounds on their capacity and a unit cost assigned to each of them, and

  – nodes, with supply/demand numbers assigned to them (typically we denote demand by
    positive numbers and demand by negative ones);

- decision variables, which most often correspond to flows along the arcs/edges, but there
  can also be node-related decision variables, for example, in problems related to product
  manufacturing at nodes, which gets distributed further along the arcs.

This is a very convenient framework for modeling and thinking about problems, but it is not
the only advantage of network problems.

As already seen in Example 4.1, under certain conditions often seen in network problems,
we can obtain integer solutions to a MILO problem *without* requiring that explicitly and
without needing to resort to the branch-and-bound or cutting-plane algorithms. Just as
in Example 4.1 we solve the linear relaxation of the problem at hand and still obtain an
integer solution. Furthermore, for some network problems, we can come up with specialized
algorithms that work even faster than solving the problem using an LO formulation, which
was also illustrated in Example 4.1.

For these reasons, it makes sense to exploit the network structure, and very often a problem
that can be modeled as a network can be transformed into an equivalent problem that belongs
to one of the well-studied network problems. The rest of the chapter is structured as follows.
First, in Section 4.2 we introduce the notion of total unimodularity – a property of the set
of equality constraints that allows us to ignore the integrality constraints without losing the
integrality of the optimal solutions. Next, in Section 4.3 we build a catalog of well-known
network problems, studying their computational difficulty, and showing how their structure
often implies total unimodularity holds.

4.2 Totally Unimodular Matrices and Integrality of the Solutions

We start by introducing the essential notion of a totally unimodular matrix.

Definition 4.1 An $m \times n$ matrix A is said to be **totally unimodular (TU)** if every $k \times k$
submatrix of A has determinant equal to -1, 0, or 1 for $k \in \{1, \ldots, \min\{n, m\}\}$.

This property turns out to be enough to immediately conclude whether the optimal solution
of a given LO problem is integer or not, as illustrated by the following proposition.

Proposition 4.1 *Consider a linear optimization problem of the form*

$$\begin{aligned} \min \quad & c^\top x \\ \text{s.t.} \quad & Ax \geq b \\ & x \geq 0. \end{aligned}$$

*If the constraint matrix A is TU and the vector b has integer entries, then the entries of any
optimal solution x^* are all integers.*

The advantage of recognizing that the MILO problem at hand has a TU constraint
matrix is that we can solve it as an LO problem, ignoring the integrality constraints,

because we are guaranteed that the optimum solution (if it exists) has integer entries.

Of course, checking the TU-ness of a matrix using the definition is often a tedious task. However, there exist simple criteria to check whether the constraint matrix A is totally unimodular. We now provide, without proof, two examples of such sufficient (but not necessary) conditions.

Criterion 1 for TU matrix: An $m \times n$ matrix A is TU if *all* the following statements hold:

- $a_{ij} \in \{-1,0,1\}$ for all $i \in \{1,\ldots,m\}$ and $j \in \{1,\ldots,n\}$;
- every column of A contains at most two nonzeros;
- there exists a partition (P,Q) of the rows $\{1,\ldots,m\}$ such that $\sum_{i \in P} a_{ij} = \sum_{i \in Q} a_{ij}$ for all $j \in \{1,\ldots,n\}$.

Criterion 2 for TU matrix: An $m \times n$ matrix A is TU if all its entries are in $\{-1,0,1\}$ and each column has at most one -1 and one 1.

Note that we stated Proposition 4.1 for problems written in canonical form, which requires the objective to be a "min," the constraints to be $\leq$, and all variables to be nonnegative. We already know that all LO problems can be rewritten into an equivalent problem that looks like this. It is then helpful to know that typical operations used in such a "rewriting" of the problem are also preserving the TU property. In other words, the fact that the final, canonical problem will yield an integer solution without requiring it, the same will hold for the "original" problem, regardless of its form.

Proposition 4.2 *If A is a totally unimodular matrix, then the following operations on it preserve the TU property:*

- *deleting rows or columns from A;*
- *rearranging rows or columns of A;*
- *introducing in A a copy of an existing row or column;*
- *introducing in A a unit vector or a zero vector as a row or column;*
- *multiplying an entire row or column of A by -1;*
- *taking the transpose of A;*
- *taking the inverse of A (only when A is a square matrix and its determinant is nonzero).*

Based on all those results one could be tempted to think that, for example, the combined matrix $[A_1|A_2]$ (obtained by placing the columns of two matrices next to each other into a single matrix) is TU if both A_1 and A_2 are. However, this is not the case in general, and it is rather simple to come up with counterexamples.

Example 4.3 (Dinner seating arrangement – Total unimodularity of the matrix) Let us apply the operations outlined in Proposition 4.2 to verify the TU property for the problem in (4.1). To do so, we need to analyze the matrix A and the vector x used to formulate this problem in the canonical form. Let us group all decision variables into a long vector:

$$x = [x_{1,1},\ldots,x_{1,n_T},x_{2,1},\ldots,x_{2,n_T},\ldots,x_{n_F,1},\ldots,x_{n_F,n_T}]^{\top}.$$

Moving to the constraints, we can rewrite the "table capacity constraints" as

$$-\sum_f x_{ft} \geq -c_t, \quad \forall t \in T$$

and the "everybody from a given family must go somewhere" equality constraints as two inequality constraints each, which together force the equality to hold:

$$\sum_t x_{ft} \geq m_f, \quad \forall f \in F$$

$$-\sum_t x_{ft} \geq -m_f, \quad \forall f \in F.$$

Since the nonnegativity constraints are treated separately in the canonical form, the only constraints left to rewrite are the "not more than k persons from a given family to each table constraints" which can be rewritten as

$$-x_{ft} \geq -k$$

By putting 0s, -1s, and 1s in the right entries of the matrix A, we can rewrite all constraints of the problem as

$$
\underbrace{\begin{bmatrix}
-1 & -1 & \cdots & -1 & & & & & & \cdots & \\
& & & & -1 & -1 & \cdots & -1 & \cdots & & \\
\vdots & & & & & & & & & & \vdots \\
& & & & & & \cdots & -1 & -1 & \cdots & -1 \\
1 & & & 1 & & & & \cdots & 1 & & \\
\vdots & & & & & & & & & & \vdots \\
& & 1 & & & 1 & \cdots & & & \cdots & 1 \\
-1 & & & -1 & & & \cdots & -1 & & & \\
\vdots & & & & & & & & & & \vdots \\
& & -1 & & & -1 & \cdots & & \cdots & & -1 \\
-1 & & & & & & \cdots & & & & \\
\vdots & & & & & & & & & & \vdots \\
& & & & & & \cdots & & \cdots & & -1
\end{bmatrix}}_{A}
\, x \geq
\underbrace{\begin{bmatrix}
c_1 \\ c_2 \\ \vdots \\ c_{n_T} \\ m_1 \\ \vdots \\ m_{n_F} \\ -m_1 \\ \vdots \\ -m_{n_F} \\ -k \\ \vdots \\ -k
\end{bmatrix}}_{b}.
$$

The integrality of the vector b follows immediately from the problem data. Therefore, we simply need to show that the matrix A verifies the TU property, which seems not obvious at first sight.

However, the last $n_F \times n_T$ rows are unit vectors (vectors consisting of all zeros except for a single 1), multiplied by -1. Due to this fact and Proposition 4.2, the rules about "introducing a unit vector as a row" and "multiplying an entire row or column by -1" as being TU-preserving operations, we know that we can reduce proving the TU-ness of A to proving the TU-ness of a smaller matrix without these rows (if that matrix turns out to be TU, then the larger matrix is TU as well because adding such rows does not change the TU-status):

$$
\begin{bmatrix}
-1 & -1 & \cdots & -1 & & & & & & \cdots & & \\
 & & & & -1 & -1 & \cdots & -1 & \cdots & & & \\
\vdots & & & & & & & & & & & \vdots \\
 & & & & & & & \cdots & -1 & -1 & \cdots & -1 \\
1 & & & & 1 & & & \cdots & 1 & & & \\
\vdots & & & & & & & & & & & \vdots \\
 & & & 1 & & & 1 & \cdots & & & \cdots & 1 \\
-1 & & & & -1 & & & \cdots & -1 & & & \\
\vdots & & & & & & & & & & & \vdots \\
 & & & -1 & & & -1 & \cdots & & & \cdots & -1
\end{bmatrix}.
$$

Now, the last sequence of rows (the ones with -1 entries) consists of copies of the sequence of rows just before them, multiplied by -1. Using again Proposition 4.2 and specifically the rules about "introducing a copy of a given row" and "multiplying an entire row or column by -1," we can limit ourselves to studying the TU property of the smaller matrix obtained removing these rows, that is,

$$
\begin{bmatrix}
-1 & -1 & \cdots & -1 & & & & & & \cdots & & \\
 & & & & -1 & -1 & \cdots & -1 & \cdots & & & \\
\vdots & & & & & & & & & & & \vdots \\
 & & & & & & & \cdots & -1 & -1 & \cdots & -1 \\
1 & & & & 1 & & & \cdots & 1 & & & \\
\vdots & & & & & & & & & & & \vdots \\
 & & & 1 & & & 1 & \cdots & & & \cdots & 1
\end{bmatrix}.
$$

All entries in this smaller matrix are in $\{-1, 0, 1\}$ and each column has at most one -1 and at most one 1, so by virtue of Criterion 2 it is TU, and hence the original matrix A is also TU. This guarantees that an LO algorithm is sufficient to find the integral solution to the dinner seating problem.

The reasoning in the last example may seem rather involved at first encounter. The good news is that it is rather uncommon to prove the TU-ness of a newly encountered network problem in this way. Indeed, it is far more common to first try to check whether a newly encountered problem can be rewritten as a well-understood established network problem whose TU property is already established. This reformulation often involves a series of techniques, such as the addition of nodes and arcs, which we will elaborate on in Section 4.3.9.

4.3 Modeling and Special Network Problems

Earlier in this chapter, we have shown that the dinner seating problem can be easily reformulated as a max-flow problem. While this transformation might have initially seemed superfluous and unnecessary, it offers significant advantages. The max-flow problem is extremely well studied in the academic literature, with numerous established properties. Consequently, leveraging this existing body of knowledge allows us to build on the work of others, thereby enhancing the efficiency and robustness of our problem-solving approach. In particular, for well-established network problems, specialized algorithms may already be available that outperform LO solvers in terms of computational time. The art of modeling and solving network problems often involves reducing the problem at hand to a well-known problem for which efficient algorithms exist.

In this section, we present a collection of notable classic network problems, accompanied by real-world situations where they are commonly encountered. This "gallery" will include both tractable problems that satisfy the TU property and more challenging ones that are known to lack this property and are much harder to solve. In each example, we will emphasize whether, after formulating the problem as an LO problem, the corresponding matrices are TU. Furthermore, we will indicate whether the problem at hand can be solved using more efficient adaptations of standard LO algorithms or through the application of specialized algorithms.

Finally, in Section 4.3.9, we will introduce a set of operations and modeling tricks designed to facilitate the transformation of one network problem into another.

4.3.1 Minimum-Cost Flow

We begin with the classic **minimum-cost flow problem**, which we already introduced in Example 4.2. This is a typical mathematical "prototypical problem" to which, as we shall see, many other problems can be reduced.

To analyze the total unimodularity of this problem, note that the flow conservation constraints in (4.3b) can be written in a matrix form as $Ax = b$. The matrix A is of size $|E| \times |V|$ and each of its columns corresponds to a node and each of its rows to an edge. Given the precise formulation of the constraint (4.3b), every column contains one -1 and one $+1$, corresponding to the two endpoints (the source and the target nodes) of that directed arc. Therefore, it readily follows from Criterion 2 that the matrix A is TU.

For that specific problem, there are also other (even more efficient) algorithms, such as the **network simplex algorithm**, the **cycle canceling algorithm**, and the **cut canceling algorithm**.

4.3.2 Shortest Path

There are several network optimization problems that can be reduced to the minimum-cost flow problem. The first example is the problem of finding the shortest path between two nodes in a given network.

Assume that we are given a network $G = (V, E)$ in which a length/cost of travel c_{ij} is associated with each edge $(i, j) \in E$. For any pair of origin $s \in V$ and destination $t \in V$ nodes, the **shortest path problem** aims to determine the path from s to t through the network that minimizes the total length of the edges traversed. It is illustrated in Figure 4.4.

As a linear optimization model, the shortest path problem is a special case of the minimum-cost flow problem. Indeed, it can be formulated as (4.3) using unit edge capacities, that is, $u_{ij} = 1$ for every $(i, j) \in E$, and picking

$$
b_j = \begin{cases} 1, & \text{for } j = s \\ -1, & \text{for } j = t \\ 0, & \text{otherwise.} \end{cases}
$$

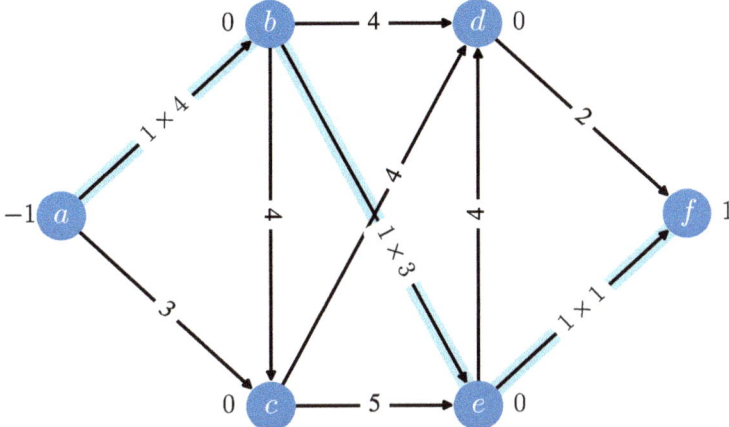

Figure 4.4 Small instance of a shortest path problem reformulated as a minimum-cost flow problem in which we send one unit of flow from the origin node a to the destination node f. The optimal solution is highlighted in blue.

This fact readily implies that the shortest path problem inherits the TU of the resulting constraint matrix from the minimum-cost flow problem and thus also the solution of a shortest path problem is guaranteed to be integer.

As already mentioned earlier, there are many network models whose optimal solution can be found by special-purpose algorithms. For the shortest path problem with nonnegative edge weights, **Dijkstra's algorithm** is the go-to computational method [Dij22]. It is based on the simple and intuitive fact that shortest paths concatenate and can be summarized as follows:

1. Let the starting node be node 1 and the target node be node n. Initialize a list of nodes with their temporary labels (label = shortest distance found so far) $s_1 = 0$ and $s_i = +\infty$, $i \neq 1$.
   Put all node indices $1, \ldots, n$ in a set $\mathcal{S} = \{1, \ldots, n\}$.
2. While the node i from $\mathcal{S}$ with the lowest s_i is not node n repeat:
   Remove i from $\mathcal{S}$. From now on, the label s_i becomes permanent – the length of the path from 1 to i cannot be further shortened. For all j for which an arc (i,j) exists update their corresponding temporary labels as

$$s_j := \min\{s_j, s_i + c_{ij}\}.$$

3. The shortest path from 1 to n has length s_n. A simple backtracking algorithm recovers the path by initializing it with n and prepending with the node i for which $s_i + c_{ij} = s_j$ being j the last node prepended to the path, until node 1 is finally added.

This algorithm can find the shortest path from 1 to any node in a number of operations that scales quadratically with the number n of nodes in the graph, which we denote as $\mathcal{O}(n^2)$. This notation means that there exists a constant C such that the total number of steps is less than or equal to Cn^2 for any graph.

Example 4.4 (Illustration of Dijkstra's algorithm) We show Dijkstra's algorithm in action while finding the shortest path from node a to node c in the small network instance depicted in Figure 4.5.

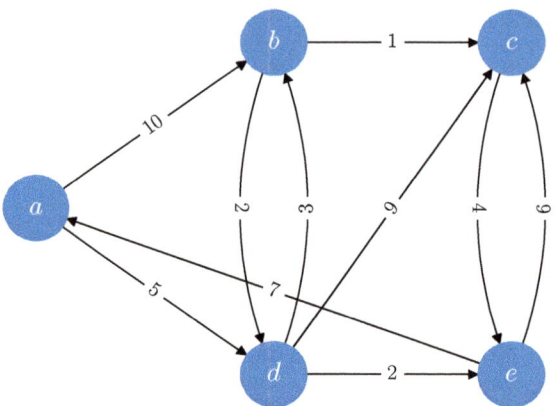

Figure 4.5 Small instance of network with its associated costs on each arc.

We begin with Step 1 of the algorithm. Since we start at a, we may give a the label 0, which means that the shortest path from a to a has length 0, and we give to all the other nodes the label $+\infty$, which means that we do not have an estimate of the shortest distance from a to them.

We use Figure 4.6 to describe Step 2 of the algorithm.

Since a now has the lowest temporary label, we make it permanent and scan the nodes that can be reached from a as in Figure 4.6a. For each such node, we update the label to the minimum between the present value and the sum of the (permanent) label of the current node with the length of the corresponding arc. Due to this, b receives label $10 = 0 + 10$ and d receives label $5 = 0 + 5$. The other nodes retain their $+\infty$ labels.

Next, the node with the lowest temporary label is d. We make it permanent and scan the successors, namely $\{b,c,e\}$, updating their labels; see Figure 4.6b.

In the next iteration of Step 2 of the algorithm, the node with the lowest temporary label is e. As Figure 4.6c shows, we make this label permanent and scan the successors of e, updating their labels. Obviously, nodes that already have a permanent label, like a, are skipped, as their label is necessarily lower than it would become.

Proceeding in this way, we finally label node c with a permanent label in Figure 4.6d indicating a shortest distance of 9.

The values of the labels in the subsequent steps of the algorithm are summarized in Table 4.3, where each row adds one permanent label and lists the labels at that moment.

The last row of the table lists the final label values. In our small example all labels became permanent, that is not necessarily the case as the algorithm terminates when the destination receives a permanent label.

Table 4.3 *Summarizing the label updates.*

node	a	b	c	d	e
	0				
a	0	10		5	
d		8	14	5	7
e		8	13		7
b		8	9		
c			9		
	0	8	9	5	7

From the last label values and the original network as in Figure 4.6 we can recover the path as being $[a, d, b, c]$ with length $5 + 3 + 1 = 9$ (see Figure 4.7).

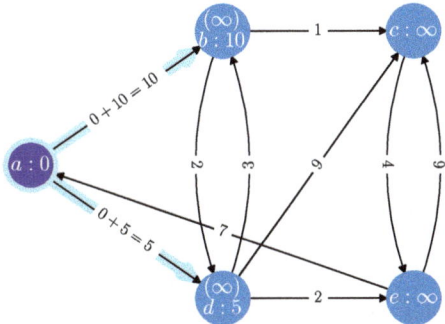

(a) Scan the successors of a and update the labels.

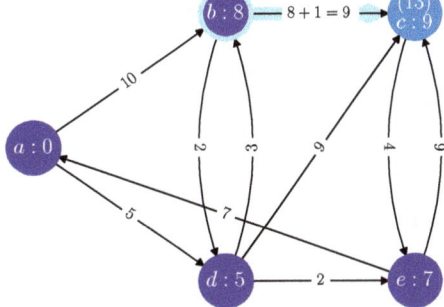

(b) Scan the successors of d and update the labels.

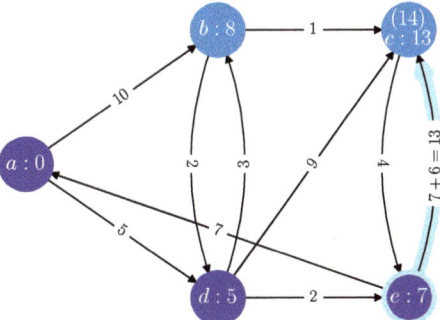

(c) Scan the successors of e and update the labels.

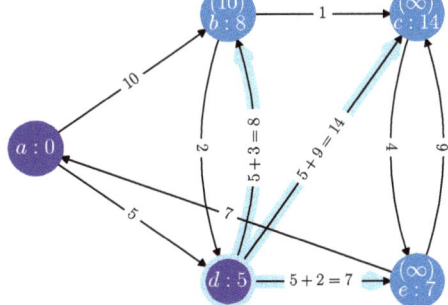

(d) Scan the successors of b and update the labels.

Figure 4.6 The four repetitions of Step 2 needed to label permanently node c and hence find the shortest path from a to c. Darker nodes have permanent labels and improved labels are shown in brackets.

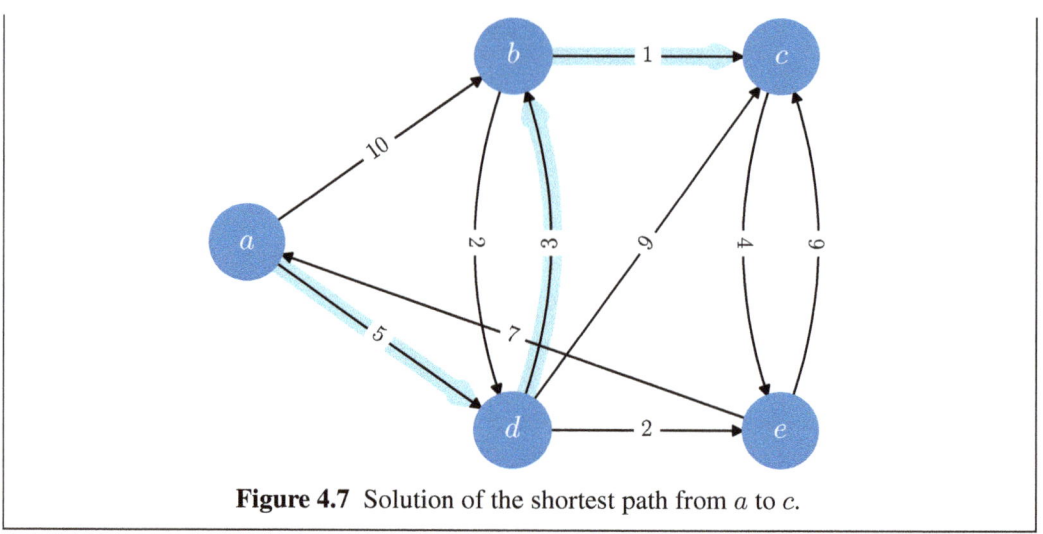

Figure 4.7 Solution of the shortest path from a to c.

4.3.3 Transportation

A simpler version of the min-cost flow problem from Example 4.2 is the basic transportation problem with direct arcs only from supply nodes to demand nodes. Consider n suppliers with inventory amounts $s_1, \ldots, s_n \geq 0$ and m demand sites with inventory demands $d_1, \ldots, d_m \geq 0$. The transport of one unit of inventory from supplier i to demand site j costs $c_{ij} \geq 0$. The transportation problem aims to allocate the available inventory to satisfy the demand in the most cost-effective way.

This **transportation problem** can be reformulated as a minimum-cost flow problem on a weighted bipartite network $G = (V, E)$ depicted in Figure 4.8. More specifically, we

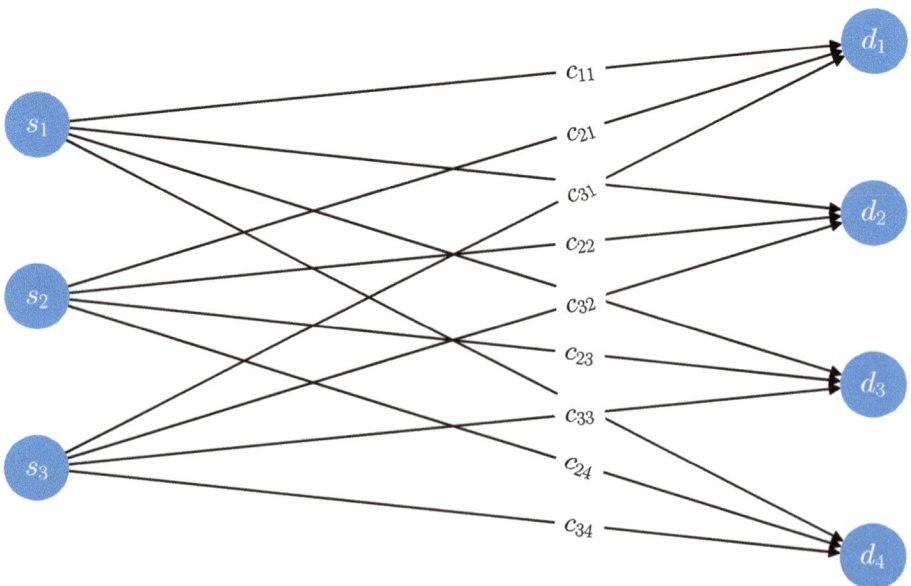

Figure 4.8 Visualization of a transportation problem with three supply nodes (on the left) and four demand nodes (on the right).

consider the two types of nodes, supply nodes and demand nodes, and define the set of nodes as $V = \{s_1, \ldots, s_n, d_1, \ldots, d_m\}$. Then, we draw directed edges only between each supply node and each demand node, obtaining the set of edges $E = \{(s_i, d_j) : i \in \{1, \ldots, n\}, j \in \{1, \ldots, m\}\}$.

This transportation problem can be recovered as an instance of the minimum-cost flow problem on the network $G = (V, E)$ for which we set $b_{s_i} = s_i$ for $i \in \{1, \ldots, n\}$, $b_{d_j} = -d_j$ for $j \in \{1, \ldots, m\}$ and capacities $u_{ij} = \infty$ for all $(i, j) \in E$ (if there are no transport limits). Consequently, the transportation problem inherits the total unimodularity property.

Due to its special bipartite structure, the transportation problem can also be solved with ad hoc algorithms such as the (α, β) algorithm, which is equivalent to the simplex method but exploits the bipartite structure at each step to achieve maximum efficiency.

Example 4.5 (Gasoline distribution) YaYa Gas-n-Grub is the franchisor and operator of a network of regional convenience stores that sell gasoline and convenience items in the United States. Each store is individually owned by a YaYa Gas-n-Grub franchisee who pays fees to the franchisor for services. Gasoline is delivered by truck from regional distribution terminals by the current supplier. Each truck delivers 8000 gallons at a fixed charge of $700 per delivery or $0.0875 per gallon.

Franchise owners are eager to reduce delivery costs to boost profits. YaYa Gas-n-Grub decides to accept proposals from other distribution terminals, A and B, to supply the franchise operators. Rather than a fixed fee per delivery, they proposed pricing based on location. Since they already have existing customers, A and B can only provide a limited amount of gasoline to new customers totaling 100,000 and 80,000 gallons, respectively. The only difference between the new suppliers and the current supplier is the delivery charge.

The operator of YaYa Gas-n-Grub wants to allocate gasoline delivery in such a way that the costs to franchise owners are minimized (see Table 4.4).

Table 4.4 *The demand recorded by each franchisee and their individual prices of gasoline delivery in dollar cents/gallon.*

Franchisee	Demand	Current supplier (500,000 available)	Terminal A (100,000 available)	Terminal B (80,000 available)
Alice	30,000	8.75	8.3	10.2
Badri	40,000	8.75	8.1	12.0
Cara	50,000	8.75	8.3	100.0
Dan	80,000	8.75	9.3	8.0
Emma	30,000	8.75	10.1	10.0
Fujita	45,000	8.75	9.8	10.0
Grace	80,000	8.75	100.0	8.0
Helen	18,000	8.75	7.5	10.0

We introduce the decision variables $x_{d,s} \geq 0$, where subscript $d \in 1, \ldots, n_d$ refers to the destination of the delivery and subscript $s \in 1, \ldots, n_s$ to the source. The value of $x_{d,s}$ is the volume of gasoline shipped to destination d from source s.

Given the cost rate $r_{d,s}$ for delivering one unit of gasoline from d to s, the objective is to minimize the total cost of transporting gasoline from the sources to the destinations subject to meeting the demand requirements, D_d, at all destinations, and satisfying the supply constraints, S_s, at all sources. In mathematical terms, we can write the full problem as

$$\min \quad \sum_{d=1}^{n_d} \sum_{s=1}^{n_s} r_{d,s} x_{d,s}$$

$$\text{s.t.} \quad \sum_{s=1}^{n_s} x_{d,s} = D_d, \qquad \forall\, d = 1, \ldots, n_d \quad \text{(demand constraints)}$$

$$\sum_{d=1}^{n_d} x_{d,s} \leq S_s, \qquad \forall\, s = 1, \ldots, n_s \quad \text{(supply constraints)}$$

$$x_{d,s} \geq 0, \qquad \forall\, d = 1, \ldots, n_d,\; s = 1, \ldots, n_s.$$

The Pyomo model below is an implementation of this mathematical model, with the solution displayed in Table 4.5 and Figure 4.9. Sets and indices have been designated with more descriptive symbols for readability.

Table 4.5 *Details of the optimal gasoline distribution and corresponding savings.*

	Terminal A	Terminal B	Curr. Supplier	Curr. costs	New costs	Savings
Alice	30,000	0	0	2625.0	2490.0	135.0
Badri	40,000	0	0	3500.0	3240.0	260.0
Cara	12,000	0	38,000	4375.0	4321.0	54.0
Dan	0	20,000	0	1750.0	1600.0	150.0
Emma	0	0	30,000	2625.0	2625.0	0.0
Fujita	0	0	45,000	3937.5	3937.5	0.0
Grace	0	60,000	20,000	7000.0	6550.0	450.0
Helen	18,000	0	0	1575.0	1350.0	225.0

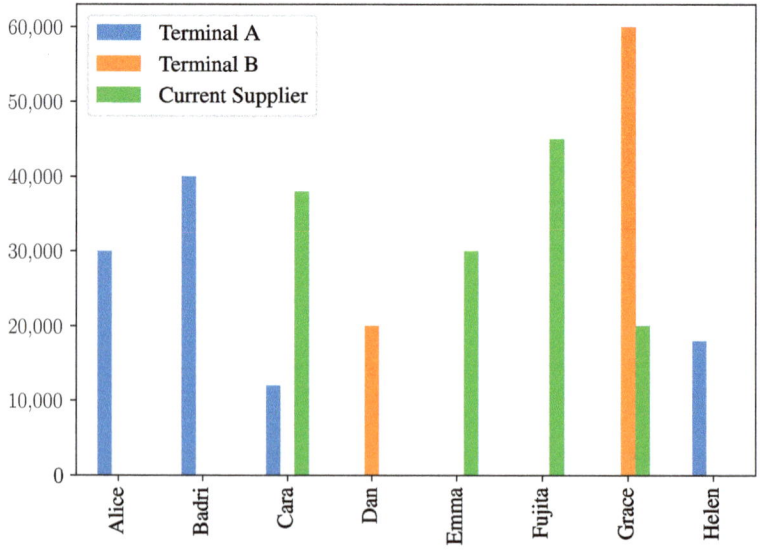

Figure 4.9 Visualization of the optimal plan for gasoline delivery.

```python
rates = pd.DataFrame(
    [
        ["Alice", 8.3, 10.2, 8.75],
        ["Badri", 8.1, 12.0, 8.75],
        ["Cara", 8.3, 100.0, 8.75],
        ["Dan", 9.3, 8.0, 8.75],
        ["Emma", 10.1, 10.0, 8.75],
        ["Fujita", 9.8, 10.0, 8.75],
        ["Grace", 100, 8.0, 8.75],
        ["Helen", 7.5, 10.0, 8.75],
    ],
    columns=["Destination", "Terminal A", "Terminal B", "Current
    ↪ Supplier"],
).set_index("Destination")

demand = pd.Series(
    {
        "Alice": 30000,
        "Badri": 40000,
        "Cara": 50000,
        "Dan": 20000,
        "Emma": 30000,
        "Fujita": 45000,
        "Grace": 80000,
        "Helen": 18000,
    },
    name="demand",
)

supply = pd.Series(
    {"Terminal A": 100000, "Terminal B": 80000, "Current Supplier":
    ↪ 500000},
    name="supply",
)


def transport(supply, demand, rates):
    m = pyo.ConcreteModel("Gasoline distribution")

    m.SOURCES = pyo.Set(initialize=rates.columns)
    m.DESTINATIONS = pyo.Set(initialize=rates.index)

    m.x = pyo.Var(m.DESTINATIONS, m.SOURCES, domain=pyo.NonNegativeReals)

    @m.Param(m.DESTINATIONS, m.SOURCES)
    def Rates(m, dst, src):
        return rates.loc[dst, src]
```

```
@m.Objective(sense=pyo.minimize)
def total_cost(m):
    return sum(
        m.Rates[dst, src] * m.x[dst, src] for dst, src in
        ↪ m.DESTINATIONS * m.SOURCES
    )


@m.Expression(m.DESTINATIONS)
def cost_to_destination(m, dst):
    return sum(m.Rates[dst, src] * m.x[dst, src] for src in m.SOURCES)


@m.Expression(m.DESTINATIONS)
def shipped_to_destination(m, dst):
    return sum(m.x[dst, src] for src in m.SOURCES)


@m.Expression(m.SOURCES)
def shipped_from_source(m, src):
    return sum(m.x[dst, src] for dst in m.DESTINATIONS)


@m.Constraint(m.SOURCES)
def supply_constraint(m, src):
    return m.shipped_from_source[src] <= supply[src]


@m.Constraint(m.DESTINATIONS)
def demand_constraint(m, dst):
    return m.shipped_to_destination[dst] == demand[dst]


m.dual = pyo.Suffix(direction=pyo.Suffix.IMPORT)


return m


m = transport(supply, demand, rates / 100)
SOLVER.solve(m)
```

The old delivery total cost is $313,000/8000 \cdot 700 = \$27,387.50$, while with the new contract it would be only $\$26,113.50$.

Example 4.6 (Rebalancing a bike-sharing system) Consider the following simplified problem motivated by rebalancing issues in a bike-sharing network. Consider a directed graph $G = (V, E)$, where the nodes are bike-sharing stations and a directed edge between two stations i and j indicates the possibility of relocating bikes from station i to station j at a unit cost c_{ij}. The goal of the bike rebalancing problem is to relocate the bikes to meet the demand in the most cost-effective way. Let x_{ij} be the decision variable describing the number of bikes to be transported from bike station i to bike station j. Assume that each bike station i currently has $b_i \geq 0$ bikes and forecast demand for $d_i \geq 0$ bikes. If the total bike availability is equal to the total forecast bike demand, that is, $\sum_{i \in V} b_i = \sum_{i \in V} d_i$, then we can formulate the following MILO problem:

$$\min \quad \sum_{(i,j) \in E} c_{ij} x_{ij} \tag{4.4}$$

$$\text{s.t.} \quad b_i + \sum_{j:(j,i) \in E} x_{ji} - \sum_{j:(i,j) \in E} x_{ij} = d_i, \qquad \forall i \in V$$

$$x_{ij} \in \mathbb{Z}_+, \qquad \forall (i,j) \in E.$$

The equality constraints prescribe that at each node the demand is met by calculating the "net bike balance" at that node, which takes into account the initial number of bikes and both the inflow and outflow of bikes. In this formulation, the bike-sharing rebalancing problem is a special instance of a transportation problem.

If the total forecast demand is larger than the total bike availability, we can introduce a nonnegative integer variable δ_i for every node i describing the unmet demand at that node. We then try to find a solution that satisfies as much of the forecast demand as possible by adding an additional penalty term in the objective function. This variant can be formulated as follows:

$$\min \quad \sum_{(i,j) \in E} c_{ij} x_{ij} + \sum_{i \in V} \alpha_i \delta_i \tag{4.5}$$

$$\text{s.t.} \quad b_i + \sum_{j:(j,i) \in E} x_{ji} - \sum_{j:(i,j) \in E} x_{ij} = d_i - \delta_i, \qquad \forall i \in V$$

$$x_{ij} \in \mathbb{Z}_+, \qquad \forall (i,j) \in E$$

$$\delta_i \in \mathbb{Z}_+, \qquad \forall i \in V.$$

4.3.4 Max-Weight Bipartite Matching

The **maximum-weight bipartite matching problem** considers two sets I and J with the same cardinality and wants to pair each $i \in I$ with exactly one $j \in J$, and the cost of pairing i and j. Two examples of assignment problems are: (i) assigning workers/agents with heterogeneous rates/skills to different tasks or (ii) assigning customers requesting a taxi to specific taxi drivers.

This problem can be modeled as a minimum-cost flow problem by taking again a bipartite network with $V = I \cup J$ and $E = \{(i,j) : i \in I, j \in J\}$, and by setting $b_i = 1$ for $i \in I$ and $b_j = -1$ for $j \in J$, and $u_{ij} = 1$ for all $(i,j) \in E$. This conversion is illustrated in Figure 4.10. Because the resulting matrix is TU, there is an optimal integer solution consisting only of 0s and 1s, which gives us the desired assignment.

4.3.5 Max-Flow and Min-Cut

The **maximum-flow problem (max-flow problem)** is another foundational "mother problem" to which many other network optimization problems can be reduced. In this particular problem,

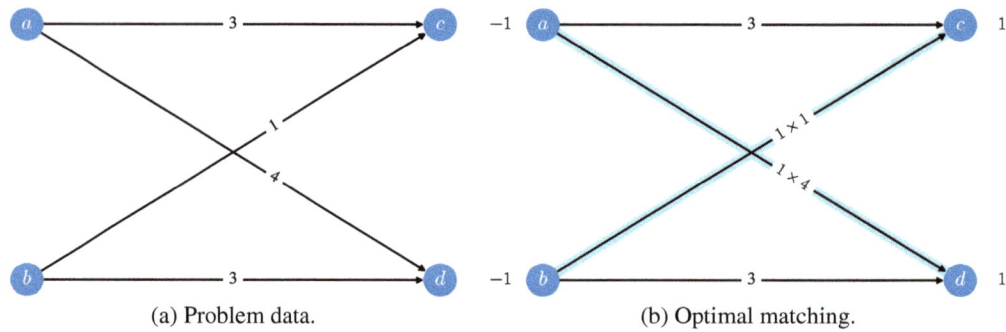

(a) Problem data.   (b) Optimal matching.

Figure 4.10 Max-weight bipartite matching between the nodes on the left and on the right, and its conversion to a min-cost flow problem.

the objective is not to minimize the cost associated with the flow but rather to maximize the magnitude of the flow itself.

Consider a network $G = (V, E)$ with a source $s \in V$ and a target $t \in V$, and such that edge $(i, j) \in E$ has capacity $u_{ij} \geq 0$. The goal is to send as much flow from the source to the target as possible, without exceeding the capacities of the edges. This problem can be formulated as a minimum-cost flow problem by adding the edge (t, s) to E, obtaining a new edge set $\tilde{E}$, setting $b_i = 0$ for all $i \in V$, and taking $c_{ts} = -1$ and $c_{ij} = 0$ for $(i, j) \in \tilde{E} \setminus \{(t, s)\}$. For completeness, the formulation can be rewritten as

$$
\begin{aligned}
\max \quad & x_{ts} \\
\text{s.t.} \quad & \sum_{i:(j,i)\in\tilde{E}} x_{ji} - \sum_{i:(i,j)\in\tilde{E}} x_{ij} = 0, && \forall j \in V \\
& 0 \leq x_{ij} \leq u_{ij}, && \forall (i,j) \in \tilde{E},
\end{aligned}
$$

where we assume u_{ts} to be some very large number M so that capacity of that edge is not an obstacle to maximizing the flow.

The constraint matrix corresponding to this problem is TU as the constraints can be rewritten as

$$
\begin{bmatrix} A & 0_{|E|^2} \\ I_{|E|} & I_{|E|} \end{bmatrix} \begin{bmatrix} x \\ y \end{bmatrix} = \begin{bmatrix} 0 \\ u \end{bmatrix},
$$

with y being an extra vector of $|E|$ nonnegative decision variables to enforce the capacity constraints. We use the notation 0_n and I_n to denote, respectively, the matrix of all zeros of size n and the identity matrix of size n; see Appendix A. The TU-ness of such an extended matrix follows from the argument for A in the case of the shortest path problem and then by the large matrix being obtained by introducing unit vectors as additional rows and columns (which are all TU-preserving operations).

The "mirror" of the max-flow problem is the **minimum-cut problem (min-cut problem)**, which aims to find the best particular partition of V, called **cut**, into two disjoint sets V_s and V_t such that $s \in V_s, t \in V_t, V_s \cap V_t = \emptyset$, and $V_s \cup V_t = V$. The capacity of a cut (V_s, V_t) is then defined as the sum of capacities of edges passing from V_s to V_t, that is,

$$c(V_s, V_t) = \sum_{(i,j) \in E \, : \, i \in V_s, j \in V_t} u_{ij}.$$

The goal is to find a cut with minimum capacity and the min-cut problem can be formulated as follows:

$$\min \quad \sum_{(i,j) \in E} u_{ij} x_{ij}$$

$$\text{s.t.} \quad x_{ij} \geq y_j - y_i, \qquad\qquad \forall \, (i,j) \in E$$

$$y_s = 0$$

$$y_t = 1$$

$$x_{ij} \geq 0, \qquad\qquad \forall \, (i,j) \in E$$

$$y_i \in \mathbb{B}, \qquad\qquad \forall \, i \in V.$$

In the optimal solution, $y_i = 0$ if $i \in V_s$ and $y_i = 0$ if $i \in V_t$, while the variable x_{ij} is equal to 1 if there is an edge from $i \in V_s$ to $j \in V_t$.

It can be shown that without integrality constraints the min-cut problem is (a reformulation of) the dual of the max-flow problem. Also, it is well known that both problems have the same optimal objective value for a given network and capacities. Efficient dedicated algorithms, such as the Ford–Fulkerson algorithm, exist for both problems.

Finally, the following result is a well-known theorem, which is a special case of the strong duality theorem that holds for this linear network optimization problem.

Theorem 4.1 (Max-flow min-cut theorem) *The maximum amount of flow passing from the source to the sink is equal to the total weight of the edges in a minimum cut, that is, the smallest total weight of the edges that, if removed, would disconnect the source from the sink.*

4.3.6 Maximum Matching

Consider now the special case of the max-flow problem, known as the **maximum matching problem**, in which the goal is to identify the largest possible number of distinct node pairs that are connected by an edge in a given graph. Maximum matching problems arise, for example, when we try to find as many matchings as possible between potential kidney donors and recipients while taking into account suitable donors/recipients compatibility and willingness to donate.

First, we introduce the simpler version of this problem, in which we deal with bipartite graphs, as illustrated in Figure 4.11. In the bipartite maximum matching problem, we are given two sets I and J, and the edges connect only nodes in I with nodes in J, that is, $E \subseteq I \times J$. A matching $M \subseteq E$ is a subset of edges such that no two edges in M have a shared node. The objective is to find a matching with maximum cardinality. In the aforementioned example in healthcare, this corresponds to matching kidney donors and suitable recipients.

The bipartite maximum matching problem can be modeled as a max-flow problem by adding a source node s connected to I, a sink node t from J, and setting all edge capacities to 1. On the other hand, its MILO formulation is

$$\max \quad \sum_{(i,j) \in E} x_{ij}$$

$$\text{s.t.} \quad \sum_{i \in I : (i,j) \in E} x_{ij} \leq 1$$

$$\sum_{j \in J : (i,j) \in E} x_{ij} \leq 1$$

$$x_{ij} \in \mathbb{B}, \qquad\qquad\qquad \forall\, (i,j) \in E.$$

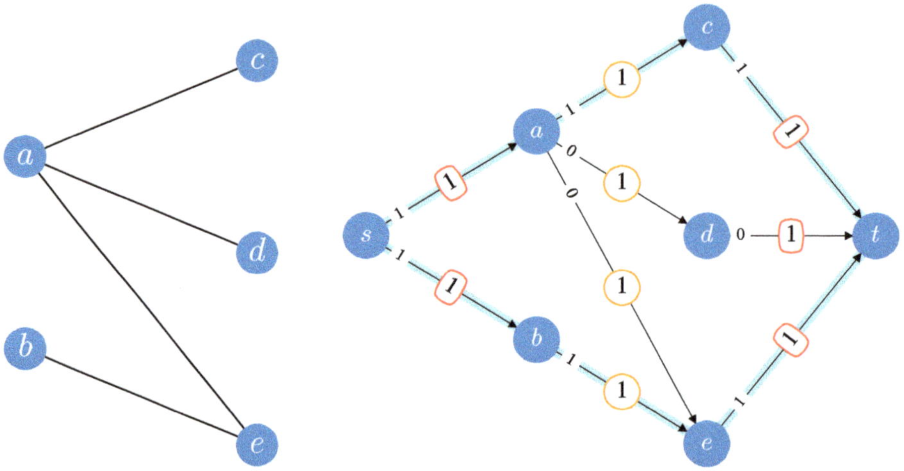

(a) Problem data.   (b) Min-cost flow reformulation and the corresponding solution.

Figure 4.11 Bipartite maximum matching between left-hand and right-hand side nodes, and its conversion to a max-flow problem (arc capacities in ovals on the arcs, optimal solution as numbers).

The constraint matrix of bipartite matching is TU, implying that the solution of the linear relaxation is integer. Hence, we can solve this problem with linear optimization. There also exists a dedicated algorithm called the **Hungarian method**.

A more difficult general version of this problem is the one in which the nodes cannot be partitioned into disjoint sets I and J such that $E \subseteq I \times J$, as illustrated in Figure 4.12.

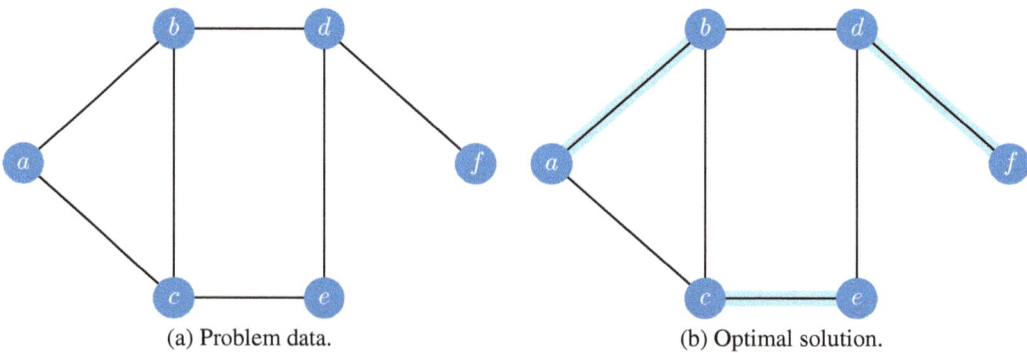

(a) Problem data.   (b) Optimal solution.

Figure 4.12 Maximum-weight nonbipartite matching.

Let A be the **incidence matrix** of the network, that is, $A \subseteq \mathbb{B}^{|V| \times |E|}$ with $a_{k,(i,j)} = 1$ if $k \in \{i,j\}$ and $a_{k,(i,j)} = 0$ otherwise. The MILO formulation is

$$\max \quad \sum_{e \in E} x_e$$

$$\text{s.t.} \quad A\boldsymbol{x} \leq 1,$$

$$\boldsymbol{x} \in \mathbb{B}^{|E|}.$$

In this more general setting, the constraint matrix is *not* TU and the linear relaxation does not have integer optimal solutions. However, the general matching problem can still be efficiently solved using a dedicated algorithm like the **blossom algorithm** [Sch03].

4.3.7 Traveling Salesman

We now present an important network problem for which the resulting constraint matrix is, in general, not TU, and which is one of the "holy grail" problems in network-based optimization. Given a set of cities V and distances $c_{ij} \in \mathbb{R}$ between cities $i, j \in V$, the **traveling salesman problem (TSP)** determines the shortest route to visit all cities exactly once. A formulation for the TSP is

$$\min \quad \sum_{i \in V} \sum_{j \in V} c_{ij} x_{ij}$$

$$\text{s.t.} \quad \sum_{i \in V} x_{ij} = 1, \qquad\qquad \forall\, j \in V$$

$$\sum_{j \in V} x_{ij} = 1, \qquad\qquad \forall\, i \in V$$

$$\sum_{i \in S, j \notin S} x_{ij} \geq 1, \qquad\qquad \forall\, S \subseteq V \qquad\qquad (4.6)$$

$$x_{ij} \in \mathbb{B}, \qquad\qquad \forall\, i, j \in V.$$

Note that this formulation has exponentially many constraints, due to the connectivity constraints (4.6) (a form of subtour elimination).

The TSP is NP-hard and there are no known algorithms to solve the problem efficiently in general. For more details on the TSP, its history, and current state of the art, we refer the interested reader to the book [App+07] and the website [Coo].

4.3.8 Graph Coloring

Consider an undirected graph $G = (V, E)$ consisting of a set V of nodes and a collection $E \subseteq V \times V$ of edges. A **graph coloring** of G is an assignment of colors to the vertices of G such that no two adjacent vertices share the same color. The **graph coloring problem** aims to find any such coloring. In addition to finding a coloring, one may be interested in using as few different colors as possible. This seemingly theoretical problem can be used to model real-life situations, such as scheduling problems like the one illustrated in the following example.

Example 4.7 (Exam schedule as graph coloring) In a small university, there is a single lecture hall that is suitable for exams. Consider the problem of choosing the time slots for each exam among a finite number of available time slots during the same week. Assume that the lecture hall is large enough to accommodate any number of exams in parallel. Since many courses are taken by the same students, exams for two courses that share at least one student cannot be scheduled in the same time slot. Every time slot the lecture hall is allocated for an exam has some indirect costs related to the hiring of temporary staff to invigilate the exams. The goal is thus to use as few time slots as possible while obeying the above restriction.

This situation can be modeled as a graph coloring problem. Consider an undirected graph $G = (V, E)$, in which the various courses are represented by nodes and two nodes are connected whenever at least one student takes the exams for both courses. Each node must be assigned a color, each color representing a different time slot. Solving the scheduling problem above corresponds to finding a coloring that uses the smallest number of colors possible.

We can formulate this problem using MILO as follows. Consider K possible colors (time slots) and define the following decision variables:

- $w_k \in \mathbb{B}$, $k = 1, \ldots, K$ is equal to 1 if the kth color (time slot) is used;
- $x_{ik} \in \mathbb{B}$, $i \in V$, $k = 1, \ldots, K$ is equal to 1 if the node (course) i is assigned the kth color (time slot).

Using these variables, we can formulate the minimum graph coloring problem as

$$\min \quad \sum_{k=1}^{K} w_k$$

$$\begin{aligned}
\text{s.t.} \quad & x_{ik} + x_{jk} \leq 1, && \forall (i,j) \in E, \forall k = 1, \ldots, K \\
& x_{ik} \leq w_k, && \forall i \in V, \forall k = 1, \ldots, K \\
& \sum_{i \in V} x_{ik} = 1, && \forall k = 1, \ldots, K \\
& w_k \in \mathbb{B}, && \forall k = 1, \ldots, K \\
& x_{ik} \in \mathbb{B}, && \forall i \in V, \forall k = 1, \ldots, K.
\end{aligned}$$

In this formulation, the objective function tries to minimize the number of colors used, while the constraints ensure that (i) two connected nodes do not share the same color, (ii) only a color k that is being used can be assigned to a node i, and (iii) each node has to be assigned exactly one color. The optimal solution to this problem is known as the **chromatic number** of the graph G and corresponds to the minimum number of time slots needed to schedule all exams (see Figure 4.13 for example problem and solution).

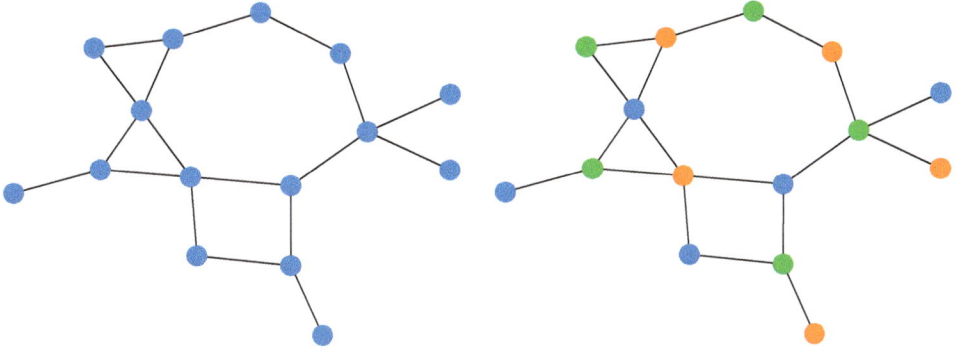

(a) Original network instance.   (b) Optimal solution using the
                                      minimum number of colors, that is, 3.

Figure 4.13 Example of a graph coloring problem solved using $K = 7$.

```python
def graph_coloring(G, K=7):
    m = pyo.ConcreteModel("Graph Coloring")

    m.colors = pyo.RangeSet(0, K - 1)
    m.nodes = pyo.Set(initialize=list(G.nodes))
    m.edges = pyo.Set(initialize=list(G.edges), dimen=2)

    m.x = pyo.Var(m.nodes, m.colors, domain=pyo.Binary)
    m.w = pyo.Var(m.colors, domain=pyo.Binary)

    @m.Constraint(m.edges, m.colors)
    def edge_constraint(m, i, j, k):
        return m.x[i, k] + m.x[j, k] <= 1

    @m.Constraint(m.nodes)
    def node_constraint(m, i):
        return pyo.quicksum(m.x[i, k] for k in m.colors) == 1

    @m.Constraint(m.nodes, m.colors)
    def used_color_constraint(m, i, k):
        return m.x[i, k] <= m.w[k]

    @m.Objective(sense=pyo.minimize)
    def number_used_colors(m):
        return pyo.quicksum(m.w[k] for k in m.colors)

    return m


# assumes G is a networkx graph
m = graph_coloring(G, K)
SOLVER.solve(m)
```

The graph coloring problem is known to be NP-hard and, hence, certainly does not admit a TU-matrix constraint formulation. As mentioned above, it can be used to model a variety of problems where nonoverlapping constraints are used, and tricks developed to solve the graph coloring problem more efficiently can also support the solution of such problems.

4.3.9 Modeling Techniques

As we have seen already in the previous examples, to "tweak" a new problem into an already known one, several gimmicks have been devised. The most common of them are:

- *Modifying/adding edges.* In a network, edges/arcs can be reversed/added and their capacities and/or costs can be set/adjusted. We introduced additional edges, for instance, in Example 4.1 and Figure 4.4.
- *Splitting nodes.* A node $v \in V$ can be split into two nodes v^+ and v^-. All edges $(v,w) \in E$ now originate from v^-, that is, $(v^-,w) \in \hat{E}$ for all $(v,w) \in E$. The edges $(w,v) \in E$ point to v^+: $(w,v^+) \in \hat{E}$ for all $(w,v) \in E$. Finally, a new edge (v^+,v^-) is added. By setting a capacity on (v^+,v^-), we can limit the amount of flow sent through node v (when translating the flow back to the original graph). We have seen this trick applied in Example 4.1.
- *Adding nodes.* Nodes can be added to model specific features of the network. An important special case is the one in which a source node s and a sink node t can be added. The source has edges to all original nodes: $(s,v) \in \hat{E}$ for all $v \in V$. The sink has edges from all original nodes: $(v,t) \in \hat{E}$ for all $v \in V$. We have seen this trick being applied in Figure 4.4.

4.4 A Complete Example: Arbitrage Search in Cryptocurrency Markets

Cryptocurrency exchanges are web services that enable the purchase, sale, and exchange of cryptocurrencies. These exchanges provide liquidity to owners and establish the relative value of these currencies. Joining an exchange enables a user to maintain multiple currencies in a digital wallet, buy and sell currencies, and use cryptocurrencies for financial transactions.

In this example, we explore the efficiency of cryptocurrency exchanges by testing for arbitrage opportunities. Arbitrage exists if a customer can realize a net profit through a sequence of risk-free trades. The efficient market hypothesis assumes that arbitrage opportunities are quickly identified and exploited by investors. As a result of their trading, prices reach a new equilibrium, so that any arbitrage opportunities would be small and fleeting in an efficient market. The question here is whether it is possible, with real-time data and rapid execution, for a trader to profit from these fleeting arbitrage opportunities.

In addition to the standard packages, in this example, we also need the packages `networkx` for graph-related functionalities and `ccxt`, an open-source library that supports the real-time APIs of the largest and most common exchanges on which cryptocurrencies are traded.

First, we need some terminology. Trading between two specific currencies is called a market, and each exchange hosts multiple markets; `ccxt` labels each market with a symbol common across exchanges. The market symbol is an upper-case string with abbreviations for a pair of traded currencies separated by a slash (/). The first abbreviation is the base currency, the second is the quote currency. Prices for the base currency are denominated in units of the quote currency. As an example, ETH/BTC refers to a market for the base currency

Table 4.6 *Example exchange order book.*

	base	quote	bid_price	bid_volume	ask_price	ask_volume
BTC/USD	BTC	USD	22,802.31	0.305724	22,806.98	1.06225
BUSD/USD	BUSD	USD	1.0	305,903.93	1.0001	526,363.54
ETH/USD	ETH	USD	1,821.78	0.06938	1,822.12	3.7
USDC/USD	USDC	USD	1.0001	496,371.76	1.0002	369,691.6
USDT/USD	USDT	USD	1.0	301,354.9	1.0001	1,043,381.46
UST/USD	UST	USD	–	–	–	–
BTC/USDT	BTC	USDT	22,800.85	0.15	22,807.99	0.159694
BUSD/USDT	BUSD	USDT	0.9998	249,813.73	0.9999	174,993.84
ETH/USDT	ETH	USDT	1,821.31	1.0	1,822.08	1.0
USDC/USDT	USDC	USDT	0.9998	264,185.0	0.9999	168,052.0
UST/USDT	UST	USDT	–	–	–	–
ETH/BTC	ETH	BTC	0.079848	1.266	0.079884	0.365
BTC/BUSD	BTC	BUSD	22,800.39	0.04	22,810.05	0.002462
ETH/BUSD	ETH	BUSD	1,821.14	0.305	1,822.18	0.305
USDC/BUSD	USDC	BUSD	0.9999	339,672.09	1.0001	365,694.2
BTC/USDC	BTC	USDC	22,801.97	0.095	22,808.93	0.095
ETH/USDC	ETH	USDC	1,821.25	0.295	1,822.13	0.295
BTC/UST	BTC	UST	–	–	–	–

Ethereum (ETH) quoted in units of the Bitcoin (BTC). The same market symbol can refer to an offer to sell the base currency (a "bid") or to an offer to sell the base currency (an "ask"). For example, x ETH/BTC means we can buy x units of BTC with one unit of ETH.

An exchange can be represented by a directed graph constructed from the market symbols available on that exchange, as in Figure 4.14. In such a graph currencies correspond to nodes on the directed graph. Market symbols correspond to edges in the directed graph, with the source indicating the quote currency and the destination indicating the base currency.

The order book for a currency exchange is the real-time inventory of trading orders. A *bid* is an offer to buy up to a specified amount of the base currency at a price not exceeding the "bid price" in the quote currency. An *ask* is an offer to sell up to a specified amount of the base currency at a price no less than a value specified given in the quote currency. The exchange attempts to match the bid to ask order at a price less than or equal to the bid price. If a transaction occurs, the buyer will receive an amount of base currency less than or equal to the bid volume and the ask volume, at a price less than or equal to the bid price and no less than the specified value.

The order book for currency exchange is the real-time inventory of orders, for example, in Table 4.6. The exchange order book maintains a list of all active orders for symbols traded on the exchange. Incoming bids above the lowest ask or incoming asks below the highest bid will be immediately matched, and transactions executed following the rules of the exchange.

Modeling the Arbitrage Search Problem as a Graph

A bid appearing in the order book for the market symbol b/q is an order from a prospective counterparty to purchase an amount of the base currency b at a bid price given in a quote currency q. For a currency trader, a bid in the order book is an opportunity to convert the base currency b into the quote currency q.

The order book can be represented as a directed graph where the nodes correspond to individual currencies. A directed edge $b \to q$ from node b to node q describes an opportunity for us to convert currency b into units of currency q. Let V_b and V_q denote the amounts of each currency held by us, and let $x_{b \to q}$ denote the amount of currency b exchanged for currency q. Following the transaction $x_{b \to q}$ we have the following changes to the currency holdings:

$$\Delta V_b = -x_{b \to q}$$
$$\Delta V_q = a_{b \to q} x_{b \to q},$$

where $a_{b \to q}$ is a *conversion coefficient* equal to the price of b expressed in terms of currency q. The capacity $c_{b \to q}$ of a trading along edge $b \to q$ is specified by a relationship

$$x_{b \to q} \le c_{b \to q}.$$

Because the arcs in our graph correspond to two types of orders – bid and ask – we need to build a consistent way of expressing them in our $a_{b \to q}$, $c_{b \to q}$ notation. Now, imagine that we are the party that accepts the buy and ask bids existing in the graph.

For bid orders, we have the opportunity to convert the base currency b into the quote currency q, for which we will use the following notation:

$$a_{b \to q} = \text{bid price}$$
$$c_{b \to q} = \text{bid volume}.$$

An ask order for symbol b/q is an order to sell the base currency at a price not less than the ask price given in terms of the quote currency. The ask volume is the amount of base currency to be sold. For us, a sell order is an opportunity to convert the quoted currency into the base currency such that

$$a_{q \to b} = \frac{1}{\text{ask price}}$$
$$c_{q \to b} = \text{ask volume} \times \text{ask volume}.$$

In the directed graph shown in Figure 4.14, created using data from an exchange order book, we distinguish between different order types, highlighting the bid orders with green color, and ask orders with red color.

Searching for a Negative Cycle: Dedicated Network Algorithms

With this unified treatment of bid and ask orders, we are ready to formulate the mathematical problem. An arbitrage exists if it is possible to find a closed path and a sequence of transactions in the directed graph, resulting in a net increase in currency holdings. Given a path

$$i_0 \to i_1 \to i_2 \to \cdots \to i_{n-1} \to i_n$$

the path is closed if $i_n = i_0$. The path has finite capacity if each edge in the path has a nonzero capacity. For a sufficiently small holding w_{i_0} of currency i_0 (because of capacity constraints), a closed path with $i_0 = i_n$ represents an arbitrage opportunity if

$$\prod_{k=0}^{n-1} a_{i_k \to i_{k+1}} > 1. \tag{4.7}$$

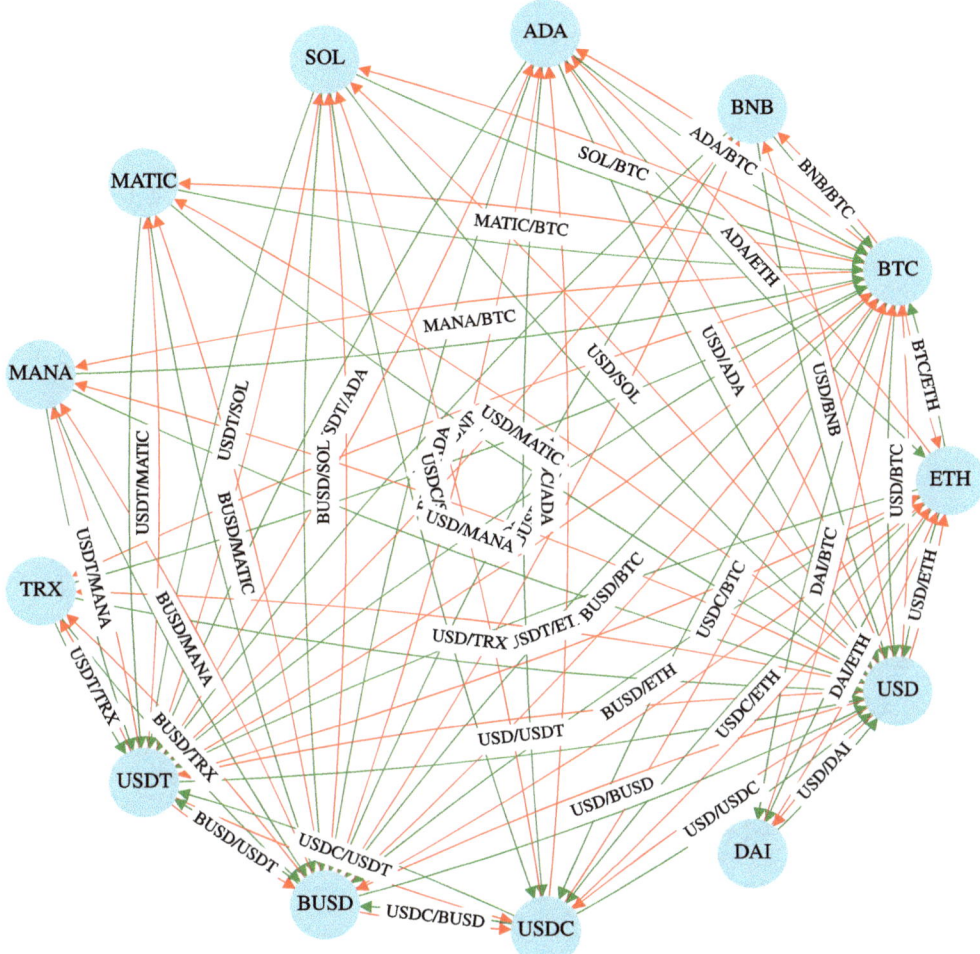

Figure 4.14 An example of an exchange order book.

If all we care about is simply finding an arbitrage cycle, regardless of the volume traded, we can use one of the many shortest path algorithms from the `networkx` library. To convert the problem of finding a path meeting (4.7) into a sum-of-terms to be minimized, we can take the negative logarithm of both sides to obtain the condition:

$$-\log\left(\prod_{k=0}^{n-1} a_{i_k \to i_{k+1}}\right) = -\sum_{k=0}^{n-1} \log(a_{i_k \to i_{k+1}}) < 0,$$

In other words, if we assign the negative logarithm as the weight of arcs in a graph, then our problem just becomes translated into the problem of searching for a cycle with a total sum of weights along it to be negative.

A simple cycle is a closed path in which no node appears twice. Simple cycles are distinct if they are not cyclic permutations (essentially, rewriting the same path but with a different

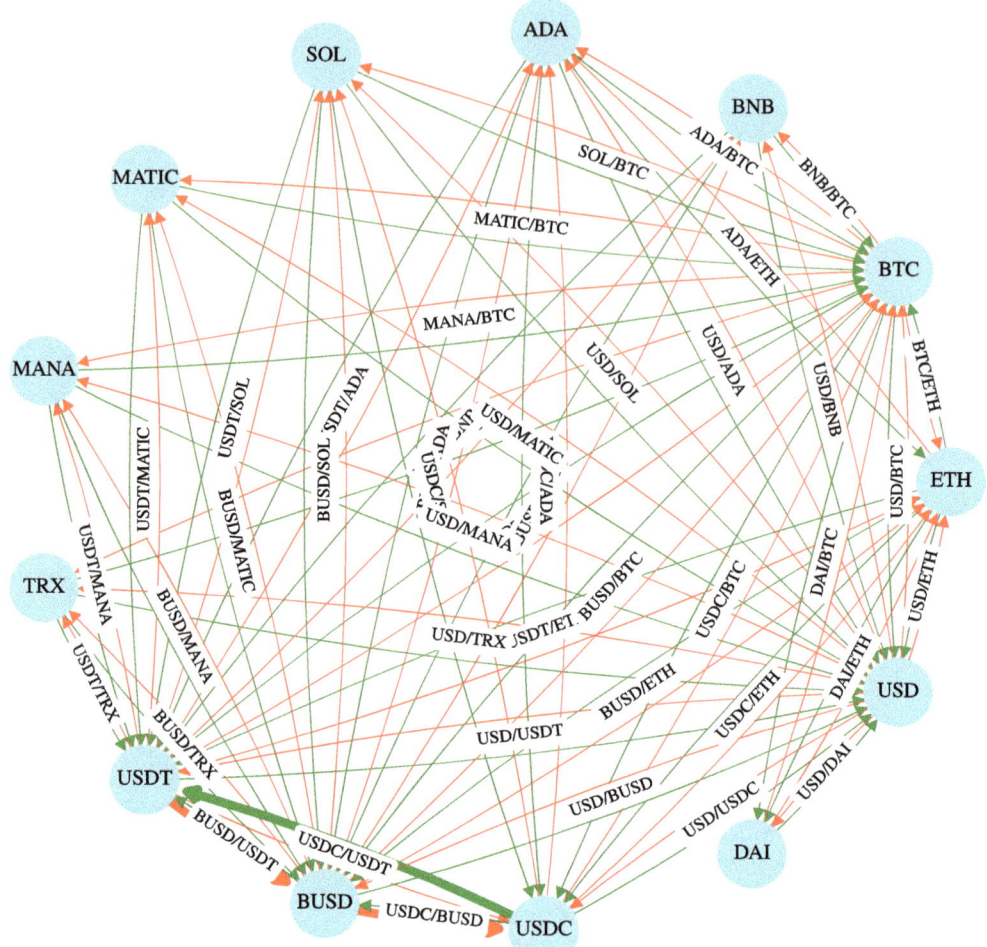

Figure 4.15 An arbitrage opportunity with a return of 0.05% identified as a negative cycle. The arcs corresponding to the involved transactions are displayed as thicker lines.

start=end point) of each other. One could check for arbitrage opportunities by checking if there are any negative simple cycles in the graph.

However, searching for a negative-weight cycle through searching for an arbitrage opportunity can be a daunting task – a brute-force search over all simple cycles has complexity $(n + e)(c + 1)$, which is impractical for large-scale applications.

A more efficient search based on the Bellman–Ford algorithm is embedded in the `networkx` function `negative_edge_cycle` that checks whether a negative cycle exists in a directed graph. This function is fast, but it only indicates if there is a negative cycle or not, and we do not even know what kind of cycle it is, so it would be hard to use that information to perform arbitrage. Luckily, the `networkx` library also includes the function `find_negative_cycle`, which returns a single negative edge cycle, if it exists. In Figure 4.15, we show the arbitrage opportunity found in this way, reporting the corresponding trading return.

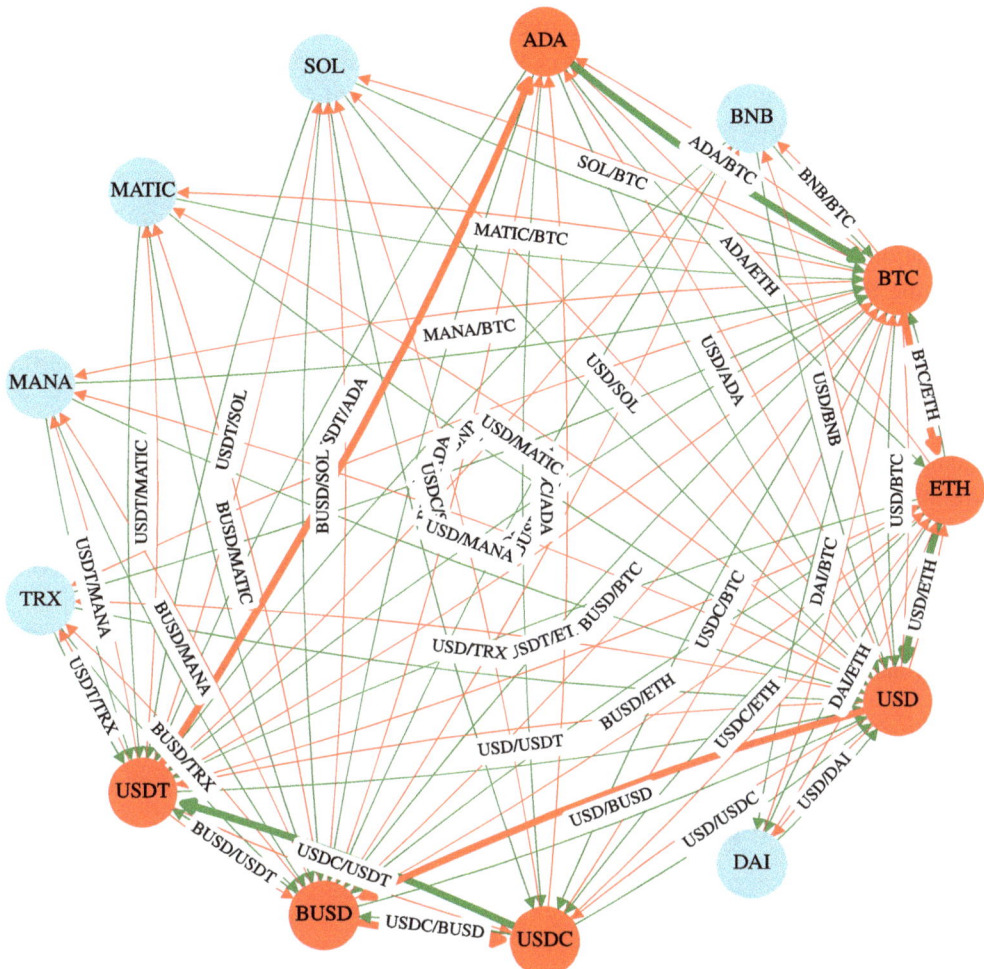

Figure 4.16 The rate-wise best single-cycle arbitrage opportunity with a return of 0.148% identified as a negative cycle. The arcs corresponding to the involved transactions are displayed as thicker lines.

This cycle has a return of 0.05% and it may not be the trading cycle with the maximum return. There may be other cycles with higher or lower returns that allow higher or lower trading volumes. A brute-force search over all simple cycles (i.e., the closed paths where no node appears twice) shows that there is a cycle of length 8 with a return of 0.148%, see Figure 4.16.

Optimizing the Arbitrage Under Capacity Constraints Using LO

The preceding analysis demonstrates some practical limitations of relying on generic implementations of network algorithms:

- First, more than one negative cycle may exist, so more than one arbitrage opportunity may exist, that is an optimal strategy consists of a combination of cycles.

- Second, simply searching for a negative cycle using shortest path algorithms does not account for capacity constraints, that is, the maximum size of each of the exchanges. For that reason, one may end up with a cycle on which a good "rate" of arbitrage is available, but where the absolute gain need not be large due to the maximum amounts that can be traded.

Instead, we can formulate the problem of searching for a maximum-gain arbitrage via linear optimization.

Assume that we are given a directed graph where each edge $i \to j$ is labeled with a "multiplier" $a_{i \to j}$ indicating how many units of currency j will be received for one unit of currency i, and a "capacity" $c_{i \to j}$ indicating how many units of currency i can be converted to currency j.

We break the trading process into steps indexed by $t = 1, 2, \ldots, T$, where currencies are exchanged between two adjacent nodes in a single step. We denote by $x_{i \to j}(t)$ the amount of currency traded from node i to j in step t. In this way, we start with the amount $w_{USD}(0)$ at time 0 and aim to maximize the amount $w_{USD}(T)$ at time T. Denote by O_j the set of nodes to which outgoing arcs from j lead, and by I_j the set of nodes from which incoming arcs lead.

A single transaction converts $x_{i \to j}(t)$ units of currency i to currency j. Following all transactions at event t, the trader will hold $v_j(t)$ units of currency j, where

$$v_j(t) = v_j(t-1) + \sum_{i \in I_j} a_{i \to j} x_{i \to j}(t) - \sum_{k \in O_j} x_{j \to k}(t).$$

For every directed edge $(j,k) \in E$, the sum of all transactions must satisfy

$$\sum_{t=1}^{T} x_{j \to k}(t) \le c_{j \to k}.$$

The objective of the optimization model is to find a sequence of currency transactions that increase the holdings of a reference currency. The solution is constrained by assuming that the trader cannot short-sell any currency. The resulting model formulation then is

$$\max \quad w_{USD}(T)$$
$$\text{s.t.} \quad w_{USD}(0) = w_0 \tag{4.8}$$
$$w_j(t) = w_j(t-1) + \sum_{i \in I_j} a_{i \to j} x_{i \to j}(t) - \sum_{k \in O_j} x_{j \to k}(t), \quad \forall j \in N, \forall t > 0 \tag{4.9}$$
$$w_j(t-1) \ge \sum_{k \in O_k} x_{j \to k}(t), \quad \forall t > 0 \tag{4.10}$$
$$\sum_{t=1}^{T} x_{j,k}(t) \le c_{j \to k}, \quad \forall t > 0, \forall (j,k) \in E \tag{4.11}$$
$$w_j(t), x_{i \to j}(t) \ge 0, \tag{4.12}$$

where the constraints are

- (4.8): the initial amount condition;
- (4.9): the balance equations linking the state of the given node in the previous and subsequent time periods;
- (4.10): the constraint capturing the fact we cannot trade at time step t more of a given currency than we had in this currency from time step $t - 1$. This constraint "enforces" the time order of trades, that is, we cannot trade in time period t units that have been received in the same time period;
- (4.11): the capacity constraints related to the maximum allowed trade volumes;
- (4.12): the nonnegativity constraints.

The following Python code implements this formulation and Figure 4.17 illustrates the optimal arbitrage opportunity found in this way.

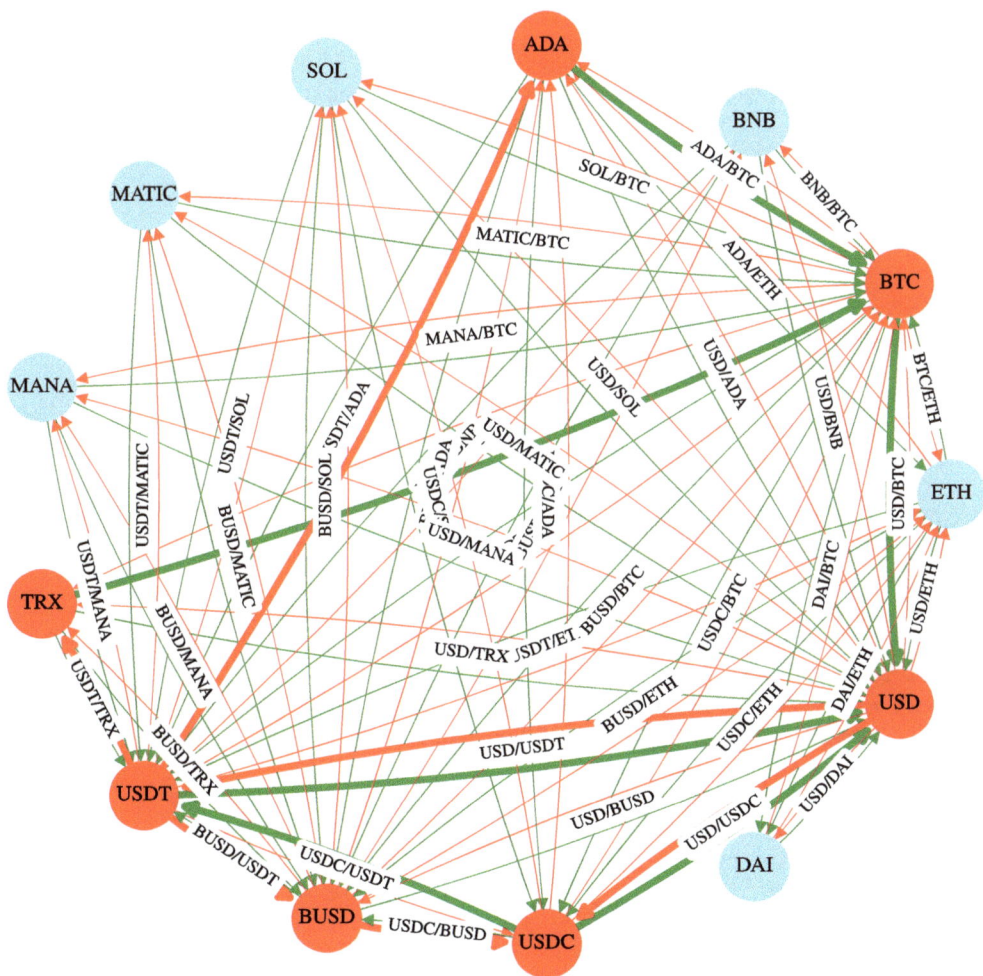

Figure 4.17 The optimal arbitrage opportunity with a return of 0.09% found in the case of capacity constraints with an initial wealth of $10,000 and using $T = 8$ as the maximum transaction chain length.

```python
def crypto_model(order_book_dg, T=10, v0=100.0):
    m = pyo.ConcreteModel("Cryptocurrency arbitrage")

    # length of the trading chain
    m.T0 = pyo.RangeSet(0, T)
    m.T1 = pyo.RangeSet(1, T)

    # currency nodes and trading edges
    m.NODES = pyo.Set(initialize=list(order_book_dg.nodes))
    m.EDGES = pyo.Set(initialize=list(order_book_dg.edges))

    # currency on hand at each node
    m.v = pyo.Var(m.NODES, m.T0, domain=pyo.NonNegativeReals)

    # amount traded on each edge at each trade
    m.x = pyo.Var(m.EDGES, m.T1, domain=pyo.NonNegativeReals)

    # total amount traded on each edge over all trades
    m.z = pyo.Var(m.EDGES, domain=pyo.NonNegativeReals)

    # "multiplier" on each trading edge
    @m.Param(m.EDGES)
    def a(m, src, dst):
        return order_book_dg.edges[(src, dst)]["a"]

    @m.Param(m.EDGES)
    def c(m, src, dst):
        return order_book_dg.edges[(src, dst)]["capacity"]

    @m.Objective(sense=pyo.maximize)
    def wealth(m):
        return m.v["USD", T]

    @m.Constraint(m.EDGES)
    def total_traded(m, src, dst):
        return m.z[src, dst] == sum([m.x[src, dst, t] for t in m.T1])

    @m.Constraint(m.EDGES)
    def edge_capacity(m, src, dst):
        return m.z[src, dst] <= m.c[src, dst]

    # initial assignment of 100 units on a selected currency
    @m.Constraint(m.NODES)
    def initial(m, node):
        if node == "USD":
            return m.v[node, 0] == v0
        return m.v[node, 0] == 0
```

```
@m.Constraint(m.NODES, m.T1)
def no_shorting(m, node, t):
    out_nodes = [dst for src, dst in m.EDGES if src == node]
    return m.v[node, t - 1] >= sum(m.x[node, dst, t] for dst in out_nodes)

@m.Constraint(m.NODES, m.T1)
def balances(m, node, t):
    in_nodes = [src for src, dst in m.EDGES if dst == node]
    out_nodes = [dst for src, dst in m.EDGES if src == node]
    return m.v[node, t] == m.v[node, t - 1] + sum(
        m.a[src, node] * m.x[src, node, t] for src in in_nodes
    ) - sum(m.x[node, dst, t] for dst in out_nodes)

    return m


# assume order_book_dg is the order book in the form of a directed weighted
↪    graph
v0 = 10000.0
T = 8
m = crypto_model(order_book_dg, T=T, v0=v0)
SOLVER.solve(m)
```

Exercises

Exercise 4.1 On a given day, five events take place at specified locations, with known starting hours and duration (also in hours) as listed below on the left.

event	location	start	duration		A	B	C	D	E
a	A	10	2	A	0	2	3	4	3
b	B	11	1	B	2	0	1	3	4
c	C	13	1	C	3	1	0	1	3
d	D	15	1	D	4	3	1	0	1
e	E	16	2	E	3	4	3	1	0

Each event needs a jury to be present and the same jury may be present at different events, provided that they can travel from one to the other on time. The travel times, in hours, between the event locations are also known, as listed above on the right.

(a) Establish a network of feasible connections, where each node is an event, and find the optimal solution to the problem of minimizing the number of required juries. Could you come up with a general procedure for larger problems that would always determine the optimal solution?

(b) Consider now a different model, where each event is an arc connecting its start node to its end node. In this network, end nodes connect to start nodes of events that can follow. Formulate the problem of the previous question as an MILO problem and solve it. Is it possible to solve it using only an LO solver?

(c) How could one modify the LO model of the previous question if the goal is to minimize the total number of hours spent by the juries, starting from the start of the first event to the end of the last event each of them attends? Is it possible to solve it using only an LO solver?

(d) What happens if we limit the total working hours (from the start of their first events until the end of their last event) spent by all juries involved to less than or equal to 10, but still want to minimize the number of juries involved? Is it possible to solve it using only an LO solver?

(e) What happens if the work of each individual jury must be at most 6 hours while minimizing either the number of juries involved or the total working time?

Exercise 4.2 Consider a network of computers connected such that for each pair of computers there exists a path between them, despite not all computers being directly connected. It is assumed that the time taken by a message to travel between two adjacent computers is known. Computers may forward messages, and therefore messages may traverse a path before reaching their destination. The goal is to send a message to all computers so that each computer receives the message at the earliest possible moment. How should such a message be routed in the network? Assume that the time spent processing the message before forwarding can be neglected.

Exercise 4.3 Consider an individual who just purchased (at time 0) a new car for $12,000. The cost of maintaining a car for one year depends on its age at the beginning of the year, as described in the table below. To avoid the high maintenance costs associated with an older car, it is possible to trade-in the car and purchase a new car. The price received on a trade-in depends on the age of the car at the time of trade-in (see the table below). We assume that at any time, it costs $12,000 to purchase a new car. The goal is to minimize the net cost (purchasing costs + maintenance costs − money received in trade-ins) incurred during the next five years.

Age of car (years)	Annual maintenance costs ($)	Trade-in price ($)
0	2,000	7,000
1	4,000	6,000
2	5,000	2,000
3	9,000	1,000
4	12,000	0

Formulate this problem as a shortest path problem and solve it using an algorithm of your choice.

Exercise 4.4 Consider a group of n_S students who need to be matched with n_P research projects. Each student has preferences for the projects they want to work on, denoted by a number $w_{sp} \in \{1, \ldots, 5\}$ where the higher the number, the higher the preference of student s for project p. At the same time, each project has a limited number c_p of available spots for students. Your goal is to maximize the satisfaction of both students and project supervisors by matching students with their preferred projects while respecting project capacity constraints.

Explain how to create a network optimization model to solve the problem of maximizing student satisfaction while respecting project capacity constraints.

Exercise 4.5 Consider organizing a social event where attendees need to be matched into pairs for activities. Each attendee can specify a list of people with whom they would like to be paired. Assuming the number of attendees is even, formulate the problem of maximizing the number of preferred matches.

Exercise 4.6 Assume that you are the curator of a prestigious art exhibition and you have an extensive collection of art pieces from different artists, each with its unique style and a finite set of colors used. The challenge is to arrange these artworks in a way that maximizes the visual appeal of the exhibition.

To create a visually stunning experience, you decide to hang the artwork on the walls of a large gallery. The catch is that you want to organize them in such a way that no two adjacent pieces share the same colors used.

Additionally, the artists have provided you with a list of constraints indicating which artworks should not be placed next to each other due to thematic or stylistic clashes.

Formulate the problem of finding the most aesthetically pleasing arrangement of the artworks on the walls, considering the constraints provided.

Exercise 4.7 In the city of Techville, a group of engineers and data scientists faces a unique and complex challenge. They are tasked with creating an efficient cable network using n strategically positioned connectors. Assume that the cost of building a direct cable between two connectors i and j is given by $w_{ij} > 0$. Formulate a mathematical model to determine the cheapest possible network under the assumption that it is possible to traverse between any two connectors along the cables (possibly via other connectors). How would you efficiently solve this problem?

Exercise 4.8 Consider again the problem of the previous exercise. The municipality of Techville adds a "simple" extra constraint: In order to have a serial connection of the terminals, each connector can at most be connected to at most two others. Formulate a mathematical model that incorporates this constraint.

Exercise 4.9 Consider a directed graph $G = (V, E)$ and two distinct vertices s and t in it. Two paths from s to t are called edge-disjoint if each edge of the graph appears at most in one of these paths. How would you determine the highest number of edge-disjoint paths?

Exercise 4.10 In 1927 Karl Menger proved that the maximum number of edge-disjoint s–t paths is equal to the minimum cardinality of a set of arcs whose removal disconnects s from t. This is a special case of a result discussed in Chapter 4; which one?

5

Convex Optimization

5.1 Introduction

Our aim so far has been to formulate a real-life decision problem as a (mixed integer) linear optimization problem. The reason was clear: Linear functions are simple, so problems formulated with them should also be simple. However, two questions arise. First, is the world of linear functions flexible enough to model all real-life problems? Second, are MILOs the only simple problems we can solve quickly?

Regarding the first question, an example will show that numerous real-life problems are naturally nonlinear, which might lead to difficulties. Regarding the second question, we will see that the real separation between simple and difficult problems luckily is not linear vs. nonlinear, but instead, convex vs. nonconvex. Convexity is a desirable property of an optimization problem that makes the optimization problem provably solvable to optimality.

We now begin with an example, showing what nonlinearity might lead to.

Example 5.1 (Milk pooling and blending) A bulk distributor supplies custom milk blends to several customers, denoted as 1, 2, and 3. Each customer has specified a minimum fat content, a maximum price, and a maximum amount of milk they wish to buy. The distributor sources raw milk from local farms denoted as A and B. Each farm produces milk with a known fat content and cost.

The distributor has recently identified more affordable sources of raw milk from two remote farms, denoted as C and D. These remote farms produce milk grades that can be blended with milk from local farms. However, the distributor only has one truck with a single tank to transport milk from the remote farms. As a result, milk from the remote farms must be mixed in the tank before being transported to the blending station. This creates a "pool" of uniform composition to be blended with local milk to meet the customer's requirements.

The problem data are summarized in Table 5.1. The fat content and cost of raw milk are given for each farm. For each customer, data is provided for the required milk fat content, price, and maximum demand. Milk can be transported from each farm A or B to each customer and also pooled from the remote farms C and D for subsequent delivery at each customer.

What is the optimal way to combine the transport of milk from the local farms A and B, and to pool the milk from the remote farms C and D, for the distributor to maximize revenue?

Table 5.1 *Customer data (top) and supplier data (bottom) for the milk pooling problem.*

	fat	price	demand
Customer 1	0.045	52.0	6000.0
Customer 2	0.030	48.0	2500.0
Customer 3	0.040	50.0	4000.0

	fat	cost	location
Farm A	0.045	45.0	local
Farm B	0.03	42.0	local
Farm C	0.033	37.0	remote
Farm D	0.05	45.0	remote

This problem is known as a **pooling problem** and several equivalent mathematical formulations exist for it. Here, we will use a formulation called p-parameterization, where the pool composition is a new continuous decision variable $p \geq 0$. The other decision variables are:

- $z_{l,c}$: the amount of milk transported from a local farm $l \in L = \{A,B\}$ to customer $c \in K$;
- x_r: the amount of raw milk purchased from remote farm $r \in R = \{C,D\}$;
- y_c: the amount delivered to customer $c \in K = \{1,2,3\}$ from the milk pooled from the remote farms.

The objective includes the cost of purchasing raw milk from the remote farms and the income received for selling the pooled milk:

$$\text{Profit} = \sum_{(l,c)\in L\times K} (\text{price}_c - \text{cost}_l)\, z_{l,c} + \sum_{c\in K} \text{price}_c y_c - \sum_{r\in R} \text{cost}_r x_r.$$

The first constraint states that the product delivered to each customer from local farms and the pool cannot exceed demand:

$$\sum_{l\in L} z_{l,c} + y_c \leq \text{demand}_c, \qquad \forall c \in K.$$

Additionally, purchases from remote farms and amounts delivered to customers from the pool must balance:

$$\sum_{r\in R} x_r = \sum_{c\in K} y_c.$$

Next, we assume that the milk fat composition of the milk pools is linear, that is, the average milk fat composition of the pool, p, must satisfy an overall balance on the milk fat entering the pool from remote farms and the milk fat delivered to customers:

$$\sum_{r\in R} \text{fat}_r \cdot x_r = p \underbrace{\sum_{c\in K} y_c}_{\text{nonlinear}}.$$

Finally, the milk fat required by each customer $c \in K$ satisfies a blending constraint:

$$\underbrace{p \cdot y_c}_{\text{nonlinear}} + \sum_{l \in L} \text{fat}_l \, z_{l,c} \geq \text{fat}_c^{\min} \cdot \left(\sum_{l \in L} z_{l,c} + y_c \right), \qquad \forall c \in K.$$

The last two constraints include nonlinear (to be precise, **bilinear**) terms, each consisting of the product $p \cdot y_c$ of the decision variable p with decision variables y_c for all $c \in K$.

The full mathematical model is as follows:

$$\max \quad \sum_{(l,c) \in L \times K} (\text{price}_c - \text{cost}_l) \, z_{l,c} + \sum_{c \in K} \text{price}_c y_c - \sum_{r \in R} \text{cost}_r x_r$$

$$\text{s.t.} \quad \sum_{l \in L} z_{l,c} + y_c \leq \text{demand}_c, \qquad\qquad\qquad \forall c \in K$$

$$\sum_{r \in R} x_r = \sum_{c \in K} y_c$$

$$\sum_{r \in R} \text{fat}_r \cdot x_r = p \sum_{c \in K} y_c$$

$$p \cdot y_c + \sum_{l \in K} \text{fat}_l \cdot z_{l,c} \geq \text{fat}_c^{\min} \left(\sum_{l \in L} z_{l,c} + y_c \right), \qquad \forall c \in K$$

$$z_{l,c}, x_r, y_c \geq 0, \qquad\qquad\qquad\qquad\qquad\qquad \forall r \in R, c \in K, l \in L$$

$$p \geq 0.$$

The products of p with the other variables have a profound consequence on the nature of the optimization problem, which is no longer an LO problem. This is not a modeling mistake, as there is no linear formulation of this pooling problem.

How can we proceed in this situation? For fixed p, the problem becomes linear, suggesting that a possible strategy could be solving the problem for a "dense grid" of possible p values to see how the optimal value changes. The resulting plot is given in Figure 5.1. This plot demonstrates how the nonconvex bilinear constraints can result in

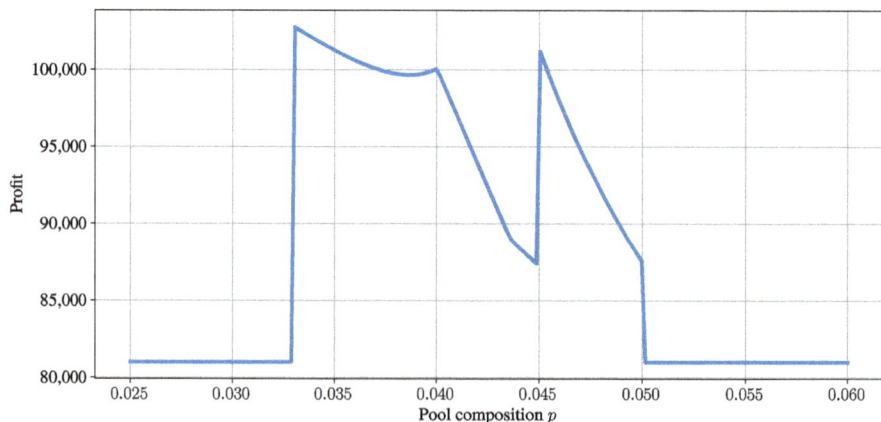

Figure 5.1 The profit as a function of the continuous variable p describing the average milk fat composition of the pool.

multiple local maxima – values $p = 0.04$ and $p = 0.045$ are both local maxima, that is, they yield higher profit than the values close to them, but they are not as good as the global maximum $p = 0.033$, which produces the largest profit $102,833.33$.

In this specific case, this shortcut was possible as there was only one trouble-making variable p, and we could search through the possible values. However, with more variables causing the nonlinearity, this is no longer easily done. We need an optimization technique that can solve for the optimal pool composition p and profit in one go.

The following Pyomo code solves this problem using a generic nonlinear solver `ipopt` [WB06], arriving at the optimal solution with $p = 0.033$ with a profit equal to $102,833.33$.

```python
def milk_pooling_bilinear():
    m = pyo.ConcreteModel("Milk pooling bilinear model")

    # define sources
    m.L = pyo.Set(initialize=suppliers.index[suppliers["location"] ==
    ↪    "local"].tolist())
    m.R = pyo.Set(
        initialize=suppliers.index[suppliers["location"] ==
        ↪    "remote"].tolist()
    )

    # define customers
    m.C = pyo.Set(initialize=customers.index.tolist())

    # define flowrates
    m.x = pyo.Var(m.R, domain=pyo.NonNegativeReals)
    m.y = pyo.Var(m.C, bounds=lambda m, c: (0, customers.loc[c, "demand"]))
    m.z = pyo.Var(m.L * m.C, domain=pyo.NonNegativeReals)
    m.p = pyo.Var(bounds=(remote_suppliers["fat"].min(),
    ↪    remote_suppliers["fat"].max()))

    @m.Objective(sense=pyo.maximize)
    def profit(m):
        return (
            +sum(
                m.z[l, c] * (customers.loc[c, "price"] - suppliers.loc[l,
                ↪    "cost"])
                for l, c in m.L * m.C
            )
            + sum(m.y[c] * customers.loc[c, "price"] for c in m.C)
            - sum(m.x[r] * suppliers.loc[r, "cost"] for r in m.R)
        )

    @m.Constraint(m.C)
    def customer_demand(m, c):
        return sum(m.z[l, c] for l in m.L) + m.y[c] <= customers.loc[c,
        ↪    "demand"]
```

```python
    @m.Constraint()
    def pool_balance(
        m,
    ):
        return sum(m.x[r] for r in m.R) == sum(m.y[c] for c in m.C)

    @m.Constraint()
    def pool_quality(m):
        return sum(suppliers.loc[r, "fat"] * m.x[r] for r in m.R) == m.p *
        ↪    sum(
            m.x[r] for r in m.R
        )

    @m.Constraint(m.C)
    def customer_quality(m, c):
        return m.p * m.y[c] + sum(
            suppliers.loc[l, "fat"] * m.z[l, c] for l in m.L
        ) >= customers.loc[c, "min_fat"] * (sum(m.z[l, c] for l in m.L) +
        ↪    m.y[c])

    return m


m_global = milk_pooling_bilinear()
pyo.SolverFactory("ipopt").solve(m_global)
```

In this case, however, the convergence to the optimal point was pure luck. The way general nonlinear optimization solvers work is that they begin from an initial solution and improve from then on, stopping when a set of conditions (known as the KKT conditions, to be explained later in Section 5.3.3), is met. If we start from a different initial point, we can end up in one of the suboptimal local minima. We can trigger this by changing the way `ipopt` picks the initial point using the optional argument of the function `milk_pooling_bilinear`. We then obtain as the solution the local maximum $p = 0.045$, corresponding to a profit of 101,392.15, which is roughly 1.4% worse than the global optimum. Although this might appear to be a slight difference, it can cost a lot of money in high-volume pooling applications.

As Example 5.1, the problem of minimizing a nonlinear function can easily have multiple **local optima**; see Figure 5.2 for the illustration of the concept. As the number of decision variables and product-of-variables terms increases, the number of local optima may also increase. Because nonlinear optimization solvers are essentially "local-optimum hunters," each of them will end up in one of the local optima, not knowing if it is the "best" one among them, the **global optimum**.

Does this mean we are doomed if a problem can only be expressed using nonlinear functions? No, as there is a broad class of optimization problems for which we can show that every local optimum is also global. Practically, it means that any nonlinear solver will give us the genuinely best solution.

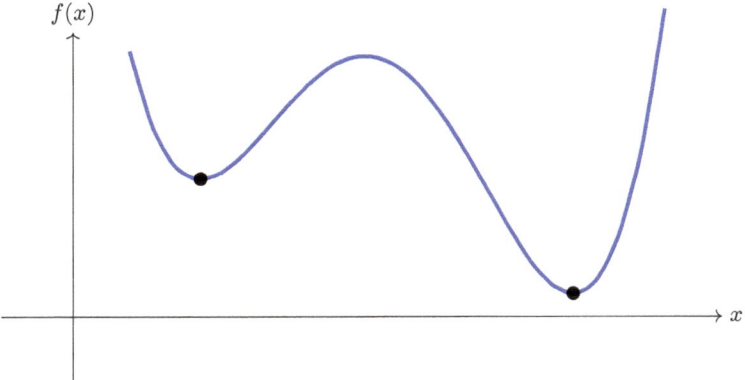

Figure 5.2 Illustration of a local minimum (on the left), a point at least as good as all the points sufficiently close to it, and a global minimum (on the right), a point at least as good as any other point) for a minimization problem.

Example 5.2 (Center of mass of a glass of water) Consider the following problem where we need to minimize a nonlinear function over an interval, which originally stems from determining the optimal height x of water in a glass (of cylindrical shape with diameter 4, height 20, and total mass 100) to ensure its maximum stability by minimizing the height of the center of mass:

$$\min \quad \frac{\pi x^2 + 500}{2\pi x + 50}$$
$$\text{s.t.} \quad 0 \le x \le 20.$$

Although nonlinear, this problem is easy to solve both analytically (searching for the point where the derivative equals 0) and numerically to global optimality.

To understand the essence of the difference between Example 5.1 and Example 5.2, we best see both of them as special cases of a general optimization problem:

$$\min_{\boldsymbol{x} \in S} f(\boldsymbol{x}). \tag{5.1}$$

In this context, we introduce convexity of functions and sets.

Definition 5.1 (Convex sets) A set $S \subseteq \mathbb{R}^n$ is **convex** if

$$\theta\boldsymbol{x} + (1 - \theta)\boldsymbol{y} \in S, \quad \forall \boldsymbol{x}, \boldsymbol{y} \in S, \theta \in [0,1].$$

A first example of a convex set is the feasible region of an LO in Figure 2.1. This set, called a polyhedron, is the intersection of half-spaces, each of which is a convex set. Additional examples of sets, one convex and two nonconvex, are shown in Figure 5.3. The rightmost set is an example of a feasible set of a MILO problem, which partly explains the difficulty of solving such problems.

Next, we define the concept of a **convex function**.

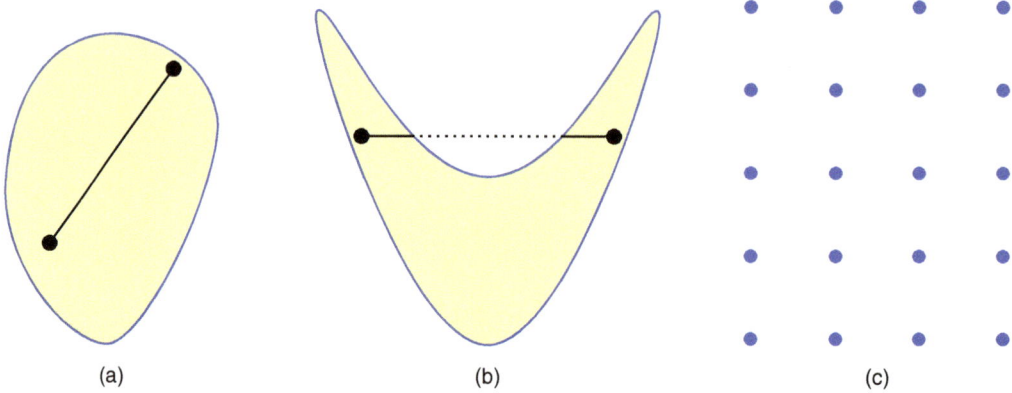

Figure 5.3 A convex set (a) and two nonconvex ones (b and c).

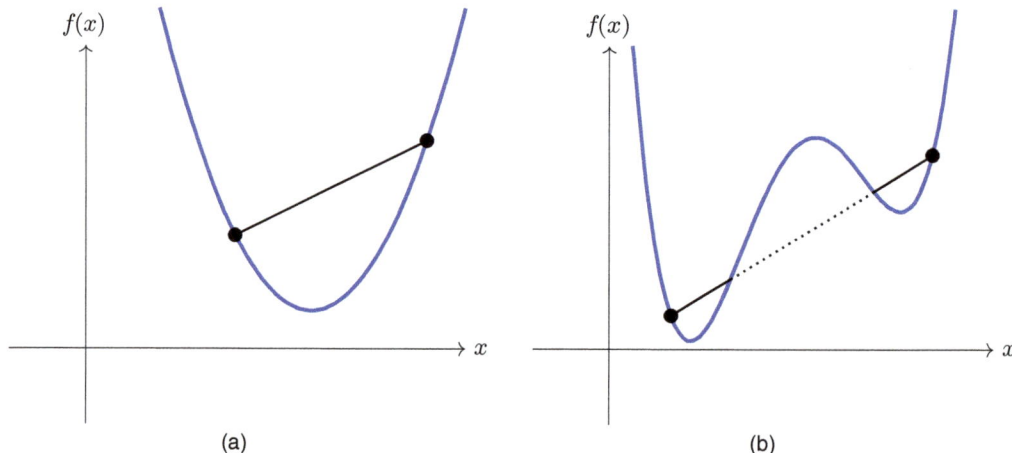

Figure 5.4 Examples of convex (a) and nonconvex (b) functions.

Definition 5.2 (Convex functions) We say that a function $f: S \subseteq \mathbb{R}^n \to \mathbb{R}$ is convex if its domain S is convex and for all points $x, y \in S$ and any scalar $0 \leq \theta \leq 1$

$$f\big(\theta x + (1-\theta)y\big) \leq \theta f(x) + (1-\theta)f(y).$$

We say that a function f is concave if $-f$ is convex. Finally, f is strictly convex (concave) if the above inequality is strict for all points $x, y \in S$.

In short, convex functions are the ones for which any line segment connecting two points on the graph of the function lies entirely above or on the graph. Examples of convex and nonconvex functions are given in Figures 5.4 and 5.5.

Returning to the optimization problem (5.1), our interest in the convexity of sets and functions is motivated by the following result.

Theorem 5.1 *The local minimum (maximum) of a convex (concave) function $f: S \subseteq \mathbb{R}^n \to \mathbb{R}^n$ over a convex set S is also its global minimum (maximum).*

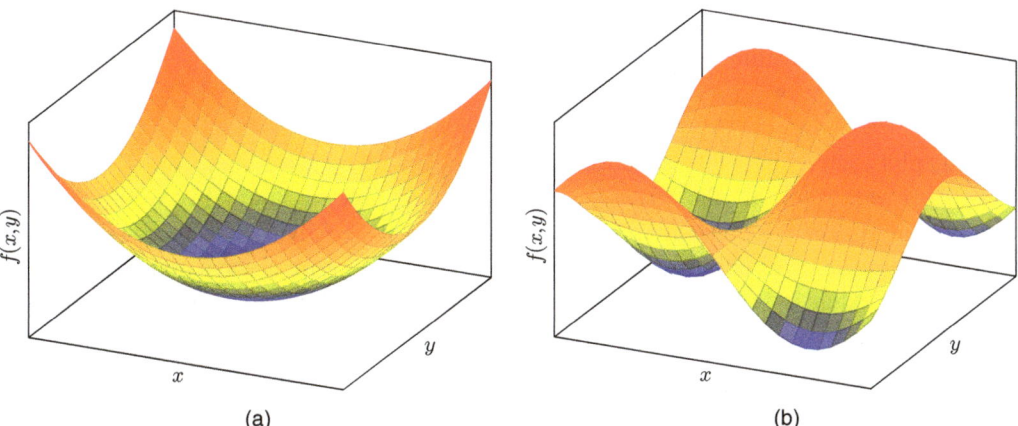

Figure 5.5 Examples of surface plots of convex (a) and nonconvex (b) functions of two variables (x, y).

If f is a convex function and S is a convex set, we refer to (5.1) as a **convex optimization problem**.

Now, let us get back to our examples and see them through the lens of the above definitions. In Example 5.1, when considering the problem of optimizing over p, the feasible set for p was convex. Still, the objective function was not (since it had multiple local minima), thus making the problem not convex. On the contrary, in Example 5.2, both the feasible set and the objective function were convex, which is why we can hope that any solver will give us a globally optimal solution.

For the above examples, we could verify the convexity of the functions and sets involved "with the naked eye." However, for large problems, we need to be able to model them in a convex way or verify their convexity using more scalable techniques. In the remainder of the chapter, we first focus on this aspect. Next, we extend the concept of duality introduced in Chapter 2 for LO to the convex case and briefly explain the most common solution algorithms.

5.2 Convex Optimization Problems: Modeling and Verification

Verifying whether an optimization problem of the form (5.1) is convex can be reduced to check one by one whether the functions used to formulate it are convex. This is because convex sets can be formulated using convex functions. More specifically, following a particular set of rules known as **disciplined convex programming**, we are able to build very complex convex problems using relatively simple convex functions. Thus, we begin with simple convex functions and a few conditions to verify convexity.

5.2.1 Convex Functions

Example 5.3 The following are examples of single-variable convex functions $\mathbb{R} \to \mathbb{R}$ are:

- any linear function $x \mapsto ax + b$;
- the exponential $x \mapsto e^x$;

- the power functions $x \mapsto x^a$ for fixed $a \geq 1$ or $a \leq 0$;
- the negative logarithm $x \mapsto -\log(x)$;
- the negative entropy $x \mapsto x \log(x)$.

One can plot each of the above functions to become convinced about their convexity, but how can we prove that for "harder" functions without being able to plot them? We need a more convenient criterion for convexity. The above functions are easily seen to be convex because they are all twice-differentiable and their second derivatives are always nonnegative. As it turns out, an extension of this criterion applies to functions of more variables. To formulate this extension, we need the concept of definiteness of a square matrix.

Definition 5.3 Let $S^n := \{Q \in \mathbb{R}^{n \times n} : Q = Q^\top\}$ be the set of all real symmetric $n \times n$ matrices. A matrix $Q \in S^n$ is called:

- **positive (semi)definite** if $t^\top Q t > 0$ ($t^\top Q t \geq 0$, respectively) for all vectors $t \in \mathbb{R}^n$ with $t \neq 0$;
- **negative (semi)definite** if $-Q$ is positive (semi)definite or, equivalently, if $t^\top Q t < 0$ ($t^\top Q t \leq 0$, respectively) for all vectors $t \in \mathbb{R}^n$ with $t \neq 0$;
- **indefinite** if it is neither positive semidefinite nor negative semidefinite.

We write $Q \succ 0$ (respectively, $Q \succeq 0$) if Q is a positive (semi)definite matrix and use the analogous notation for negative (semi)definite matrices.

Note that an equivalent condition for a symmetric matrix to be definite (semidefinite) positive is to have only strictly positive (nonnegative) eigenvalues of the matrices. An indefinite matrix has eigenvalues of both signs.

Example 5.4 Consider the following matrices:

$$Q_1 = \begin{bmatrix} 2 & -1 \\ -1 & 2 \end{bmatrix}, \quad Q_2 = \begin{bmatrix} -2 & -2 \\ 2 & -5 \end{bmatrix}, \quad Q_3 = \begin{bmatrix} 1 & -1 \\ -1 & 1 \end{bmatrix}, \quad Q_4 = \begin{bmatrix} -3 & 2 \\ -4 & 3 \end{bmatrix}.$$

The matrix Q_1 is positive definite, since $t^\top Q_1 t = 2\left(t_1^2 - t_2 t_1 + t_2^2\right) = \left(t_1 - \frac{t_2}{2}\right)^2 + \frac{3t_2^2}{4} > 0$ for every $t = (t_1, t_2) \in \mathbb{R}^2$ such that $t \neq 0$. The matrix Q_2 is negative definite, since $t^\top Q_2 t = -t_1^2 + 4t_1 t_2 - 4t_2^2 = -(t_1 - 2t_2)^2 < 0$ for every $t = (t_1, t_2) \in \mathbb{R}^2$ such that $t \neq 0$. The matrix Q_3 is positive semidefinite, since $t^\top Q_3 t = (t_1 - t_2)^2 \geq 0$, and zero can be attained by a nontrivial pair, for example, $(1, 1)^\top Q_3 (1, 1) = 0$. The matrix Q_4 is indefinite since $(0, 1)^\top Q_4 (0, 1) = -2 < 0$ and $(1, 0)^\top Q_4 (1, 0) = 1 > 0$.

We could have also reached the same conclusions by calculating the eigenvalues. We have $Q_1 \succ 0$, having all strictly positive eigenvalues, namely 1 and 3. Similarly, $Q_2 \prec 0$, since it has strictly negative eigenvalues (-6 and -1), while $Q_3 \succeq 0$ because it has nonnegative eigenvalues (2 and 0). Q_4 is indefinite because it has eigenvalues of both signs, namely -3 and 2.

Equipped with the above, we can state two general convexity criteria.

Proposition 5.1 *Let* $f \colon S \to \mathbb{R}$ *be a function with a convex domain* $S \subseteq \mathbb{R}^n$. *The following statements hold:*

(i) If f *is differentiable, then* f *is convex if and only if*

$$f(\boldsymbol{y}) \geq f(\boldsymbol{x}) + \nabla f(\boldsymbol{x})^\top (\boldsymbol{y} - \boldsymbol{x}), \quad \forall \, \boldsymbol{x}, \boldsymbol{y} \in S.$$

(ii) If f *is twice differentiable, then* f *is convex if and only if its Hessian* $\nabla^2 f(\boldsymbol{x})$ *is positive semidefinite for all* $\boldsymbol{x} \in S$.

Example 5.5 This criterion is sufficient to verify the convexity, for example, of a quadratic function

$$f(x_1, x_2) = x_1^2 + 4x_2^2,$$

since its Hessian is given by

$$\nabla^2 f(x_1, x_2) = \begin{bmatrix} 2 & 0 \\ 0 & 8 \end{bmatrix},$$

which is positive semidefinite since $\boldsymbol{t}^\top \nabla^2 f(x_1, x_2) \boldsymbol{t} = 2t_1^2 + 8t_2^2 \geq 0$.

Example 5.6 (Ordinary least squares regression part 1) In Chapter 2, we introduced linear regression with least absolute deviations (LAD); see Example 2.5. Here, we consider the same problem setting but slightly change the underlying optimization problem, in particular its objective function, obtaining the classical **ordinary least squares (OLS)** regression.

Like in Example 2.5, suppose that we have a finite dataset consisting of n points $\{(\boldsymbol{X}^{(i)}, y^{(i)})\}_{i=1,\dots,n}$ with $\boldsymbol{X}^{(i)} \in \mathbb{R}^k$ and $y^{(i)} \in \mathbb{R}$. We want to fit a linear model with intercept, whose error or deviation term e_i is equal to

$$e_i = y^{(i)} - \boldsymbol{m}^\top \boldsymbol{X}^{(i)} - b = y^{(i)} - \sum_{j=1}^{k} X_j^{(i)} m_j - b, \tag{5.2}$$

for some real numbers $b, m_1, \dots, m_k$. The OLS is a possible statistical optimality criterion for such a linear regression, which tries to minimize the sum of the squared errors, that is, $\sum_{i=1}^{n} e_i^2$. The OLS regression can thus be formulated as an optimization problem with coefficients b and m_i and errors e_i as the decision variables, namely:

$$\min \quad \sum_{i=1}^{n} e_i^2 \tag{5.3}$$

$$\text{s.t.} \quad e_i = y^{(i)} - \boldsymbol{m}^\top \boldsymbol{X}^{(i)} - b, \quad \forall \, i = 1, \dots, n$$

$$\boldsymbol{m} \in \mathbb{R}^k$$

$$b \in \mathbb{R}$$

$$\boldsymbol{e} \in \mathbb{R}^n.$$

Let $\boldsymbol{\theta} = (b, \boldsymbol{m}) \in \mathbb{R}^{k+1}$ be the vector comprising all decision variables. Denote the vector $\boldsymbol{y} = (y^{(1)}, \ldots, y^{(n)})$ and define the so-called **design matrix** $\tilde{\boldsymbol{X}} = \mathbb{R}^{n \times (k+1)}$ associated with the dataset as

$$\tilde{\boldsymbol{X}} = \begin{bmatrix} 1 & \boldsymbol{X}^{(1)} \\ 1 & \boldsymbol{X}^{(2)} \\ \vdots & \vdots \\ 1 & \boldsymbol{X}^{(n)} \end{bmatrix}.$$

We can then rewrite (5.3) as an unconstrained optimization problem in the vector of variables $\boldsymbol{\theta}$, namely,

$$\min_{\boldsymbol{\theta}} f(\boldsymbol{\theta}),$$

with $f : \mathbb{R}^{k+1} \to \mathbb{R}$ defined as $f(\boldsymbol{\theta}) := \|\boldsymbol{y} - \tilde{\boldsymbol{X}}\boldsymbol{\theta}\|_2^2$. Note that here $\boldsymbol{y}$ and $\boldsymbol{X}^{(i)}$, $i = 1, \ldots, n$ are not vectors of variables, but rather of known parameters. The Hessian of the objective function can be calculated to be

$$\nabla^2 f(\theta) = 2\tilde{\boldsymbol{X}}^\top \tilde{\boldsymbol{X}}.$$

In particular, it is a constant matrix that does not depend on the variables $\boldsymbol{\theta}$ and is always positive semidefinite, since

$$\boldsymbol{t}^\top \nabla^2 f(\theta)\boldsymbol{t} = 2\boldsymbol{t}^\top \tilde{\boldsymbol{X}}^\top \tilde{\boldsymbol{X}}\boldsymbol{t} = 2\|\tilde{\boldsymbol{X}}\boldsymbol{t}\|_2^2 \geq 0, \qquad \forall \boldsymbol{t} \in \mathbb{R}^{k+1}.$$

The OLS optimization problem is then always convex.

The OLS problem can be implemented in Pyomo as follows. Leveraging the problem's convexity and thus the solution's uniqueness, we can resort to the generic nonlinear solver `ipopt`. Figure 5.6 shows the two linear fits obtained using the OLS and LAD regression.

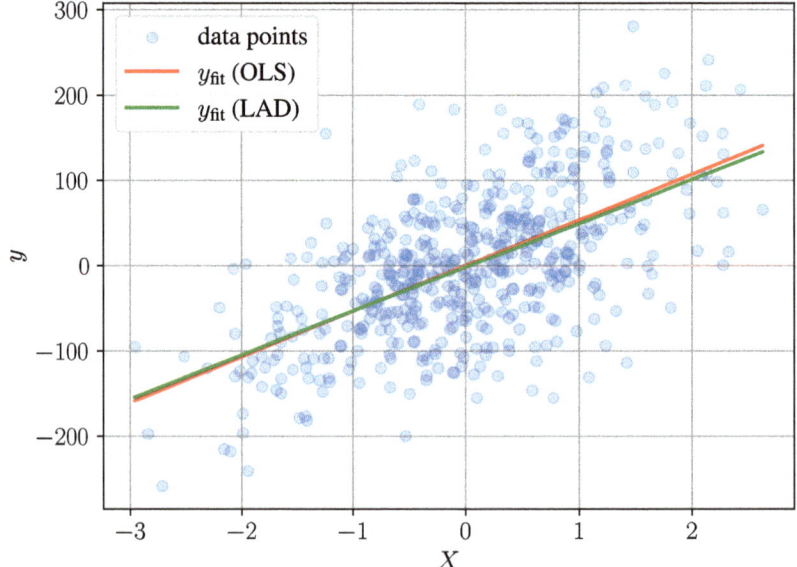

Figure 5.6 Comparison of the OLS regression and LAD regression for a synthetic dataset of 500 points.

```python
def ols_regression(X, y):
    m = pyo.ConcreteModel("OLS Regression")

    n, k = X.shape

    m.I = pyo.RangeSet(0, n - 1)
    m.J = pyo.RangeSet(0, k - 1)

    m.e = pyo.Var(m.I, domain=pyo.Reals)
    m.m = pyo.Var(m.J)
    m.b = pyo.Var()

    @m.Constraint(m.I)
    def residuals(m, i):
        return m.e[i] == y[i] - sum(X[i][j] * m.m[j] for j in m.J) - m.b

    @m.Objective(sense=pyo.minimize)
    def sum_of_square_errors(m):
        return sum((m.e[i]) ** 2 for i in m.I)

    return m


m = ols_regression(X, y)
pyo.SolverFactory("ipopt").solve(m)
```

In the above example, we could use the differentiability-based convexity criteria. Not every function is differentiable though, and in real-world problems we encounter mostly functions consisting of many subfunctions for which even a simple criterion like the one of Proposition 5.1 is difficult to verify, due to, for example, nondifferentiability. However, just as we have rules on computing derivatives of complicated functions based on the derivatives of their simple subparts, for verifying convexity, we have a set of operations on convex functions that preserve convexity. Examples of such operations are:

- **nonnegative weighted sums**: if $f_1, \ldots, f_m$ are convex and $w_1, \ldots, w_m \geq 0$, then $g = w_1 f_1 + \cdots + w_m f_m$ is convex;
- **linear transformations**: if f is convex, then the function $g(\boldsymbol{x}) := f(A\boldsymbol{x} + b)$ is convex;
- **point-wise maximum**: if $f_1, \ldots, f_m$ are convex, then the function $g(\boldsymbol{x}) = \max\{f_1(\boldsymbol{x}), \ldots, f_m(\boldsymbol{x})\}$ is convex;
- **minimization**: if f is convex in $(\boldsymbol{x}, \boldsymbol{y})$ and C is a nonempty convex set, then $g(\boldsymbol{x}) = \inf_{\boldsymbol{y} \in C} f(\boldsymbol{x}, \boldsymbol{y})$ is convex in $\boldsymbol{x}$ if $g(\boldsymbol{x}) > -\infty$ for all $\boldsymbol{x}$;
- **composition**: if f and g are convex and g is nondecreasing, then their composition $h(\boldsymbol{x}) = g(f(\boldsymbol{x}))$ is convex.

We refer the reader to [BV04, chapter 3] for more examples. One can verify/-construct complicated multivariate convex functions using the above criteria and ruleset.

Example 5.7 Examples of multidimensional convex functions $\mathbb{R}^n \rightarrow \mathbb{R}$ are as follows:

- every norm is convex, so in particular any p-norm on $\mathbb{R}^n$ with $p \geq 1$ is, that is, $\boldsymbol{x} \mapsto \|\boldsymbol{x}\|_p := \left(\sum_{i=1}^n |x_i|^p \right)^{1/p}$;
- the maximum $\boldsymbol{x} \mapsto \max\{x_1, \ldots, x_n\}$;
- the quadratic-over-linear function $(x,y) \mapsto \frac{x^2}{y}$ with $y > 0$;
- the log-sum-exp function $\boldsymbol{x} \mapsto \log(e^{x_1} + \cdots + e^{x_n})$;
- the geometric mean $\boldsymbol{x} \mapsto \left(\prod_{i=1}^n x_i \right)^{1/n}$;
- the log-determinant $X \mapsto \log \det X$ (X is a square matrix).

5.2.2 Convex Sets Built With Convex Functions

When solving real-world optimization problems, directly visualizing the feasible set S is often impossible. Consequently, we need to study its convexity focusing exclusively on the constraints that define this set and specifically on the properties of the functions that appear in those constraints.

Convex sets are linked to convex functions by the fact that for every $t \in \mathbb{R}$ the **sublevel set** $S(t)$ of the function $f : \mathbb{R}^n \rightarrow \mathbb{R}$, that is, the set

$$S(t) = \{\boldsymbol{x} \in \mathbb{R}^n \; : \; f(\boldsymbol{x}) \leq t\},$$

is convex whenever f is convex.

Example 5.8 If an optimization problem only has a single constraint of the form

$$x^2 - 4 \leq 0,$$

we can conclude that its feasible set, that is, the set of points satisfying this condition $\{\boldsymbol{x} \in \mathbb{R}^n : \; f(\boldsymbol{x}) \leq 0\}$, is convex from the convexity of the function $f(x) = x^2 - 4$.

Similarly to functions, we can rely on a set of "rules" by which we can construct complicated convex sets out of simpler convex sets. The most important one for us is that the intersection of an arbitrary number of convex sets is convex. For that reason, if we write all the constraints of an optimization problem as

$$X = \{x \in \mathbb{R}^n : \; g_1(x) \leq 0, \; \ldots, \; g_m(x) \leq 0\}$$

then as long as each of the sets $X_i = \{x \in \mathbb{R}^n : \; g_i(x) \leq 0\}$ is convex, it will automatically follow that $X = X_1 \cap \cdots \cap X_m$ is convex as well. Combined with our earlier statement, it means that it is sufficient to check separately if each of the functions $g_1, \ldots, g_m$ is convex! This is usually how the convexity of optimization problems is verified in practice.

Example 5.9 (Linear equalities) Feasible sets determined by linear equality constraints of the form

$$\boldsymbol{a}^\top \boldsymbol{x} = b$$

are also convex, since they can be rewritten equivalently to be the intersection of two inequality constraints involving linear (hence convex) functions, namely

$$\begin{cases} \boldsymbol{a}^\top \boldsymbol{x} - b \leq 0 \\ -\boldsymbol{a}^\top \boldsymbol{x} + b \leq 0. \end{cases}$$

Another example of a convex set that arises from the intersection of simpler convex sets is the feasible set of an LO problem, called **polyhedron** (plural: polyhedra) $\{\boldsymbol{x} \in \mathbb{R}^n : A\boldsymbol{x} \leq \boldsymbol{b}\}$ for given matrix $A \in \mathbb{R}^{m \times n}$ and vector $\boldsymbol{b} \in \mathbb{R}^m$. A **polytope** is a bounded polyhedron.

Other examples of a convex set are singletons, each orthant of a linear space, linear subspaces (also called hyperplanes), cones, spheres, etc. See [BV04, chapter 2] for more examples of convexity-preserving operations.

5.2.3 Convex Optimization Problems

Knowing how to verify the convexity of a function and of a set defined using functions, we can build a general convex optimization problem:

$$\begin{align} \min \quad & f(\boldsymbol{x}) \tag{5.4} \\ \text{s.t.} \quad & g_i(\boldsymbol{x}) \leq 0, \quad \forall i = 1,\ldots,m \\ & h_i(\boldsymbol{x}) = 0, \quad \forall i = m+1,\ldots,p, \\ & \boldsymbol{x} \in \mathbb{R}^n, \end{align}$$

where $f, g_1, \ldots, g_m$ are convex functions from $S \subseteq \mathbb{R}^n$ to $\mathbb{R}$ and $h_i(\boldsymbol{x}) = \boldsymbol{a}_i^\top \boldsymbol{x} - b_i$ are linear functions. Being the intersection of convex sets, the feasible set of (5.4) is convex. Maximizing a concave function is equivalent to minimizing a convex function.

Example 5.10 (Markowitz portfolio optimization part 1) A canonical stochastic optimization problem is the so-called portfolio selection problem, also known as **Markowitz portfolio optimization**. Assume that an investor has an initial capital C that she wants to invest in n possible risky assets, each of them with an unknown return rate r_i, $i = 1,\ldots,n$, or in another risk-free asset with a guaranteed return rate R. Let $\boldsymbol{x}$ be the vector whose ith component x_i describes the amount invested in asset i and $\tilde{x}$ the amount invested in the risk-free asset. We consider a stochastic model in which the return of the n risky assets is then a random vector $\boldsymbol{r}$ with known expected values $\boldsymbol{\mu} = \mathbb{E}\boldsymbol{r}$ and covariance

$$\boldsymbol{\Sigma} = \mathbb{E}\left[(\boldsymbol{r} - \boldsymbol{\mu})(\boldsymbol{r} - \boldsymbol{\mu})^\top\right].$$

The investment return $y = R\tilde{x} + \boldsymbol{r}^\top \boldsymbol{x}$ then becomes also a random variable with mean

$$\mathbb{E}y = R\tilde{x} + \mathbb{E}\boldsymbol{r}^\top \boldsymbol{x} = R\tilde{x} + \boldsymbol{\mu}^\top \boldsymbol{x}$$

and variance

$$\text{Var}(y) = \mathbb{E}(y - \mathbb{E}y)^2 = \boldsymbol{x}^\top \boldsymbol{\Sigma} \boldsymbol{x}.$$

The variance of the investment return is a possible way to quantify the risk of the investment $\boldsymbol{x}$.

The investor needs to select a portfolio that achieves a good compromise between risk and expected return. One could try to maximize the expected return subject to an upper bound on the tolerable risk, obtaining the following optimization problem:

$$\max \quad R\tilde{x} + \boldsymbol{\mu}^\top \boldsymbol{x} \tag{5.5}$$

$$\text{s.t.} \quad \sum_{i=1}^{n} x_i + \tilde{x} = C$$

$$\boldsymbol{x}^\top \boldsymbol{\Sigma} \boldsymbol{x} \leq \gamma$$

$$\tilde{x} \geq 0$$

$$\boldsymbol{x} \geq 0.$$

The first constraint describes the fact that the total amount invested must be equal to the initial capital. The second constraint ensures that the variance of the chosen portfolio is upper bounded by a parameter γ, which captures the risk the investor is willing to take on. The last nonnegativity constraint excludes the possibility of short-selling.

One can easily show that the quadratic constraint $\boldsymbol{x}^\top \boldsymbol{\Sigma} \boldsymbol{x} \leq \gamma$ is convex in x because $\boldsymbol{\Sigma}$ is positive semidefinite, being a covariance matrix. Therefore, the optimization problem (5.5) is convex. Due to the convexity of the problem, there are no local maxima; thus, a generic nonlinear optimization solver like `ipopt` is enough to find the unique global maximum.

```python
def markowitz(gamma, mu, Sigma):
    m = pyo.ConcreteModel("Markowitz portfolio optimization")

    m.xtilde = pyo.Var(domain=pyo.NonNegativeReals)
    m.x = pyo.Var(range(n), domain=pyo.NonNegativeReals)

    @m.Objective(sense=pyo.maximize)
    def objective(m):
        return mu @ m.x + R * m.xtilde

    @m.Constraint()
    def bounded_variance(m):
        return (m.x @ (Sigma @ m.x)) <= gamma

    @m.Constraint()
    def total_assets(m):
        return sum(m.x[i] for i in range(n)) + m.xtilde == C

    return m


# Specify the initial capital, the risk threshold, and the guaranteed
↪   return rate.
C = 1
gamma = 1
R = 1.01
```

```
# Specify the number of assets, their expected return, and their covariance
↪   matrix.
n = 3
mu = np.array([1.2, 1.1, 1.3])
Sigma = np.array([[1.5, 0.5, 2], [0.5, 2, 0], [2, 0, 5]])

# Check that the matrix Sigma is semi-definite positive
assert np.all(np.linalg.eigvals(Sigma) >= 0)

# When changing the matrix Sigma, make sure to input a semi-definite
↪   positive matrix.
# The easiest way to generate a random covariance matrix is as follows:
# first generate a random m x m matrix A and than take the matrix A A^T
# which is always semi-definite positive by construction.
#
# m = 3
# A = np.random.rand(m, m)
# Sigma = A.T @ A

m = markowitz(gamma, mu, Sigma)
pyo.SolverFactory("ipopt").solve(m)
```

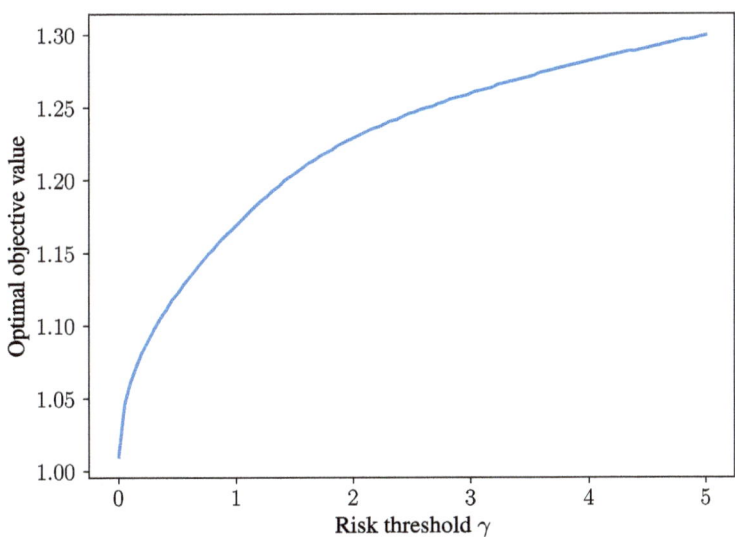

Figure 5.7 The optimal investment return changes as the risk threshold parameter γ varies.

Figure 5.7 shows how the return on investment changes when the risk threshold γ increases.

Example 5.11 (Convex quadratic optimization) The previous problem was a specific case of a very important category – **convex quadratic problems**. More specifically, for P being a positive semidefinite matrix and $g_1, \ldots, g_m$ convex functions, the problem can be stated as

$$\min \quad \frac{1}{2} \boldsymbol{x}^\top P \boldsymbol{x} + \boldsymbol{q}^\top \boldsymbol{x} + r$$
$$\text{s.t.} \quad g_i(\boldsymbol{x}) \leq 0, \qquad \forall\, i = 1, \ldots, m,$$
$$\boldsymbol{a}_i^\top \boldsymbol{x} = b_i, \qquad \forall\, i = m+1, \ldots, p.$$

An important thing about modeling problems is that the same problem, depending on how it is formulated, may or may not meet the simple criterion for convexity (all functions being convex). Therefore, we should always seek maximal simplicity in formulating problem constraints while building an optimization model.

Example 5.12 Consider the optimization problem

$$\min \quad x_1^2 + x_2^2$$
$$\text{s.t.} \quad x_1/(1 + x_2^2) \leq 0,$$
$$(x_1 + x_2)^2 = 0.$$

It has a feasible convex set, but the function used in the inequality constraint is not convex in general, and the equality constraints involve a nonlinear function. However, the problem can be rewritten in an equivalent form, namely

$$\min \quad x_1^2 + x_2^2$$
$$\text{s.t.} \quad x_1 \leq 0,$$
$$x_1 + x_2 = 0,$$

where all the functions involved are convex.

5.3 Duality for Convex Optimization

In Section 2.3, we learned that every LO problem has its associated dual linear optimization problem. That was extremely useful because it was the way to provide optimality certificates. The power of convex optimization lies in the fact that precisely the same property holds, with some minor changes. This section shows how the optimal solution–optimality certificate solution can be seen as a solution to the so-called Karush–Kuhn–Tucker (KKT) conditions. This is important because, under the hood, every nonlinear solver is essentially a numerical procedure to solve the KKT conditions.

5.3.1 Deriving the Dual: Analogy to LO

While the dual for LO problems could be derived by playing "manually" with the constraints and the objective, for convex problems a more general approach is needed. To build this more general approach, we first rederive the dual problem of an LO problem using the so-called **Lagrangian relaxation**.

Example 5.13 (Lagrangian dual for an LO problem) Consider the LO problem

$$v(P) = \min \quad c^\top x$$
$$\text{s.t.} \quad Ax \leq b,$$
$$x \geq 0.$$

We define a special function called the Lagrangian associated with this problem as follows:

$$L(x, \lambda) = c^\top x + \lambda^\top (Ax - b), \quad \lambda \geq 0.$$

The interpretation of this function is: Instead of enforcing the constraints in a "hard way" in the optimization problem, we move them to a new augmented objective function in such a way that the larger the violation of the original constraint $Ax \leq b$, the lower the value of the objective function – note the nonnegativity of the "penalty parameter" λ.

For every value of $\lambda \geq 0$, we can obtain a lower bound on the optimal value of the LO problem by maximizing the value of $L(x, \lambda)$. Indeed, for every $x \geq 0$ that satisfies the constraint $Ax \leq b$, we have that

$$L(x, \lambda) = c^\top x + \lambda^\top (Ax - b) \leq c^\top x.$$

Consider the Lagrangian dual function $\ell(\lambda)$ defined as

$$\ell(\lambda) := \inf_{x \geq 0} L(x, \lambda). \tag{5.6}$$

As argued above, this provides a lower bound, since

$$\ell(\lambda) = \inf_{x \geq 0} L(x, \lambda) \leq v(P).$$

The infimum on the left-hand side is a function of λ and, since this is itself an optimization problem, we can derive its value as

$$\ell(\lambda) = \min_{x \geq 0} c^\top x + \lambda^\top (Ax - b)$$
$$= -\lambda^\top b + \min_{x \geq 0} \left(c^\top + \lambda^\top A \right) x$$
$$= \begin{cases} -\lambda^\top b, & \text{if } \left(c + A^\top \lambda \right) \leq 0 \\ -\infty, & \text{otherwise.} \end{cases}$$

Because $\ell(\lambda) \leq v(P)$, the search for the tightest lower bound on $v(P)$ can be cast as the problem of maximizing $\ell(\lambda)$:

$$\max_{\lambda \geq 0} \ell(\lambda). \tag{5.7}$$

Because, when we minimize, we do not want to obtain a value of $-\infty$, we can restrict λ only to the values for which $\ell(\lambda)$ is finite to obtain the following optimization problem:

$$\max \quad -\boldsymbol{\lambda}^\top \boldsymbol{b}$$
$$\text{s.t.} \quad -\boldsymbol{\lambda}^\top A \le \boldsymbol{c}^\top$$
$$\boldsymbol{\lambda} \ge 0.$$

After the substitution $\boldsymbol{\lambda}' = -\boldsymbol{\lambda}$, this turns out to be precisely the dual problem derived in Section 2.3.

The above example shows that LO duality is a special case of Lagrange duality, which we now introduce formally. Consider a general convex optimization problem, which we will refer to as the **primal problem** (P):

$$v(P) = \min \quad f(\boldsymbol{x})$$
$$\text{s.t.} \quad \boldsymbol{g}(\boldsymbol{x}) \le 0$$
$$\boldsymbol{x} \in X \subseteq \mathbb{R}^n,$$

where $\boldsymbol{g}$ is a vector of convex functions and X is a convex subset of $\mathbb{R}^n$. The **Lagrange dual problem** (D) is

$$v(D) = \max \quad \ell(\boldsymbol{\lambda})$$
$$\text{s.t.} \quad \boldsymbol{\lambda} \ge 0,$$

where the value of $\ell(\boldsymbol{\lambda})$ is given by solving another subproblem, also called **Lagrangian relaxation** or the **Lagrangian dual function**, that is,

$$\ell(\boldsymbol{\lambda}) := \inf_{\boldsymbol{x} \in X} \ f(\boldsymbol{x}) + \boldsymbol{\lambda}^\top \boldsymbol{g}(\boldsymbol{x}). \tag{5.8}$$

Is this function always as easy to derive as in the case of LO problems? Not necessarily, however, for many important problem classes such as convex quadratic optimization, it is indeed possible and the resulting dual problem can help tremendously in solving the original problem. In Chapter 6, we will show that for a special subclass of convex optimization called **conic optimization** (which includes convex quadratic as well), the derivation of the dual function is just as easy as for the LO problems.

5.3.2 Weak and Strong Duality

Knowing how a dual is formulated for a convex optimization problem, we are ready to state the "optimality certificate" results for such problems. The first theorem states formally that we can use the dual problem to obtain bounds on the optimal value of the primal problem.

Theorem 5.2 (Weak duality theorem – Convex case) *For all feasible solutions $\boldsymbol{x}$ of the primal problem and all feasible solutions $\boldsymbol{\lambda}$ of the dual problem, the following inequality holds:*

$$f(\boldsymbol{x}) \ge \ell(\boldsymbol{\lambda}).$$

Proof Any feasible solution $\boldsymbol{x}$ provides an upper bound for the optimal value of $f(\boldsymbol{y}) + \boldsymbol{\lambda}^\top \boldsymbol{g}(\boldsymbol{y})$, therefore

$$\ell(\boldsymbol{\lambda}) = \min_{\boldsymbol{y} \in X} \{ f(\boldsymbol{y}) + \boldsymbol{\lambda}^\top \boldsymbol{g}(\boldsymbol{y}) \} \le f(\boldsymbol{x}) + \boldsymbol{\lambda}^\top \boldsymbol{g}(\boldsymbol{x}) \le f(\boldsymbol{x}),$$

where the last inequality is due to the fact that $\boldsymbol{g}(\boldsymbol{x}) \le 0$ for every $\boldsymbol{x} \in X$. $\qquad \square$

Let $v(P)$ be the optimal value of the primal and $v(D)$ that for the dual. The weak duality theorem has immediate consequences:

1. $v(P) \geq v(D)$;
2. if $\boldsymbol{x}$ is feasible for (P), $\boldsymbol{\lambda}$ is feasible for (D) and $f(\boldsymbol{x}) = \ell(\boldsymbol{\lambda})$, then these points are optimal solutions for the respective problems;
3. if (P) is unbounded, that is, $v(P) = -\infty$, then $\ell(\boldsymbol{\lambda}) = -\infty$ for all $\boldsymbol{\lambda} \geq 0$;
4. if (D) is unbounded, that is, $v(D) = +\infty$, then (P) is infeasible.

Notably, the Lagrangian dual can be used to computationally tractably obtain lower bounds on the optimal values of the original problem, even if the original problem was nonconvex. Indeed, regardless of the nature of f, g, h, and X the function $\ell(\boldsymbol{\lambda}, \boldsymbol{\mu})$ is always concave in $(\boldsymbol{\lambda}, \boldsymbol{\mu})$ since it is the pointwise minimum of (possibly infinitely many) functions $f(\boldsymbol{x}) + \boldsymbol{\lambda}^\top g(\boldsymbol{x}) + \boldsymbol{\mu}^\top h(\boldsymbol{x})$, which are linear in $(\boldsymbol{\lambda}, \boldsymbol{\mu})$.

Our interest in duality is mainly driven by the need to find, via the dual problem, an optimality certificate for a solution we find for the primal problem. We refer to the difference $v(P) - v(D)$ as the **duality gap**. If equality $v(P) = v(D)$ holds, that is, the optimal duality gap is zero, then we say that **strong duality** holds.

The question is: Does strong duality always hold? For well-behaved problems, the answer is yes, as the following example shows.

Example 5.14 Consider the convex problem

$$\min \quad x_1^2 + x_2^2$$
$$\text{s.t.} \quad x_1 + x_2 \geq 4$$
$$x_1, x_2 \geq 0.$$

We apply Lagrangian duality with $g(\boldsymbol{x}) = -x_1 - x_2 + 4$ and $X = \mathbb{R}_+$. The Lagrangian dual function is

$$\ell(\lambda) = \min_{\boldsymbol{x} \geq 0} x_1^2 + x_2^2 + \lambda(4 - x_1 - x_2) = 4\lambda + \min_{x_1 \geq 0} x_1^2 - \lambda x_1 + \min_{x_2 \geq 0} x_2^2 - \lambda x_2.$$

In the last step, we use the fact that the above problem is separable. Both minima are attained when $x_1 = x_2 = \lambda/2$ provided that $\lambda \geq 0$, while if $\lambda < 0$ then $x_1 = x_2 = 0$. We conclude that

$$\ell(\lambda) = \begin{cases} 4\lambda & \text{if } \lambda < 0 \\ 4\lambda, \, -\lambda^2/2, & \text{if } \lambda \geq 0, \end{cases}$$

which is a concave function, as confirmed by its plot in Figure 5.8.

The dual then is

$$\max \quad 4\lambda - \lambda^2/2$$
$$\text{s.t.} \quad \lambda \geq 0.$$

Note that in this example (i) the dual problem is maximizing a concave function over a convex set, (ii) the objective function of the dual is differentiable everywhere (including at $\lambda = 0$), and (iii) strong duality holds. The solution of the primal is $\boldsymbol{x} = (2,2)$ and that of the dual is $\lambda = 4$. It is easy to check that $v(P) = 8 = v(D)$.

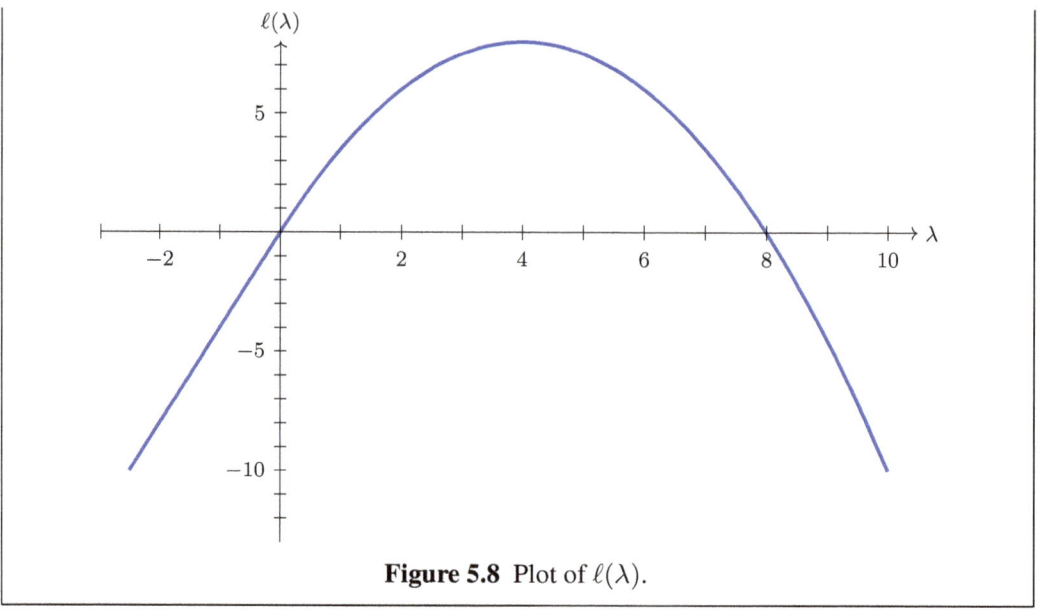

Figure 5.8 Plot of $\ell(\lambda)$.

However, in convex optimization, strong duality is not promised. In fact, an "extra" condition is needed. The next proposition states a sufficient "extra" condition for strong duality to hold.

Proposition 5.2 (Slater's condition for strong duality) *If the problem*

$$\begin{aligned}
\min \quad & f(\boldsymbol{x}) \\
\text{s.t.} \quad & \boldsymbol{g}(\boldsymbol{x}) \leq 0 \\
& \boldsymbol{h}(\boldsymbol{x}) = 0 \\
& \boldsymbol{x} \in \mathbb{R}^n
\end{aligned}$$

is convex and there exists $\hat{\boldsymbol{x}}$ such that $g_i(\hat{\boldsymbol{x}}) < 0$ for every nonlinear g_i and $g_i(\hat{\boldsymbol{x}}) \leq 0$ when g_i is linear, then strong duality holds, that is, $v(D) = v(P)$.

Roughly, it means that the set of feasible solutions $\boldsymbol{x}$ has to be "rich enough" for strong duality to hold for a given convex optimization problem.

The following example shows what might happen if Slater's condition is not satisfied.

Example 5.15 (Strong duality may not hold for convex problem) Consider the problem

$$\begin{aligned}
\min \quad & x \\
\text{s.t.} \quad & x^2 \leq 0 \\
& x \in \mathbb{R}.
\end{aligned}$$

Both the functions appearing in the objective and in the constraint are convex, so the problem is convex. We can easily solve the primal problem by hand since there is a unique feasible point $x = 0$, obtaining $v(P) = 0$. However, strong duality does not hold for this optimization problem. Indeed, by writing the Lagrangian relaxation, one can calculate that $\ell(\lambda) = -\infty$ for all $\lambda \geq 0$ and thus $v(D) = -\infty \neq 0 = v(P)$.

Except for pathological problems like those in Example 5.15 strong duality will typically hold in problems met in practice. The issue, however, is how to actually determine the optimal dual solution λ if the dual optimization problem is not defined in a closed form, but rather, the dual function is the result of the infimum in (5.8).

For this reason, the problem of finding the optimal solution together with the corresponding optimality certificate is harder. Luckily, for the situation where all the functions defining the problem are differentiable, the optimal dual solution can be determined jointly with the primal solution as a pair (x, λ) satisfying the so-called Karush–Kuhn–Tucker conditions.

5.3.3 Karush–Kuhn–Tucker (KKT) Conditions

Henceforth, we assume that all the functions defining the objective and the constraints are differentiable. Similarly to the LO case, under certain mild assumptions, the combination of a primal optimal and a dual optimal solution satisfies a certain set of conditions, which are called the **Karush–Kuhn–Tucker (KKT) conditions** and are:

$$
\begin{cases}
\nabla f(x) + \sum_{i=1}^{m} \lambda_i \nabla g_i(x) + \sum_{i=m+1}^{p} \mu_i \nabla h_i(x) = 0, & \text{(stationarity)} \\
g_i(x) \le 0, & \text{for } i = 1, \ldots, m & \text{(primal feasibility)} \\
h_i(x) = 0, & \text{for } i = m+1, \ldots, p & \text{(primal feasibility)} \\
\lambda_i \ge 0, & \text{for } i = 1, \ldots, m & \text{(dual feasibility)} \\
\lambda_i g_i(x) = 0, & \text{for } i = 1, \ldots, m. & \text{(complementary slackness)}
\end{cases}
$$

Intuitively, the first condition corresponds to the fact that the dual function $\ell(\lambda)$ is defined as the infimum of the Lagrange function over x, hence its derivative should be equal to 0. The following proposition shows in which situations it is enough to solve the KKT conditions to find the desired optimal solution.

Proposition 5.3 (When do KKT conditions return the optimal solution?) *Consider an optimization problem P with differentiable objective and constraint functions:*

(a) *If strong duality holds for P, then any x primal optimal and λ, μ dual optimal must satisfy the KKT conditions.*

(b) *If P is convex, then x, λ, μ together satisfy the KKT conditions if and only if x is primal optimal and λ, μ are dual optimal and strong duality holds.*

(c) *If P satisfies Slater's condition (hence, in particular P is convex), then x, λ, μ together satisfy the KKT conditions if and only if x is primal optimal and λ, μ are dual optimal.*

The KKT conditions play an important role in optimization. In a few special cases, it is possible to solve the KKT conditions analytically and thus solve the optimization problem. More generally, many algorithms and numerical procedures for convex optimization have been conceived (or can be interpreted) as methods for solving the KKT conditions.

5.4 Solution Methods for Convex Optimization Problems

This section provides a brief overview of solution methods for convex optimization problems. Because numerical methods, in one way or another, boil down to methods for finding solutions to the system of KKT conditions, we begin by showing how we can solve them analytically by hand in several simple problems. Later, we will give a brief idea of basic numerical methods for such problems.

5.4.1 Analytical Solution

Example 5.16 (KKT conditions for a convex problem) Consider the following convex problem with five inequality constraints:

$$
\begin{aligned}
\min \quad & f(\boldsymbol{x}) = x_1^2 + x_2^2 + x_3^2 \\
\text{s.t.} \quad & g_1(\boldsymbol{x}) = 2x_1 + x_2 - 5 \leq 0 \\
& g_2(\boldsymbol{x}) = x_1 + x_3 - 2 \leq 0 \\
& g_3(\boldsymbol{x}) = 1 - x_1 \leq 0 \\
& g_4(\boldsymbol{x}) = 2 - x_2 \leq 0 \\
& g_5(\boldsymbol{x}) = -x_3 \leq 0 \\
& x_1, x_2, x_3 \geq 0.
\end{aligned}
$$

To write the stationarity condition, we first compute the gradients $\nabla f(\boldsymbol{x}) = (2x_1, 2x_2, 2x_3)$ and

$$
\nabla \boldsymbol{g}(\boldsymbol{x}) = \begin{bmatrix} 2 & 1 & 0 \\ 1 & 0 & 1 \\ -1 & 0 & 0 \\ 0 & -1 & 0 \\ 0 & 0 & -1 \end{bmatrix}.
$$

By writing explicitly the three equations in $\nabla f(\boldsymbol{x}) + \boldsymbol{\lambda}^\top \nabla \boldsymbol{g}(x) = 0$ (one per variable), the five complementary slackness conditions (one per constraint) and all the other feasibility inequalities, the KKT conditions read

$$
\begin{cases}
2x_1 + 2\lambda_1 + \lambda_2 - \lambda_3 = 0 \\
2x_2 + \lambda_1 - \lambda_4 = 0 & \text{(stationarity)} \\
2x_3 + \lambda_2 - \lambda_5 = 0 \\
2x_1 + x_2 - 5 \leq 0 \\
x_1 + x_3 - 2 \leq 0 \\
1 - x_1 \leq 0 & \text{(primal feasibility)} \\
2 - x_2 \leq 0 \\
-x_3 \leq 0 \\
\lambda_1, \lambda_2, \lambda_3, \lambda_4, \lambda_5 \geq 0 & \text{(dual feasibility)} \\
\lambda_1(2x_1 + x_2 - 5) = 0 \\
\lambda_2(x_1 + x_3 - 2) = 0 \\
\lambda_3(1 - x_1) = 0 & \text{(complementarity slackness)} \\
\lambda_4(2 - x_2) = 0 \\
\lambda_5 x_3 = 0.
\end{cases}
$$

This system of equations can be solved, for example, using Gaussian elimination, to find that $x = (1,2,0)$ and $\lambda = (0,0,2,4,0)$. The problem is convex, and we can verify that Slater's condition is met, for instance, by the point $\hat{x} = (1.25, 2.25, 0.5)$, so Proposition 5.3(c) guarantees that x and λ are, respectively, the primal and dual optimal solutions.

5.4.2 Numerical Methods

Without going too deep into the theory, it can be said that every software used to solve a general convex optimization problem aims at finding a solution to the KKT conditions. A great book on smooth, differentiable, convex optimization is [BV04] and the algorithms described there are implemented in the solver cvxopt [ADV23]. Unfortunately, there is no interface with Pyomo for this solver.

When the functions are differentiable but not necessarily convex, differentiable interior point methods can be applied, and these exhibit very fast convergence and, especially when using self-concordant barrier functions, are also very stable. A popular solver is ipopt [WB06], which is, in fact, an interior point method. Since ipopt does not assume convexity, its applicability is quite general, at the cost of only guaranteeing local optimality, as we saw earlier in this chapter in the milk pooling example, see Example 5.1.

It is also worth mentioning CVXPY, [Dia+23a], which, while serving as a modeling language similar to Pyomo, differentiates itself by focusing specifically on convex optimization problems. It builds on the concept of **disciplined convex programming**, [Dia+23b], which is based on composition rules that ensure convexity, or quasiconvexity. CVXPY relies on external solvers and, by default, uses one quadratic and two conic solvers. Neither of those is available to Pyomo, so one may say that convex optimization is a specialized world, in a way "outside" the realm of Pyomo.

However, many of the nonlinear solvers available to Pyomo take advantage of convexity. The commercial solvers CPLEX, Gurobi, and XPRESS can deal with convex quadratic models, even including discrete variables. Gurobi can also optimize nonconvex quadratic models. The commercial solver Mosek is likely to be the most powerful nonlinear solver available to Pyomo, also in the mixed-integer case. All these solvers also exploit any conic structure of the model, which we will discuss in Chapter 6.

Although this book addresses mostly models with differentiable convex functions, it is worth mentioning that all finite-valued convex functions are differentiable almost everywhere. At points where there is no gradient, convex functions have a generalization of the gradient, the so-called subgradient, leading to simple optimization algorithms known as subgradient methods. These methods, originally devised by Shor, show slow convergence and many improvements are known. In particular, Lagrange duals are always convex and often nondifferentiable. These can ideally be solved by subgradient methods.

5.5 A Complete Example: Support Vector Machines for Binary Classification

Support vector machines (SVM) are a type of supervised machine learning model. Similarly to other machine learning techniques, training an SVM classifier involves solving an optimization problem. Such a problem is solved once using training samples with known outcomes, and the optimal solution describes the best parameters for the classifier. The resulting classifier can then be applied to classify data with unknown outcomes.

In this example, we demonstrate the process of creating an SVM for binary classification using both linear and quadratic optimization. Our implementation will initially focus on linear SVMs, which separate the feature space using a hyperplane. We will explore both primal and dual formulations. Then, using kernels, the dual formulation is extended to binary classification in higher-order and nonlinear feature spaces.

Binary Classification

A **binary classifier** is a machine learning tool to classify objects into two classes. Binary classifiers are functions designed to answer questions such as "Does this medical test indicate disease?," "Will this specific customer enjoy that specific movie?," "Does this photo include a car?," or "Is this banknote genuine or counterfeit?" These questions are answered based on the values of **feature**, which may include physical measurements or other data types collected from a representative sample with known outcomes.

In mathematical terms, a binary classifier is a function for predicting a binary outcome, say $y = \pm 1$ or $y \in \mathbb{B}$, based on the values of p features contained in a vector $x \in \mathbb{R}^p$.

For example, consider a device installed on a vending machine to detect banknotes. The classifier aims to accurately identify and accept genuine banknotes while rejecting counterfeit ones. The classifier's performance can be assessed using definitions in the following table, where "positive" refers to an instance of a genuine banknote:

	Predicted positive	Predicted negative
Actual positive	True positive (TP)	False negative (FN)
Actual negative	False positive (FP)	True negative (TN)

A vending machine user would be frustrated if a genuine banknote is incorrectly rejected as a false negative. **Sensitivity** is defined as the number of true positives (TP) divided by the total number of actual positives (TP + FN). High sensitivity indicates a low false negative rate and means that almost all genuine banknotes are accepted, making it the preferred outcome for users.

On the other hand, the vending machine owner wants the machine to avoid accepting counterfeit banknotes and would prefer a low number of false positives (FP). **Precision** is the number of true positives (TP) divided by the total number of predicted positives (TP + FP). High precision indicates a low false positive rate and implies that almost all accepted notes are genuine, making it the preferred outcome for the owner.

To achieve high sensitivity, a classifier can follow the "innocent until proven guilty" standard, rejecting banknotes only when it is certain they are counterfeit. To achieve high precision, a classifier can adopt the "guilty unless proven innocent" standard, rejecting banknotes unless absolutely sure they are genuine. The challenge in developing binary classifiers is to balance these conflicting objectives and optimize performance from both perspectives simultaneously. In this example, we will see how various SVMs strike this balance.

Problem Description and Data

Our goal is to train a binary classifier to detect counterfeit banknotes. We consider a public data set [Loh13] containing data from a collection of known genuine and known counterfeit banknote specimens. The data includes four continuous statistical measures obtained from the wavelet transform of banknote images named variance, skewness, kurtosis, and entropy, and a binary variable named class, which is 0 if genuine and 1 if counterfeit.

Following customary practices, we divide the data set into a **training set** used to train the classifier and a **test set** that will be used to evaluate the classifier's performance. To enable the plotting of the dataset for visualization purposes, we select a two-dimensional subset of the features: variance and skewness. The resulting scatter plot is available in Figure 5.9.

The SVMs are traditionally explained in a setting where the label of one class is -1 and the other is $+1$. Therefore, we rescale the class feature to have values of $+1$ for genuine banknotes and -1 for counterfeit banknotes.

Linear Support Vector Machines

A linear SVM is a binary classification method that uses a linear equation to determine the class assignment. The basic formula is expressed as

$$y^{\text{pred}} = \text{sgn}\,(\boldsymbol{w}^\top \boldsymbol{x} + b),$$

where $\boldsymbol{x} \in \mathbb{R}^p$ is a point in the feature space and sgn is a function defined as

$$\text{sgn}(x) = \begin{cases} -1 & x < 0 \\ 0 & x = 0 \\ 1 & x > 0. \end{cases}$$

Here, $\boldsymbol{w} \in \mathbb{R}^p$ represents a set of coefficients, $\boldsymbol{w}^\top \boldsymbol{x}$ is the dot product, and b is a scalar coefficient. The linear function divides the feature space using the hyperplane defined by $\boldsymbol{w}$ and b. Points on one side of the hyperplane are assigned a positive outcome ($+1$), while points on the other side are given a negative outcome (-1).

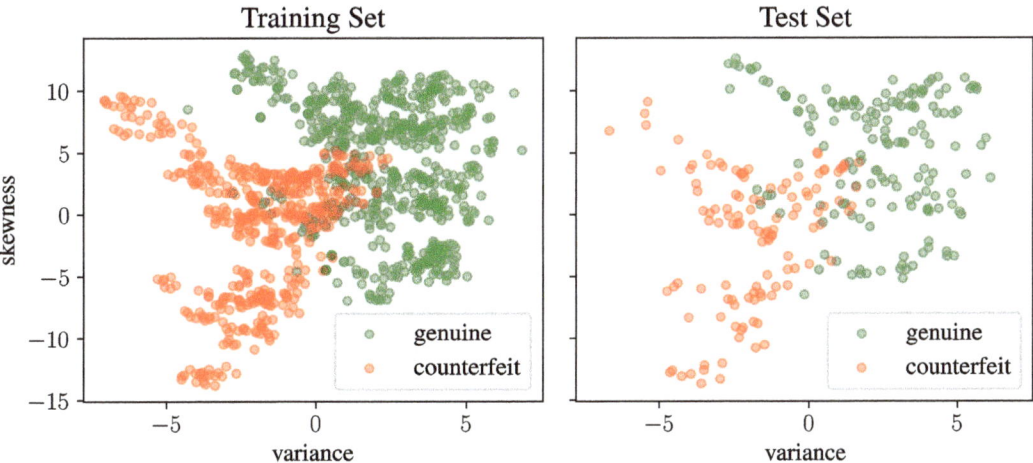

Figure 5.9 Scatter plot of the training and test sets using the two selected features.

We first implement a naive linear SVM model. The code below contains a Python implementation of a linear SVM using a class inside which to store the coefficients w and b. Indeed, a visual inspection of the training set for the banknote in Figure 5.9 shows that the two-dimensional feature set can be roughly split along the vertical axis where "variance" is zero. Most positive outcomes are on the right of the axis, and most negative outcomes are on the left. Since w is a vector normal to this surface, we manually choose

$$w = \begin{bmatrix} w_{variance} \\ w_{skewness} \end{bmatrix} = \begin{bmatrix} 1 \\ 0 \end{bmatrix}, \qquad b = 0.$$

We assess the accuracy of the linear SVM by calculating the **accuracy score**, which is the fraction of samples that were predicted accurately.

```python
# Linear Support Vector Machine (SVM) class
class LinearSVM:
    # Initialize the Linear SVM with weights w and bias b
    def __init__(self, w, b):
        self.w = pd.Series(w)
        self.b = float(b)

    # Call method to compute the decision function using the input data X
    def __call__(self, X):
        return np.sign(X.dot(self.w) + self.b)

    # String representation method for the Linear SVM class
    def __repr__(self):
        return f"LinearSvm(w = {self.w.to_dict()}, b = {self.b})"


# Assume training set is available, split into features X_train and outputs
↪   y_train
# Assume test set is available, split into features X_test and outputs y_test

# Visual estimate of w and b for a linear classifier
w = pd.Series({"variance": 1, "skewness": 0})
b = 0

# create an instance of LinearSVM
svm = LinearSvm(w, b)

# predictions for the training set
y_pred = svm(X_test)

# fraction of correct predictions
accuracy = sum(y_pred == y_test) / len(y_test)
```

The achieved accuracy is 82.9%, which is not bad, but we can certainly do better if a more sophisticated tool than a straight vertical line is used. The scatter plots in Figure 5.10 help

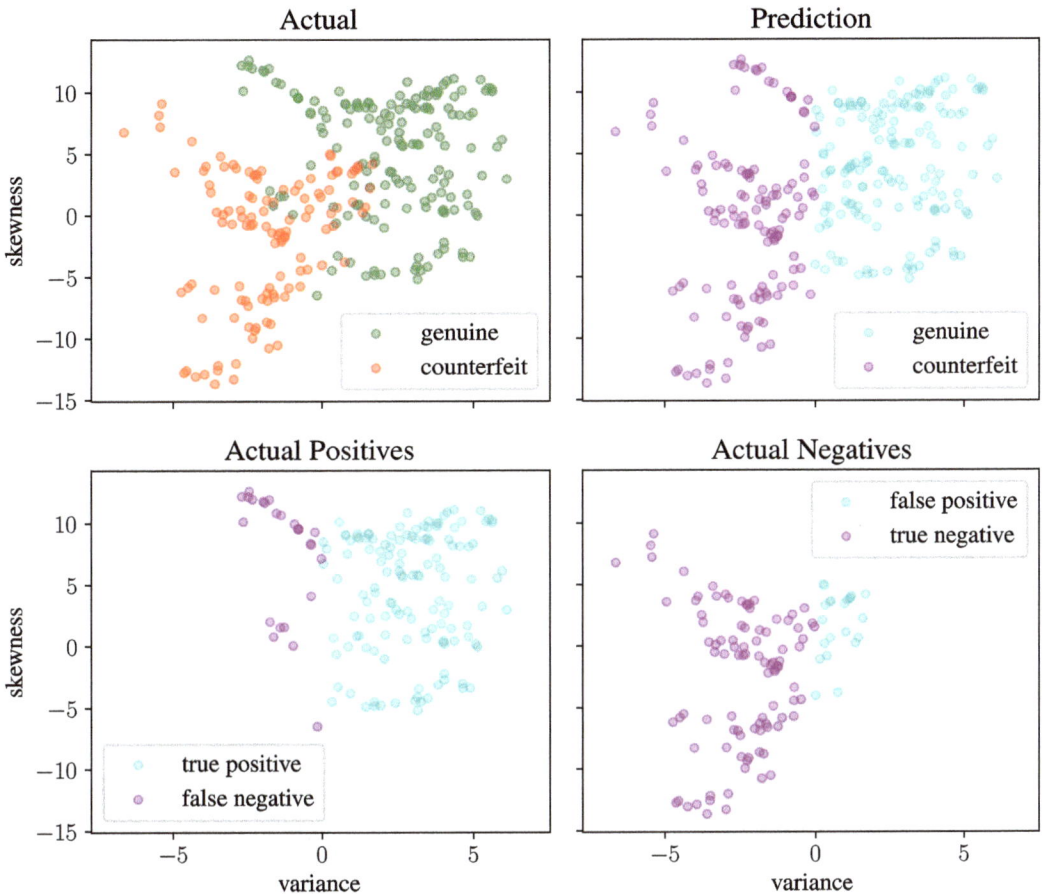

Figure 5.10 Scatter plot describing the performance of the naive linear SVM classification.

us understand when our SVM performs correctly. There are still many points that could be correctly classified and the number of false negatives and false positives is similar.

The accuracy score alone is not always a reliable metric for evaluating the performance of binary classifiers. For example, when one outcome is significantly more frequent than the other, a classifier that always predicts the more common outcome without regard to the feature vector can achieve a very high accuracy score while being completely wrong on the less common outcome.

Moreover, in many applications, the consequences of a false positive can differ from those of a false negative. For these reasons, we seek a more comprehensive set of metrics to compare binary classifiers. The **sensitivity** of a classifier measures how many of the actual positive items in the data set have been labeled as positive. The **precision**, on the other hand, counts how many of the items marked as positive are actually positive. In the literature, the **Matthews correlation coefficient (MCC)** is recommended as a reliable performance measure for binary classifiers. The above model achieves a Matthews correlation coefficient (MCC) of 0.647, a sensitivity of 85.7%, and a precision of 85.2%.

An LO Model to Find the Best Linear SVM Classifier

Our goal is to show how training a best-possible linear classifier can be posed and solved as a linear optimization problem. A training or validation set consists of n observations $(\boldsymbol{x}_i, y_i)$ where $y_i = \pm 1$ and $\boldsymbol{x}_i \in \mathbb{R}^p$ for $i = 1, \ldots, n$. The training task is to find coefficients $\boldsymbol{w} \in \mathbb{R}^p$ and $b \in \mathbb{R}$ to achieve high sensitivity and high precision for the validation set. Ideally, we would like to find w and b such that

$$\boldsymbol{w}^\top \boldsymbol{x}_i + b > 0, \qquad\qquad \forall\, i : y_i = +1$$
$$\boldsymbol{w}^\top \boldsymbol{x}_i + b < 0, \qquad\qquad \forall\, i : y_i = -1,$$

which can be wrapped into a single expression:

$$y_i \left(\boldsymbol{w}^\top \boldsymbol{x}_i + b \right) > 0, \qquad \forall\, i = 1, \ldots, n.$$

This condition does not impose a scale for $\boldsymbol{w}$ or b (i.e., if the condition is satisfied for any pair $(\boldsymbol{w}, b)$, then it is also satisfied for $(\gamma \boldsymbol{w}, \gamma b)$ where $\gamma > 0$). It is convenient, therefore, to impose a modified condition for correctly classified points where we secure a "minimum margin 1 of the inequality to hold" so that, in a way, there is a strict separation between the positive and negative data points:

$$y_i \left(\boldsymbol{w}^\top \boldsymbol{x}_i + b \right) \geq 1 \qquad \forall\, i = 1, \ldots, n.$$

This condition is known as a **hard-margin classifier**. The size of the margin (how far each of the $+1$ and -1 subsets is from the line that separates them) is determined by the scale of $\boldsymbol{w}$ and b. In practice, finding $\boldsymbol{w}$ and b that perfectly separate all the data is not always possible.

For this purpose, when fitting an SVM to the data, it is common to use a **soft-margin classifier** obtained by minimizing the so-called **hinge-loss function**, which is equal to the difference between the right- and left-hand sides in the above equation if the inequality cannot be satisfied and 0 if it works. Given parameters $\boldsymbol{w}$ and b, the hinge-loss function is defined as

$$\ell(\boldsymbol{x}, y) = \left(1 - y \left(\boldsymbol{w}^\top \boldsymbol{x} + b \right) \right)^+,$$

with the notation $z^+ = \max(0, z)$. The hinge-loss function is greater than 0 for any misclassified point and grows in proportion to how far the feature vector is from the separation plane.

With the hinge-loss function, the training problem is formulated as minimizing the hinge-loss function over all the data samples:

$$\min_{\boldsymbol{w}, b} \frac{1}{n} \sum_{i=1}^{n} \left(1 - y_i \left(\boldsymbol{w}^\top \boldsymbol{x}_i + b \right) \right)^+.$$

Practice has shown that minimizing this term alone produces classifiers with large entries for $\boldsymbol{w}$, which perform poorly on new data samples. For that reason, **regularization** adds a term to penalize the magnitude of $\boldsymbol{w}$. In most formulations, some norm $\|\boldsymbol{w}\|$ is used for regularization, commonly a sum of squares such as $\|\boldsymbol{w}\|_2^2 = \sum_i w_i^2$. Another choice is $\|\boldsymbol{w}\|_1 = \sum_i |w_i|$, which, similar to Lasso regression, may result in a sparse weighting vector w (with many zero entries) indicating which elements of the feature vector can be neglected for classification purposes.

Choosing for the latter function results in the objective function

$$\min_{\boldsymbol{w},b} \left[\lambda \|\boldsymbol{w}\|_1 + \frac{1}{n} \sum_{i=1}^{n} \left(1 - y_i \left(\boldsymbol{w}^\top \boldsymbol{x}_i + b \right) \right)^+ \right].$$

By introducing n auxiliary nonnegative variables $z_i \geq 0$ such that

$$y_i \left(\boldsymbol{w}^\top \boldsymbol{x}_i + b \right) \geq 1 - z_i, \qquad \forall i = 1, \dots, n,$$

we can formulate the problem of fitting a linear classifier as

$$
\begin{aligned}
\min \quad & \lambda \|\boldsymbol{w}\|_1 + \frac{1}{n} \sum_{i=1}^{n} z_i \\
\text{s.t.} \quad & z_i \geq 1 - y_i \left(\boldsymbol{w}^\top \boldsymbol{x}_i + b \right), && \forall i = 1, \dots, n \\
& z_i \geq 0, && \forall i = 1, \dots, n \\
& \boldsymbol{w} \in \mathbb{R}^p \\
& b \in \mathbb{R}.
\end{aligned}
$$

This is the primal optimization problem in the decision variables $\boldsymbol{w} \in \mathbb{R}^p$, $b \in \mathbb{R}$, and $\boldsymbol{z} \in \mathbb{R}^n$, a total of $n + p + 1$ unknowns with $2n$ constraints. This can be recast as a linear optimization problem with the usual technique of setting $\boldsymbol{w} = \boldsymbol{w}^+ - \boldsymbol{w}^-$ where $\boldsymbol{w}^+$ and $\boldsymbol{w}^-$ are nonnegative. Then

$$
\begin{aligned}
\min \quad & \lambda \sum_{j=1}^{p} (w_j^+ + w_j^-) + \frac{1}{n} \sum_{i=1}^{n} z_i \\
\text{s.t.} \quad & z_i \geq 1 - y_i \left(\left(\boldsymbol{w}^+ - \boldsymbol{w}^- \right)^\top \boldsymbol{x}_i + b \right), && \forall i = 1, \dots, n \\
& z_i \geq 0, && \forall i = 1, \dots, n \\
& \boldsymbol{w}^+, \boldsymbol{w}^- \geq 0 \\
& b \in \mathbb{R}.
\end{aligned}
$$

We implement the tuning of the SVM model using Pyomo. An instance of an LO problem is created with a training set. This implementation is formulated as a function that produces instances of SVM objects from training data and other specifications and returns the SVM object as we formulated it above.

```python
def svm_linear(X, y, lambd=1):
    m = pyo.ConcreteModel("Linear SVM")

    # Use dataframe columns and index to index variables and constraints
    m.P = pyo.Set(initialize=X.columns)
    m.N = pyo.Set(initialize=X.index)

    # Decision variables
    m.wp = pyo.Var(m.P, domain=pyo.NonNegativeReals)
    m.wn = pyo.Var(m.P, domain=pyo.NonNegativeReals)
    m.b = pyo.Var()
```

```
    m.z = pyo.Var(m.N, domain=pyo.NonNegativeReals)

    @m.Expression(m.P)
    def w(m, p):
        return m.wp[p] - m.wn[p]

    @m.Objective(sense=pyo.minimize)
    def lasso(m):
        return sum(m.z[i] for i in m.N) / len(m.N) + lambd * sum(
            m.wp[p] + m.wn[p] for p in m.P
        )

    @m.Constraint(m.N)
    def hingeloss(m, i):
        return m.z[i] >= 1 - y[i] * (sum(m.w[p] * X.loc[i, p] for p in m.P) +
        ↪    m.b)

    return m


m = svm_linear(X_train, y_train)
SOLVER.solve(m)
w = pd.Series([m.w[p]() for p in m.P], index=m.P)
b = m.b()
linear_svm = LinearSVM(w, b)
```

This new classifier performs better than the previous one, with a Matthews correlation coefficient (MCC) of 0.727, having a sensitivity of 95.0%, a precision of 80.7%, and an accuracy of 85.8% (see Figure 5.11 for visualization).

As we can see, the accuracy obtained with an optimal linear classifier is higher than that of the naive classifier we created initially. The pictures show that this is because the "separating hyperplane" takes a skewed shape going from the top-left to the bottom-right of our feature space. Although we managed to significantly improve the classifier accuracy with respect to the naive classifier, there are even more significant gains possible by going beyond a linear classifier and allowing more sophisticated models.

Linear SVM With L_2 Regularization Term

We will now proceed with training nonlinear classifiers that can achieve higher accuracy than nonlinear ones. For computational reasons (as explained later in the "kernel trick") it is customary to fit nonlinear classifiers by solving the dual of the tuning/fitting problem. Because it is customary to derive this dual from a problem with a **quadratic regularizer** $\|\cdot\|_2$ instead of a $\|\cdot\|_1$ one, we begin by formulating the optimization problem for the tuning of a linear SVM classifier with a quadratic penalty term.

The training objective is to minimize the total distance to misclassified data points with a regularization function $\|w\|_2^2$. The resulting optimization problem is a quadratic problem with linear constraints, namely:

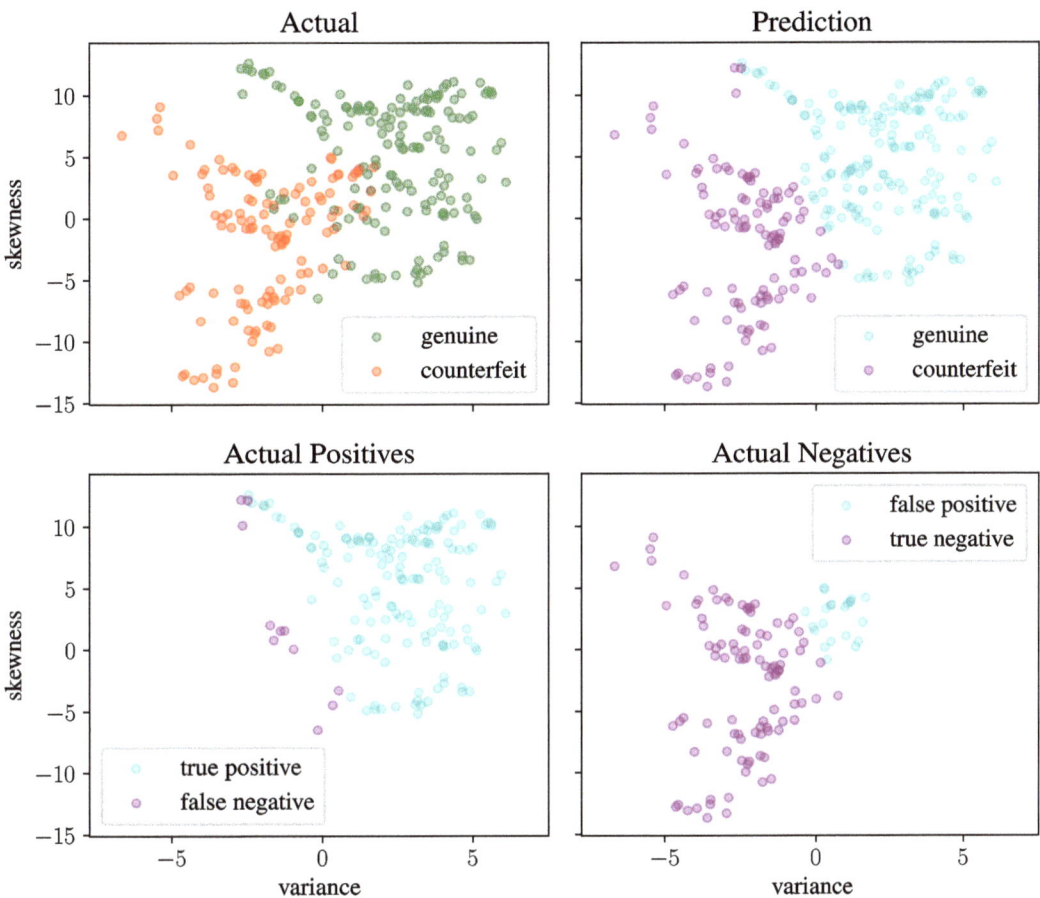

Figure 5.11 Scatter plot describing the performance of the linear SVM classification.

$$\min \quad \frac{1}{2}\|\boldsymbol{w}\|_2^2 + \frac{c}{n}\sum_{i=1}^{n} z_i$$

$$\text{s.t.} \quad z_i \geq 1 - y_i\Big(\boldsymbol{w}^\top \boldsymbol{x}_i + b\Big), \qquad \forall\, i = 1,\ldots,n$$

$$z_i \geq 0, \qquad \forall\, i = 1,\ldots,n$$

$$\boldsymbol{w} \in \mathbb{R}^p$$

$$b \in \mathbb{R}.$$

Choosing larger values of c will reduce the number and size of misclassifications. The trade-off will be higher weights $\boldsymbol{w}$ and the accompanying risk of overfitting the training data. The following code formulates this optimization problem, returning the resulting SVM as output and solves it using the general nonlinear optimization solver `ipopt`.

```python
def svm_quadratic(X, y, c=1):
    m = pyo.ConcreteModel("SVM quadratic with L2 regularization")

    # Use dataframe columns and index to index variables and constraints
    m.P = pyo.Set(initialize=X.columns)
```

```
        m.N = pyo.Set(initialize=X.index)

        # Decision variables
        m.w = pyo.Var(m.P)
        m.b = pyo.Var()
        m.z = pyo.Var(m.N, domain=pyo.NonNegativeReals)

        @m.Objective(sense=pyo.minimize)
        def qp(m):
            return sum(m.w[i] ** 2 for i in m.P) / 2 + (c / len(m.N)) * sum(
                m.z[i] for i in m.N
            )

        @m.Constraint(m.N)
        def hingeloss(m, i):
            return m.z[i] >= 1 - y[i] * (sum(m.w[p] * X.loc[i, p] for p in m.P) +
            ↪   m.b)

        return m


m = svm_quadratic(X_train, y_train)
pyo.SolverFactory("ipopt").solve(m)
w = pd.Series([m.w[p]() for p in m.P], index=m.P)
b = m.b()
quadratic_svm = LinearSVM(w, b)
```

The performance of the obtained linear classifier is very similar to that of the previous one. In fact, it has a Matthews correlation coefficient (MCC) of 0.730, a sensitivity of 93.6%, a precision of 82.0%, and an accuracy of 86.2%. Nevertheless, this helps us to move toward using nonlinear classifiers. We are now going to derive the dual of the above optimization problem.

Dual Formulation

The dual formulation for the linear SVM above provides insight into how SVM works, and it is essential for extending SVM to nonlinear classification. The dual formulation begins by creating a differentiable Lagrangian with dual variables $\alpha_i \geq 0$ and $\beta_i \geq 0$ for $i = 1, \ldots, n$:

$$\mathcal{L} = \frac{1}{2}\|\boldsymbol{w}\|_2^2 + \frac{c}{n}\sum_{i=1}^{n} z_i + \sum_{i=1}^{n} \alpha_i \left(1 - z_i - y_i(\boldsymbol{w}^\top \boldsymbol{x}_i + b)\right) + \sum_{i=1}^{n} \beta_i(-z_i).$$

To minimize this Lagrangian with respect to the original decision variables $\boldsymbol{w}$, b, and $\boldsymbol{z}$, we take the partial derivatives with respect to the primal variables and set them equal to zero, obtaining the following equations:

$$\frac{\partial \mathcal{L}}{\partial z_i} = \frac{c}{n} - \alpha_i - \beta_i = 0 \implies 0 \leq \alpha_i \leq \frac{c}{n}$$

$$\frac{\partial \mathcal{L}}{\partial \boldsymbol{w}} = \boldsymbol{w} - \sum_{i=1}^{n} \alpha_i y_i \boldsymbol{x}_i = 0 \implies \boldsymbol{w} = \sum_{i=1}^{n} \alpha_i y_i \boldsymbol{x}_i$$

$$\frac{\partial \mathcal{L}}{\partial b} = -\sum_{i=1}^{n} \alpha_i y_i = 0 \implies \sum_{i=1}^{n} \alpha_i y_i = 0.$$

These can be arranged as a standard quadratic optimization problem in n variables α_i for $i = 1, \ldots, n$:

$$\min \quad \frac{1}{2} \sum_{i=1}^{n} \sum_{j=1}^{n} \alpha_i \alpha_j y_i y_j \left(\boldsymbol{x}_i^\top \boldsymbol{x}_j \right) - \sum_{i=1}^{n} \alpha_i$$

$$\text{s.t.} \quad \sum_{i=1}^{n} \alpha_i y_i = 0$$

$$\alpha_i \in \left[0, \frac{c}{n}\right], \qquad \forall i = 1, \ldots, n.$$

The resulting dual problem is thus also a quadratic optimization problem in $\boldsymbol{\alpha}$. It is easily checked that strong duality holds between the two problems because both the primal and the dual are easily (strictly) feasible. We can formulate the dual using a symmetric $n \times n$ **Gram matrix** defined as

$$G = \begin{bmatrix} \left(y_1 \boldsymbol{x}_1^\top\right)\left(y_1 \boldsymbol{x}_1\right) & \cdots & \left(y_1 \boldsymbol{x}_1^\top\right)\left(y_n \boldsymbol{x}_n\right) \\ \vdots & \ddots & \vdots \\ \left(y_n \boldsymbol{x}_n^\top\right)\left(y_1 \boldsymbol{x}_1\right) & \cdots & \left(y_n \boldsymbol{x}_n^\top\right)\left(y_n \boldsymbol{x}_n\right) \end{bmatrix}.$$

With this matrix, the problem can be formulated as

$$\min \quad \frac{1}{2} \boldsymbol{\alpha}^\top G \boldsymbol{\alpha} - \sum_{i=1}^{n} \alpha_i$$

$$\text{s.t.} \quad \sum_{i=1}^{n} \alpha_i y_i = 0$$

$$\alpha_i \in \left[0, \frac{c}{n}\right], \qquad \forall i = 1, \ldots, n.$$

Compared to the primal, the dual formulation appears to have reduced the number of decision variables from $n + p + 1$ to n. This comes at the cost of introducing a dense matrix with n^2 coefficients and potential processing time of order n^3. This becomes a prohibitively expensive calculation for large training sets where $n \sim 10^4 - 10^6$ or even larger. Furthermore, the Gram matrix will be rank-deficient for cases $p < n$. We can avoid computing and storing the full Gram matrix G by introducing the $n \times p$ matrix F:

$$F = \begin{bmatrix} y_1 \boldsymbol{x}_1^\top \\ y_2 \boldsymbol{x}_2^\top \\ \vdots \\ y_n \boldsymbol{x}_n^\top \end{bmatrix},$$

so that $G = F F^\top$. We can then easily recover the primal optimal solution from the dual optimal solution through the relationship $\boldsymbol{w} = F^\top \boldsymbol{\alpha}$. The optimization problem becomes

$$\min \quad \frac{1}{2} \boldsymbol{w}^\top \boldsymbol{w} - \boldsymbol{1}^\top \boldsymbol{\alpha}$$
$$\text{s.t.} \quad \boldsymbol{y}^\top \boldsymbol{\alpha} = 0$$
$$\boldsymbol{w} = F^\top \boldsymbol{\alpha}$$
$$0 \le \alpha_i \le \frac{c}{n}$$
$$\boldsymbol{w} \in \mathbb{R}^p$$
$$\boldsymbol{\alpha} \in \mathbb{R}^n.$$

With respect to the other part of the primal solution (the SVM consists of two parts: $\boldsymbol{w}$ and b), we can recover the bias term b by considering the complementarity conditions on the dual variables. The slack variables z_i are zero if $\beta_i > 0$, which is equivalent to $\alpha_i < \frac{c}{n}$. If $\alpha_i > 0$ then $1 - y_i(\boldsymbol{w}^\top \boldsymbol{x}_i + b)$. Putting these facts together gives a formula for b, namely

$$b = y_i - \boldsymbol{w}^\top \boldsymbol{x}_i, \qquad \forall i = 1, \dots, n \quad \text{s.t.} \quad 0 < \alpha_i < \frac{c}{n}.$$

```python
def svm_dual(X, y, c=1):
    m = pyo.ConcreteModel("Linear SVM Dual")

    # Use dataframe columns and index to index variables and constraints
    m.P = pyo.Set(initialize=X.columns)
    m.N = pyo.Set(initialize=X.index)

    # Model parameters
    F = X.mul(y, axis=0)
    m.C = pyo.Param(initialize=c / len(m.N))

    # Decision variables
    m.w = pyo.Var(m.P)
    m.a = pyo.Var(m.N, bounds=(0, m.C))

    @m.Objective(sense=pyo.minimize)
    def dualqp(m):
        return sum(m.w[i] ** 2 for i in m.P) / 2 - sum(m.a[i] for i in m.N)

    @m.Constraint()
    def bias(m):
        return sum(y.loc[i] * m.a[i] for i in m.N) == 0

    @m.Constraint(m.P)
    def projection(m, p):
        return m.w[p] == sum(F.loc[i, p] * m.a[i] for i in m.N)

    return m


m = svm_dual(X_train, y_train)
pyo.SolverFactory("ipopt").solve(m)
```

```
# Extract the optimal values of w and b
w = pd.Series([m.w[p]() for p in m.P], index=m.P)
a = pd.Series([m.a[i]() for i in m.N], index=m.N)

# Find alpha closest to the center of [0, c/n]
i = a.index[(a - m.C / 2).abs().argmin()]
b = y_train.loc[i] - X_train.loc[i, :].dot(w)
dual_svm = LinearSVM(w, b)
```

The linear classifier trained using the dual formulation has a Matthews correlation coefficient (MCC) of 0.752, a sensitivity of 92.6%, a precision of 85.7%, and an accuracy of 87.6%.

Kernelized (Nonlinear) SVM and the Kernel Trick

We will now introduce the idea of kernelized (nonlinear) support vector machines. Although such SVMs are almost always solved by solving the dual problem (it has a much smaller dimensionality than the primal one, as the above argument has shown), we first present the derivation through the primal to give more of the intuition.

A linear SVM assumes the existence of a linear hyperplane that separates labeled sets of data points. Frequently, however, this is not possible, and some sort of nonlinear method is where the feature vector is appended with nonlinear transformations

$$\boldsymbol{x} \to \phi(\boldsymbol{x}),$$

where $\phi(\boldsymbol{x})$ is a function mapping x into a higher dimensional feature space. That is, $\phi \colon \mathbb{R}^p \to \mathbb{R}^d$ where $d \geq p$. The additional dimensions may include features such as powers of the terms in $\boldsymbol{x}$, products of those terms, or other types of nonlinear transformations. In this way, the SVM has more "room to play" to try to separate one point from another.

Corresponding to such a transformed vector, we have a binary classification tool given by

$$y^{\text{pred}} = \text{sgn}\left(\boldsymbol{w}^\top \phi(\boldsymbol{x}) + b\right).$$

As before, we wish to find a choice for $\boldsymbol{w} \in \mathbb{R}^d$ such that the soft-margin classifier

$$y_i\left(\boldsymbol{w}^\top \phi(\boldsymbol{x}_i) + b\right) \geq 1 - z_i, \quad \forall i = 1, \dots, n.$$

Using the machinery as before, we set up the Lagrangian for the $\|\cdot\|_2^2$-regularized problem:

$$\mathcal{L} = \frac{1}{2}\|\boldsymbol{w}\|_2^2 + \frac{c}{n}\sum_{i=1}^n z_i + \sum_{i=1}^n \alpha_i\left(1 - z_i - y_i\left(\boldsymbol{w}^\top \phi(\boldsymbol{x}_i) + b\right)\right) + \sum_{i=1}^n \beta_i(-z_i),$$

then take its partial derivatives to find

$$\frac{\partial \mathcal{L}}{\partial z_i} = \frac{c}{n} - \alpha_i - \beta_i = 0 \implies 0 \leq \alpha_i \leq \frac{c}{n}$$

$$\frac{\partial \mathcal{L}}{\partial \boldsymbol{w}} = \boldsymbol{w} - \sum_{i=1}^n \alpha_i y_i \phi(\boldsymbol{x}_i) = 0 \implies \boldsymbol{w} = \sum_{i=1}^n \alpha_i y_i \phi(\boldsymbol{x}_i)$$

$$\frac{\partial \mathcal{L}}{\partial b} = -\frac{c}{n}\sum_{i=1}^n \alpha_i y_i = 0 \implies \sum_{i=1}^n \alpha_i y_i = 0.$$

This is similar to the case of a linear SVM, but now with the vector of weights $w \in \mathbb{R}^d$ which can be a high-dimensional space with nonlinear features. Working through the algebra, we are once again left with a quadratic problem in n variables α_i for $i = 1, \ldots, n$:

$$\min \quad \frac{1}{2} \sum_{i=1}^{n} \sum_{j=1}^{n} \alpha_i \alpha_j y_i y_j \phi(\boldsymbol{x}_i)^\top \phi(\boldsymbol{x}_j) - \sum_{i=1}^{n} \alpha_i$$

$$\text{s.t.} \quad \sum_{i=1}^{n} \alpha_i y_i = 0$$

$$\alpha_i \in \left[0, \frac{c}{n}\right], \qquad \forall i = 1, \ldots, n,$$

where the resulting classifier for a new point x is given by

$$y^{\text{pred}} = \text{sgn}\left(\boldsymbol{w}^\top \phi(\boldsymbol{x}) + b\right) = \text{sgn}\left(\sum_{i=1}^{n} \alpha_i y_i \phi(\boldsymbol{x}_i)^\top \phi(\boldsymbol{x}) + b\right).$$

If you look at the optimization problem above and the corresponding classifier tool, this is an interesting situation where the separating hyperplane is embedded in a high-dimensional space of nonlinear features determined by the mapping $\phi(\boldsymbol{x})$, but all we need to train the classifier and to use the classifier are the inner products $\phi(\boldsymbol{x}_i)^\top \phi(\boldsymbol{x}_j)$ and $\phi(\boldsymbol{x}_i)^\top \phi(\boldsymbol{x})$, rather than the "raw" $\phi(\boldsymbol{x}_i)$ and $\phi(\boldsymbol{x})$ values. If we had a function $K(\boldsymbol{x}, \boldsymbol{z})$ that returned the value $\phi(\boldsymbol{x})^\top \phi(\boldsymbol{z})$, then we would never need to actually compute $\phi(\boldsymbol{x})$, $\phi(\boldsymbol{z})$ or their inner product.

Mercer's theorem turns the analysis on its head by specifying conditions for which a function $K(\boldsymbol{x}, \boldsymbol{z})$ can be expressed as an inner product for some $\phi(\boldsymbol{x})$. If $K(\boldsymbol{x}, \boldsymbol{z})$ is symmetric, that is, $K(\boldsymbol{x}, \boldsymbol{z}) = K(\boldsymbol{z}, \boldsymbol{x})$, and if the Gram matrix constructed for any collection of points $\boldsymbol{x}_1, \boldsymbol{x}_2, \ldots, \boldsymbol{x}_n$

$$\begin{bmatrix} K(\boldsymbol{x}_1, \boldsymbol{x}_1) & \cdots & K(\boldsymbol{x}_1, \boldsymbol{x}_n) \\ \vdots & \ddots & \vdots \\ K(\boldsymbol{x}_n, \boldsymbol{x}_1) & \cdots & K(\boldsymbol{x}_n, \boldsymbol{x}_n) \end{bmatrix}$$

is positive semi-definite, then there is some $\phi(\boldsymbol{x})$ for which $K(\boldsymbol{x}, \boldsymbol{z})$ is an inner product. We call such functions kernels. The practical consequence is that we can train and implement nonlinear classifiers using kernels without ever needing to compute the higher-dimensional features. This remarkable result is called the **kernel trick**.

To take advantage of the kernel trick, we assume that an appropriate kernel $K(\boldsymbol{x}, \boldsymbol{z})$ has been identified, then replace all instances of $\phi(\boldsymbol{x}_i)^\top \phi(\boldsymbol{x})$ with the kernel. The "kernelized" SVM is given by a solution to

$$\min \quad \frac{1}{2} \sum_{i=1}^{n} \sum_{j=1}^{n} \alpha_i \alpha_j y_i y_j K(\boldsymbol{x}_i, \boldsymbol{x}_j) - \sum_{i=1}^{n} \alpha_i$$

$$\text{s.t.} \quad \sum_{i=1}^{n} \alpha_i y_i = 0$$

$$\alpha_i \in \left[0, \frac{c}{n}\right], \qquad \forall i = 1, \ldots, n,$$

where

$$b = y_i - \sum_{j=1}^{n} \alpha_j y_j K(\boldsymbol{x}_j, \boldsymbol{x}_i), \qquad \forall\, i = 1, \ldots, n \quad \text{s.t.} \quad 0 < \alpha_i < \frac{c}{n}.$$

The resulting classifier is given by

$$y^{\text{pred}} = \text{sgn}\left(\sum_{i=1}^{n} \alpha_i y_i K(\boldsymbol{x}_i, \boldsymbol{x}) + b \right).$$

For a more compact formulation, we can define the $n \times n$ positive symmetric semidefinite Gram matrix

$$G = \begin{bmatrix} y_1 y_1 K(\boldsymbol{x}_1, \boldsymbol{x}_1) & \cdots & y_1 y_n K(\boldsymbol{x}_1, \boldsymbol{x}_n) \\ \vdots & \ddots & \vdots \\ y_n y_1 K(\boldsymbol{x}_n, \boldsymbol{x}_1) & \cdots & y_n y_n K(\boldsymbol{x}_n, \boldsymbol{x}_n) \end{bmatrix}.$$

As in the case of linear SVMs, the Gram matrix corresponding to the quadratic optimization dual problem may not be full-rank, leading to some numerical issues. To address this issue, we can factor $G = FF^\top$ where F has dimensions $n \times q$ and q is the rank of G. The factorization is not unique. As demonstrated in the Python code that follows, one suitable factorization is the spectral factorization $G = U \Lambda U^T$ where Λ is a $q \times q$ diagonal matrix of nonzero eigenvalues, and U is an $n \times q$ normal matrix such that $U^\top U = I_q$. Then

$$F = U \Lambda^{1/2}.$$

Once this factorization is complete, the optimization problem for the kernelized SVM is the same as for the linear SVM in the dual formulation

$$\begin{aligned} \min \quad & \frac{1}{2} \boldsymbol{\alpha}^\top F F^\top \boldsymbol{\alpha} - \mathbf{1}^\top \boldsymbol{\alpha} \\ \text{s.t.} \quad & \sum_{i=1}^{n} \alpha_i y_i = 0 \\ & 0 \le \alpha_i \le \frac{c}{n} \qquad \forall i = 1, \ldots, n \\ & \boldsymbol{\alpha} \in \mathbb{R}^n. \end{aligned}$$

The result is a quadratic problem for the dual coefficients $\boldsymbol{\alpha}$ and auxiliary variables $\boldsymbol{v}$:

$$\begin{aligned} \min \quad & \frac{1}{2} \boldsymbol{v}^\top \boldsymbol{v} - \mathbf{1}^\top \boldsymbol{\alpha} & (5.9) \\ \text{s.t.} \quad & \boldsymbol{y}^\top \boldsymbol{\alpha} = 1 \\ & \boldsymbol{v} = F^\top \boldsymbol{\alpha} \\ & 0 \le \alpha_i \le \frac{c}{n} \qquad \forall i = 1, \ldots, n \\ & \boldsymbol{v} \in \mathbb{R}^q \\ & \boldsymbol{\alpha} \in \mathbb{R}^n. \end{aligned}$$

In summary, the essential difference between training the linear and kernelized SVM is the need to compute and factor the Gram matrix. The result is a set of nonzero coefficients $\alpha_i > 0$ that define a set of support vectors $\mathcal{SV}$. The classifier is then given by

$$y^{\text{pred}} = \text{sgn}\left(\sum_{i \in \mathcal{SV}} \alpha_i y_i K(\boldsymbol{x}_i, \boldsymbol{x}) + b\right).$$

The following code implements this model using a polynomial kernel of order 3 (which corresponds to including in the feature vector the powers of all features up to and including degree 3) and solves it using the nonlinear solver `ipopt`.

```python
class KernelSVM:
    # Initialize the Kernel SVM with weights and bias
    def __init__(self, X, y, a, b, kernel):
        self.X = np.array(X)
        self.u = np.multiply(np.array(a), np.array(y))
        self.b = b
        self.kernel = kernel

    # Call method to compute the decision function using the input dataframe Z
    def __call__(self, Z):
        K = [
            [self.kernel(self.X[i, :], Z.loc[j, :]) for j in Z.index]
            for i in range(len(self.X))
        ]
        y_pred = np.sign((self.u @ K) + self.b)
        return pd.Series(y_pred, index=Z.index)


def svm_kernel_model(X, y, c=1, tol=1e-8, kernel=lambda x, z: x @ z):
    # Convert to numpy arrays for speed improvement
    n, p = X.shape
    X_ = X.to_numpy()
    y_ = y.to_numpy()

    # Gram matrix
    G = np.zeros((n, n))
    for i in range(n):
        for j in range(i, n):
            G[j, i] = G[i, j] = y_[i] * y_[j] * kernel(X_[i, :], X_[j, :])

    # Factor the Gram matrix
    eigvals, eigvecs = np.linalg.eigh(G)
    idx = eigvals >= tol * max(eigvals)
    F = pd.DataFrame(eigvecs[:, idx] @ np.diag(np.sqrt(eigvals[idx])),
    ↪    index=X.index)

    # Build model
    m = pyo.ConcreteModel("SVM with kernel")
```

```python
    # Use dataframe columns and index to index variables and constraints
    m.Q = pyo.Set(initialize=F.columns)
    m.N = pyo.Set(initialize=F.index)

    # Model parameters
    m.C = pyo.Param(initialize=c / len(m.N))

    # Decision variables
    m.u = pyo.Var(m.Q)
    m.a = pyo.Var(m.N, bounds=(0, m.C))

    @m.Objective(sense=pyo.minimize)
    def kernelqp(m):
        return sum(m.u[i] ** 2 for i in m.Q) / 2 - sum(m.a[i] for i in m.N)

    @m.Constraint()
    def bias(m):
        return sum(y.loc[i] * m.a[i] for i in m.N) == 0

    @m.Constraint(m.Q)
    def projection(m, q):
        return m.u[q] == sum(F.loc[i, q] * m.a[i] for i in m.N)

    return m


def svm_kernel(X, y, c=1, tol=1e-8, kernel=lambda x, z: x @ z):
    # create an optimization model for SVM binary classifier using a kernel
    m = svm_kernel_model(X, y, c=c, tol=tol, kernel=kernel)

    # Solve with the interior point method
    SOLVER_NLO.solve(m)

    # Extract solution
    sol = pd.Series([m.a[i]() for i in m.N], index=m.N)

    # Find b by locating a closest to the center of [0, c/n]
    i = sol.index[(sol - m.C / 2).abs().argmin()]
    b = y.loc[i] - sum(
        [sol[j] * y.loc[j] * kernel(X.loc[j, :], X.loc[i, :]) for j in m.N]
    )

    # Display the support vectors
    y_support = pd.Series(
        [1 if m.a[i]() > 1e-4 * m.C else -1 for i in m.N], index=X.index
    )
    scatter_labeled_data(
        X,
        y_support,
        colors=["b", "y"],
```

```
        labels=["Support Vector 1", "Support Vector 2"],
    )

    # Find support vectors
    SV = [i for i in m.N if m.a[i]() > 1e-3 * m.C]

    return KernelSVM(X.loc[SV, :], y.loc[SV], sol.loc[SV], b, kernel)
```

This new classifier performs better than the previous one, with a Matthews correlation coefficient (MCC) of 0.820, a sensitivity of 88.0%, a precision of 95.9%, and an accuracy of 90.9%.

The ability to use a nonlinear transformation of the features allowed us to gain an extra 0.1 in the value of the Matthews correlation coefficient. This improvement is also clearly visible in Figure 5.12.

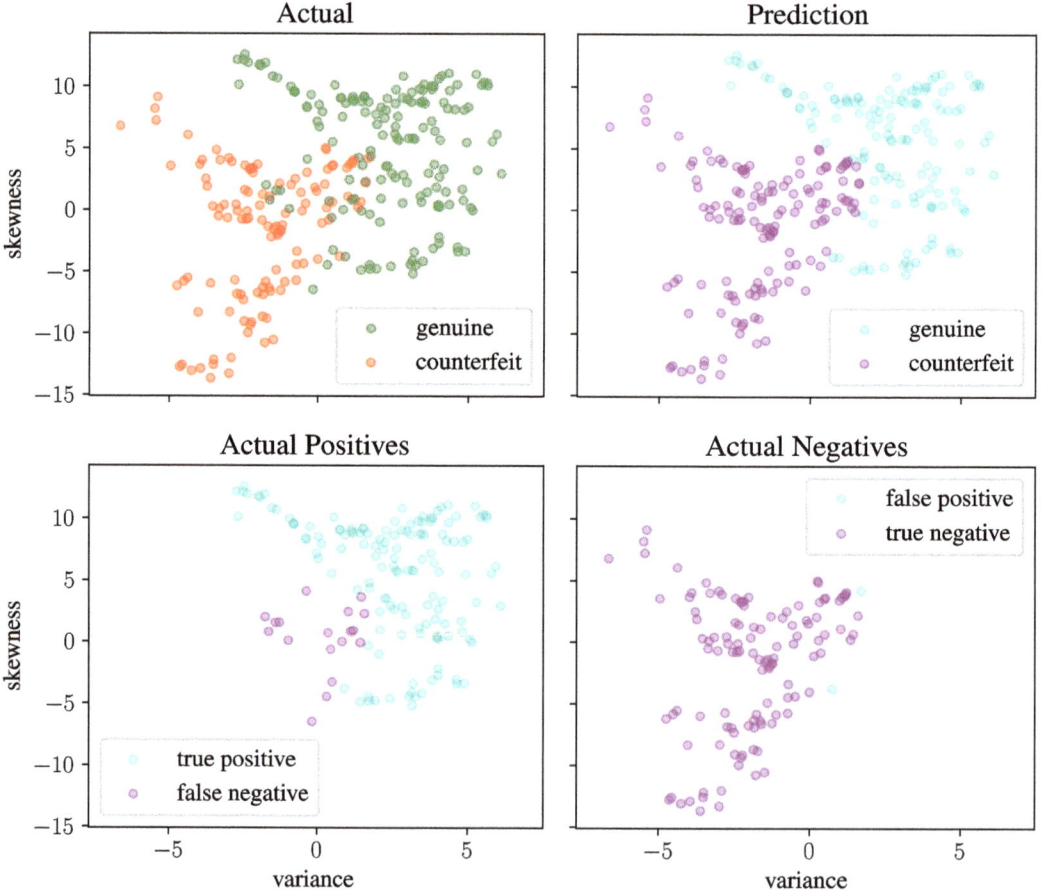

Figure 5.12 Scatter plot describing the performance of the kernelized SVM classification that uses a polynomial kernel of order 3.

Exercises

Exercise 5.1 Reformulate the following optimization problem into an equivalent linear optimization problem:

$$\min \quad \sum_{i=1}^{n} \max \left\{ i + a_1^\top y, i\left(1 + a_2^\top x\right) \right\}$$

$$\text{s.t.} \quad \frac{(1 + b_i x_i)}{1 + y_i} \leq 1 \qquad \qquad \forall i = 1, \ldots, n$$

$$\exp\left(1 + 3a_2^\top x\right) \leq 1$$

$$x, y \geq 0,$$

in which $a_1, a_2, b \in \mathbb{R}^n$ are constants.

Exercise 5.2

(a) Is the following matrix positive definite, positive semidefinite, indefinite, negative semidefinite, or negative definite?

$$A = \begin{bmatrix} 0 & 2 & 1 \\ 2 & 0 & -3 \\ 1 & -3 & 0 \end{bmatrix}$$

(b) For which values of x is the following matrix $A(x)$ positive definite? And for which values of x is the matrix $A(x)$ negative definite?

$$A(x) = \begin{bmatrix} x & -1 & 3 \\ -1 & 1 & 0 \\ 3 & 0 & 2 \end{bmatrix}$$

(c) Prove that the function $f(x,y) = e^{x-y} + x^6 - \ln(x)$ is convex by showing that its Hessian is positive semidefinite.

Exercise 5.3

(a) Show that the intersection of two convex sets is convex.
(b) Show that the intersection of any number of convex sets is convex.
(c) Give an example to show that the union of two convex sets is not always convex.

Exercise 5.4 Show that $f(x) = x^2$ is a convex function in three different ways, using the following three properties of convex functions:

(a) $f(\theta x + (1 - \theta)y) \leq \theta f(x) + (1 - \theta)f(y)$ for $0 < \theta < 1$.
(b) $f(y) \geq f(x) + \nabla_x f(x)^\mathrm{T}(y - x)$.
(c) The Hessian is positive semidefinite for all x.

Exercise 5.5 Use the Lagrangian function to find the stationary point(s) of the following problem:

$$\min \quad 5x - 3y$$
$$\text{s.t.} \quad x^2 + y^2 = 136$$
$$x, y \in \mathbb{R}.$$

Exercise 5.6 State the KKT conditions for the following problem and use them to find its global minimum:

$$\min \quad x^2 + y^2$$
$$\text{s.t.} \quad x + y \geq 4$$
$$x, y \in \mathbb{R}.$$

Exercise 5.7 Consider the problem:

$$\max \quad \ln(x+1) + y$$
$$\text{s.t.} \quad 2x + y \leq 3$$
$$x, y \geq 0.$$

(a) Show that the function $f(x,y) = \ln(x+1) + y$ is *not* convex.
(b) Reformulate the problem as an equivalent convex minimization problem.
(c) Write the KKT conditions and find the global minimum of the problem stated in (b).

Exercise 5.8 Consider the following problem (note that it is not convex):

$$\min \quad f(x) = x^3 - x^2 + 5$$
$$\text{s.t.} \quad f(x) \geq \max\{e^x, x+2\}.$$

(a) State the KKT conditions for this problem.
(b) Find the two stationary points of the *unconstrained* problem for minimizing $f(x)$ and determine for each of them if they correspond to a minimum.
(c) Show that each minimum obtained in (b) satisfies the KKT conditions. Is/are this/these solution(s) also the optimal one(s) for the constrained problem?

Exercise 5.9 Consider a product made of two ingredients, labeled A and B. We can make the product out of any combination of A and B, but the price we can charge depends on the chosen amount of ingredients. If we denote by x_A and x_B the amounts of ingredient A and B we choose for our product, then the price of our product is given by $12\sqrt{x_A} + 10\sqrt{x_B}$. The cost per unit of ingredient A is 2 and that of B is 5. Formulate and solve an optimization problem to determine the optimal composition of ingredients A and B that maximizes profit.

Exercise 5.10 A vehicle is driving for a certain period of length t. It first drives for t_1 time units at speed s_1, and after that t_1 it accelerates and drives for t_2 time units at speed $s_2 > s_1$ the rest of the time (thus $t = t_1 + t_2$). The energy per time unit used by the vehicle when driving at speed s is given by the function $u(s) = s^\alpha$, with $\alpha > 1$.

(a) Show that the function $u(s)$ is convex.

(b) The average speed of the vehicle throughout the period t is defined as

$$\bar{s} = \frac{t_1 s_1 + t_2 s_2}{t}.$$

Use the convexity of $u(s)$ to show that the energy used when driving at speed $\bar{s}$ for t time units is less than when driving first at speed s_1 for t_1 time units and then at speed s_2 for t_2 time units.

(c) Assume now that the vehicle has to drive D kilometers and must arrive at the final destination within B hours. However, it also has to pick up a customer at distance $d_1 < D$ within A hours. What is the optimal speed of the vehicle if the goal is to minimize the total energy used?

(d) Can you give an intuitive explanation for the structure of the solutions you obtained in question (c)?

6

Conic Optimization

6.1 Introduction

In Chapter 5, we claimed that the watershed between easy and difficult problems is their convexity status. Convex optimization problems are, however, a very broad class and one of their downsides is that the dual problem is not always readily available; see the discussion in Section 5.3. In view of the computational benefits of concurrently solving the primal and dual problems, a natural question arises: Is there a subclass of convex optimization problems that are expressive enough to model relevant real-life problems and, at the same time, allow us for a systematic derivation of the dual akin to linear optimization?

The answer is positive and is given by **conic optimization problems**, which are a particular class of convex optimization problems that include LO as their special case:

$$\text{linear optimization} \quad \subset \quad \text{conic optimization} \quad \subset \quad \text{convex optimization.}$$

The notation used to formulate conic problems resembles that of LO:

$$
\begin{aligned}
\min \quad & \boldsymbol{c}^\top \boldsymbol{x} \\
\text{s.t.} \quad & \boldsymbol{A}_i \boldsymbol{x} + \boldsymbol{b}_i \in \boldsymbol{K}_i, \qquad i = 1, \dots, m \\
& \boldsymbol{x} \in \mathbb{R}^n,
\end{aligned}
$$

where the sets $\boldsymbol{K}_i$ are known as **cones**. We will now introduce the idea of formulating a nonlinear problem like this through an example.

Example 6.1 (Economic order quantity) The economic order quantity (EOQ) is a classic problem in inventory management first formulated in [Har15]. This optimization model aims to identify the order quantity that minimizes the cost of maintaining a specific item in the inventory.

Let h be the annual cost of holding an item, including any financing charges, c be the fixed cost of placing and receiving an order, and d be the annual demand. If a quantity x of the product is ordered, it costs c to place the order, resulting in an order cost per item of x/c. Assuming that the demand is uniform in time, it will take x/d years for the demand to deplete the inventory entirely and, therefore, an average item will remain in the inventory for $x/2d$ years, incurring a cost $hx/2d$. To minimize the average cost per item, we minimize the function

$$\frac{hx}{2d} + \frac{c}{x},$$

which, modulo multiplying by d, is equivalent to minimizing the function

$$f(x) = \frac{hx}{2} + \frac{cd}{x},$$

which will simplify the notation later. The EOQ is the value of x minimizing $f(x)$, that is, the optimal solution of the following optimization problem:

$$EOQ = \arg\min \quad f(x) = \frac{hx}{2} + \frac{cd}{x} \tag{6.1}$$
$$\text{s.t.} \quad x > 0.$$

Figure 6.1 illustrates the nature of the problem and its optimal solution.

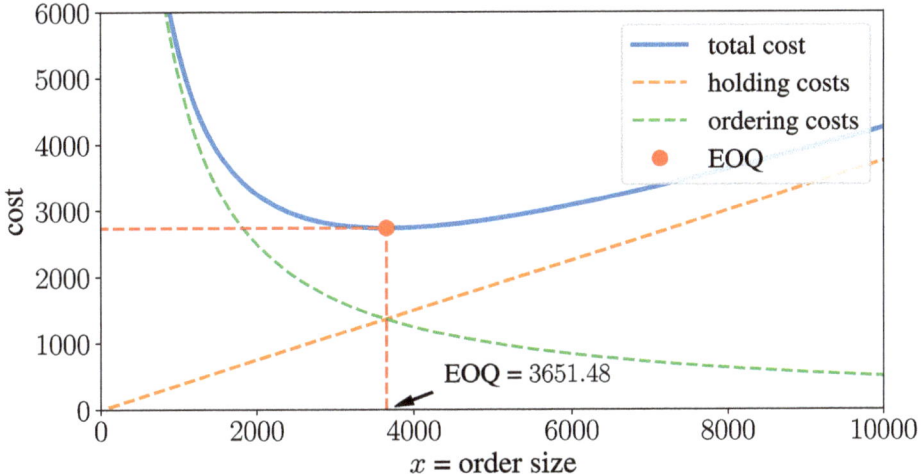

Figure 6.1 Visualization of the total cost and its two terms for an EOQ problem instance with $h = 0.75$, $c = 500$, and $d = 10{,}000$.

Given the rather simple domain, we can derive analytically the solution for the EOQ problem by setting the derivative of $f(x)$ equal to zero and solving the resulting equation, obtaining

$$EOQ = x^{opt} = \sqrt{\frac{2cd}{h}}$$
$$f^{opt} = f(x^{opt}) = \sqrt{2cdh}.$$

Using the parameters as in Figure 6.1, the optimal order size is $x^{opt} \approx 3651.48$ items, resulting in an optimal cost of $f^{opt} \approx 2738.61$.

If the problem involved multiple products for which quantities $x_1, x_2, \ldots, x_n$ would need to be ordered subject to a constraint on the total capacity of the warehouse

$$x_1 + \cdots + x_n \leq C,$$

an analytical solution would no longer be easily available. Therefore, we will now show how to formulate the above optimization problem as a simple conic one, which can be easily used to solve the multiple-product version of the same problem, subject to additional constraints.

It can be easily checked that the objective $f(x)$ is a convex function, and therefore, the problem can be solved using any convex optimization solver. However, it is a special type of convex problem, which we will make explicit in the following reformulation. The objective function of (6.1) can be linearized by adding a second auxiliary decision variable $y \geq 1/x$. The resulting optimization problem is

$$\min \quad f(x,y) = \frac{hx}{2} + cdy$$
$$\text{s.t.} \quad xy \geq 1$$
$$x,y > 0.$$

The optimal solution of this problem is always such that $y = 1/x$ because the coefficient cd in the objective is positive and, hence, y will try to be as small as possible. This problem has two decision variables, a linear objective and a hyperbolic constraint $xy \geq 1$. Figure 6.2 illustrates the structure and the optimal solution to this optimization problem.

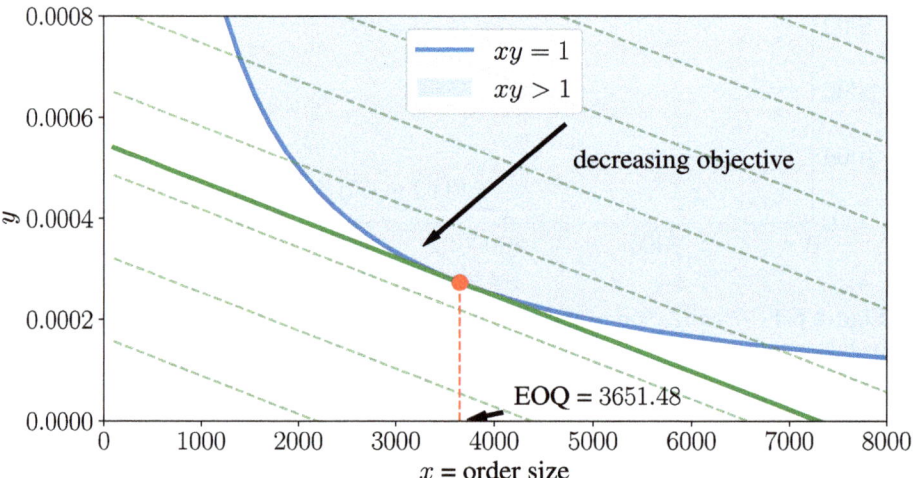

Figure 6.2 The EOQ problem reformulation as linear objective with a hyperbolic constraint.

Leveraging the nonnegativity of both variables (which implies $x + y \geq 0$), the constraint $xy \geq 1$ can be reformulated using the following trick:

$$xy \geq 1 \quad \Longleftrightarrow \quad 4xy \geq 4 \quad \Longleftrightarrow \quad (x+y)^2 - (x-y)^2 \geq 4$$
$$\Longleftrightarrow \quad \left\| \begin{bmatrix} 2 \\ x-y \end{bmatrix} \right\|_2 \leq x+y,$$

where $\|u\|_2 = \sqrt{u_1^2 + \cdots + u_n^2}$ is the 2-norm of a vector $u \in \mathbb{R}^n$. The final constraint is known as a second-order conic optimization constraint (SOCO constraint) and the EOQ problem can be rewritten as the following second-order conic optimization (SOCO) problem:

$$\min \quad f(x,y) = \frac{hx}{2} + cdy$$

$$\text{s.t.} \quad \left\| \begin{bmatrix} 2 \\ x - y \end{bmatrix} \right\|_2 \leq x + y$$

$$x, y \geq 0.$$

In general, problems with these types of constraints are more tractable and easier to solve than general convex problems. One of the reasons behind this fact is the relative ease with which one can formulate the dual problem, comparable to that of LO problems. As we demonstrate later in this chapter, the dual problem can be formulated in a more automated manner, bypassing the extensive derivations via the Lagrange function outlined in Chapter 5. In this specific case, the dual problem is

$$\max \quad -2\lambda_1$$

$$\text{s.t.} \quad \begin{bmatrix} h/2 \\ cd \end{bmatrix} = \begin{bmatrix} 0 & 1 & 1 \\ 0 & -1 & 1 \end{bmatrix} \begin{bmatrix} \lambda_1 \\ \lambda_2 \\ \lambda_3 \end{bmatrix} + \begin{bmatrix} \mu_1 \\ \mu_2 \end{bmatrix}$$

$$\left\| \begin{bmatrix} \lambda_1 \\ \lambda_2 \end{bmatrix} \right\|_2 \leq \lambda_3$$

$$\begin{bmatrix} \mu_1 \\ \mu_2 \end{bmatrix} \in \mathbb{R}_+^2$$

$$\lambda \in \mathbb{R}^3.$$

The `Pyomo kernel library` provides an experimental modeling interface for advanced application development with Pyomo. In particular, the kernel library provides direct support for conic constraints with the Mosek or Gurobi commercial solvers. The Pyomo interface to conic solvers includes six forms of conic constraints. The `conic.quadratic` constraint is expressed in the form

$$\sum_i x_i^2 \leq r^2, \quad r \geq 0,$$

where the terms x_i and r are `pyomo.kernel` variables.

The SOCO formation given above needs to be reformulated again to use with the Pyomo `conic.quadratic` constraint. The first step is to introduce rotated coordinates $t = x + y$ and $v = x - y$, and introduce a new variable with fixed value $u = 2$, to fit the Pyomo model for quadratic constraints,

$$\min \quad f(x,y) = \frac{hx}{2} + cdy$$

$$\text{s.t.} \quad t = x + y$$

$$u = 2$$

$$v = x - y$$

$$u^2 + v^2 \leq t^2$$

$$x, y, t, u, v \geq 0.$$

This version of the model with variables t, u, v, x, y could be implemented directly in Pyomo with `conic.quadratic`. However, the model can be further reduced to

$$\min \quad f(u,v) = \frac{1}{4}\left[(h + 2cd)\, t + (h - 2cd)\, v\right]$$

$$\text{s.t.} \quad u = 2$$

$$u^2 + v^2 \le t^2$$

$$t, u, v \ge 0.$$

We can now implement this EOQ model with Pyomo as follows:

```
import pyomo.kernel as pmo

h = 0.75   # cost of holding one item for one year
c = 500.0   # cost of processing one order
d = 10000.0   # annual demand

m = pmo.block()

# define variables for conic constraints
m.u = pmo.variable(lb=0)
m.v = pmo.variable(lb=0)
m.t = pmo.variable(lb=0)

# relationships for conic constraints to decision variables
m.u_eq = pmo.constraint(m.u == 2)
m.q = pmo.conic.quadratic(m.t, [m.u, m.v])

# linear objective
m.eoq = pmo.objective(((h + 2 * c * d) * m.t + (h - 2 * c * d) * m.v) / 4)

# solve with 'mosek_direct' or 'gurobi_direct'
SOLVER = pmo.SolverFactory("mosek_direct")
SOLVER.solve(m)
```

The optimal solution is $EOQ = 3654.48$, slightly different from the analytical solution we obtained earlier. This discrepancy can be attributed to numerical inaccuracies and the specific modeling approach employed. We will next explore leveraging Pyomo's capabilities to attain a more precise solution.

The `pyomo.kernel` provides additional support for conic solvers with the `.as_domain()` method that is applied to the conic solver interfaces. Adding `.as_domain()` allows the use of constants, linear expressions, or "None" in place of Pyomo variables.

`pyomo.kernel.conic.rotated_quadratic` expresses constraints in the form

$$\sum_i x_i^2 \le 2r_1 r_2, \quad r_1, r_2 \ge 0.$$

This enables a direct expression of the hyperbolic constraint $xy \geq 1$ by introducing an auxiliary variable z with fixed value $z^2 = 2$ such that

$$xy \geq 1 \iff z^2 \leq 2xy, \quad \text{where } z^2 = 2.$$

The new model formulation to be implemented in Pyomo is now

$$\min \quad f(x,y) = \frac{hx}{2} + cdy$$
$$\text{s.t.} \quad z^2 \leq 2xy$$
$$z = \sqrt{2}$$
$$x,y > 0.$$

The corresponding code is provided below, using `.as_domain()` to eliminate the need to specify a variable z.

```python
import pyomo.kernel as pmo

h = 0.75   # cost of holding one time for one year
c = 500.0  # cost of processing one order
d = 10000.0  # annual demand


m = pmo.block()

# define variables for conic constraints
m.x = pmo.variable(lb=0)
m.y = pmo.variable(lb=0)

# conic constraint
m.q = pmo.conic.rotated_quadratic.as_domain(m.x, m.y, [np.sqrt(2)])

# linear objective
m.eoq = pmo.objective(h * m.x / 2 + c * d * m.y)

# solve with 'mosek_direct'
SOLVER = pmo.SolverFactory("mosek_direct")
SOLVER.solve(m)
```

This last Pyomo implementation returns $EOQ = 3651.48$ as the optimal solution, which matches the result we obtained analytically. Note the improvement in accuracy of this calculation compared to the previous Pyomo implementation.

The problem solved in the above example was a special case of a general conic optimization problem defined as

$$\min \quad \boldsymbol{c}^\top \boldsymbol{x} \tag{6.2}$$
$$\text{s.t.} \quad \boldsymbol{A}_i \boldsymbol{x} + \boldsymbol{b}_i \in \boldsymbol{K}_i, \qquad i = 1,\dots,m$$
$$\boldsymbol{x} \in \mathbb{R}^n,$$

where the sets $\boldsymbol{K}_i$ are known as **cones**.

In this chapter, we will demonstrate the power of conic optimization. First, in Section 6.2 we show that a great variety of nonlinear constraints can be modeled in this way. Next, in Section 6.3, we will demonstrate that the construction of the dual of (6.2) is no more complicated than in the case of LO, significantly improving our ability to construct optimality certificates for the obtained solutions. In Section 6.4, we will briefly mention the positive semidefinite cone, which allows us to model as conic an even broader family of constraints. Finally, in Section 6.5, we outline the computational advantages of formulating the same problem as a general convex and as a conic problem.

6.2 Modeling

6.2.1 *Cones and the Conic-Representable Constraints*

We begin the discussion of modeling conic problems with some definitions:

- a subset $C \subseteq \mathbb{R}^n$ is a **cone** if for every $x \in C$ and $\theta \geq 0$ we have $\theta x \in C$;
- a subset $C \subseteq \mathbb{R}^n$ is a **convex cone** if it is convex and a cone, that is, if for any $x_1, x_2 \in C$ and $\theta_1, \theta_2 \geq 0$ we have $\theta_1 x_1 + \theta_2 x_2 \in C$.

There is a great variety of possible cones, but the most important ones we introduce are those for which solving a problem like (6.2) is easy and which can express the most extensive possible range of constraints encountered in applications.

Nonnegative Orthant (Linear Cone)

The first and simplest cone we introduce is the **nonnegative orthant** or **linear cone** $K_{\text{LO}}^m = \mathbb{R}_+^m$, which can be used to formulate LO problems in a conic fashion. Specifically, the canonical LO problem (2.4) can be rewritten as

$$
\begin{aligned}
\min \quad & c^\top x \quad &\text{(6.3)} \\
\text{s.t.} \quad & Ax - b \in K_{\text{LO}}^m \\
& x \in \mathbb{R}_+^n,
\end{aligned}
$$

where the first constraint means that the image of all feasible points x through the linear transformation $Ax - b$ need to belong to the cone $K_{\text{LO}}^m = \mathbb{R}_+^m$. The nonnegative orthant $K_{\text{LO}}^m = \mathbb{R}_+^m$ is trivially a convex cone.

Second-Order Cone (SOCO Cone)

The second most commonly used type of cone is the so-called **second-order cone**, which allows us to model a variety of problems with quadratic-like constraints. Formally, the second-order cone K_{SOCO}^{n+1}, also called the **SOCO cone**, **Lorentz cone**, or **ice-cream cone**, is defined as

$$
K_{\text{SOCO}}^{n+1} := \left\{ (x, t) \in \mathbb{R}^{n+1} \ : \ \|x\|_2 \leq t \right\} = \left\{ (x, t) \in \mathbb{R}^{n+1} \ : \ \sqrt{\sum_{i=1}^n x_i^2} \leq t \right\}.
$$

It is easy to show that K_{SOCO}^{n+1} is also a convex cone.

Example 6.2 (Convex quadratic constraints) For any $t \geq 0$, consider a constraint in which the squared Euclidean norm of a given vector x is upper bounded by t, that is, $x^\top x \leq t$. This constraint is equivalent to the following SOCO constraint:

$$\left\| \begin{matrix} x \\ \frac{1-t}{2} \end{matrix} \right\|_2 \leq \frac{1+t}{2}.$$

The SOCO cone can be used to formulate linear constraints and is therefore a generalization of it.

Example 6.3 (Linear constraints) Every linear constraint can be rewritten equivalently as a SOCO constraint:

$$a^\top x \leq b \quad \Longleftrightarrow \quad \begin{bmatrix} 0 \\ b - a^\top x \end{bmatrix} \in K_{\text{SOCO}}^{n+1}.$$

We provide two more examples of constraints that can be modeled using the SOCO cone.

Example 6.4 (Epigraph of a quadratic-over-linear function) Consider the constraint

$$\frac{x^\top x}{t} \leq s,$$

where both x and $t > 0$ are decision variables. Although it looks highly nonconvex, it can be reformulated as

$$\begin{cases} x^\top x \leq s\,t \\ s, t \geq 0 \end{cases} \quad \Longleftrightarrow \quad \left\| \begin{matrix} x \\ \frac{s-t}{2} \end{matrix} \right\|_2 \leq \frac{s+t}{2}.$$

Example 6.5 (Hypograph of a geometric mean) Another useful modeling trick is the following one:

$$\begin{cases} x_1, x_2 \geq 0 \\ 0 \leq t \leq \sqrt{x_1 x_2} \end{cases} \quad \Longleftrightarrow \quad \begin{cases} t \leq \tau \\ \left\| \begin{matrix} \tau \\ \frac{x_1 - x_2}{2} \end{matrix} \right\|_2 \leq \frac{x_1 + x_2}{2}. \end{cases}$$

This example is useful when modeling situations when a product of two nonnegative decision variables needs to be larger than a certain quantity, which happens, for example, when you compute the area of a rectangular space.

Example 6.6 (SVM problem revisited) Consider again the SVM problem considered in Section 5.5. By setting $v = F^\top \alpha$ and introducing an additional decision variable $r \geq 0$ to specify rotated quadratic cones (cf. Example 6.1), we have

$$\alpha^\top F F^\top \alpha \leq 2r \quad \Longleftrightarrow \quad v^\top v \leq 2r \quad \Longleftrightarrow \quad (r, 1, v) \in Q_r^{2+q}.$$

The convex problem in (5.9) then rewrites as a conic problem for the dual coefficients $\boldsymbol{\alpha}$ and auxiliary variables r and $\boldsymbol{v}$ as

$$
\begin{aligned}
\min \quad & r - 1^\top \boldsymbol{\alpha} \\
\text{s.t.} \quad & (r, 1, \boldsymbol{v}) \in \mathcal{Q}_r^{2+q} \\
& \boldsymbol{v} = F^\top \boldsymbol{\alpha} \\
& 0 \leq \alpha_i \leq \frac{c}{n} \qquad \forall i = 1, \ldots, n \\
& \boldsymbol{\alpha} \in \mathbb{R}^n \\
& \boldsymbol{v} \in \mathbb{R}^q \\
& r \geq 0.
\end{aligned}
$$

Exponential Cone

In optimization problems where logarithmic or exponential functions are used, another useful convex cone is the **exponential cone**, defined in a slightly more complex way as

$$
\begin{aligned}
\boldsymbol{K}_{\exp} = \quad & \{(x_1, x_2, x_3) \in \mathbb{R}^3 : x_1 \geq x_2 \exp(x_3/x_2),\ x_2 > 0\} \\
& \cup \{(x_1, 0, x_3) \in \mathbb{R}^3 : x_1 \geq 0,\ x_3 \leq 0\}.
\end{aligned}
$$

The second term in the union appearing in the above definition ensures that $\boldsymbol{K}$ is a closed set. The exponential cone can be used to model a variety of constraints related to probability, statistics, and other domains in which the exponential and logarithm functions appear.

Example 6.7 (Softplus function) In statistics and machine learning problems, one often needs to minimize the so-called **softplus function** defined as $f(x) = \log(1 + \exp(x))$. Minimizing a function like this is equivalent to minimizing an auxiliary variable t subject to the constraint

$$\log(1 + \exp(x)) \leq t.$$

This constraint allows for an equivalent conic representation that uses the exponential cone. More specifically, by adding two decision variables u, v, it is equivalent to the following system of constraints:

$$
\begin{cases}
u + v \leq 1 \\
(u, 1, x - t) \in \boldsymbol{K}_{\exp} \\
(v, 1, -t) \in \boldsymbol{K}_{\exp}.
\end{cases}
$$

More examples of exponential cones can be found at [Mos23a].

Example 6.8 (Growth-optimal portfolio using Kelly's criterion) The Kelly growth optimal portfolio is an investment strategy alternative to the Markowitz mean-variance portfolio presented in Example 5.10. Differently from the risk–return trade-off considerations of the mean-variance framework, the Kelly growth optimal portfolio is designed to maximize the long-term exponential growth rate of the investment capital over multiple periods.

The goal of Kelly's problem is to maximize the growth rate of an investor's wealth. At every stage $n = 1, \ldots, N$, the gross return R_n on the investor's wealth W_n is given by $W_n = W_{n-1} R_n$. After N stages, this gives

$$W_N = W_0 R_1 R_2 \cdots R_{N-1} R_N = W_0 \prod_{n=1}^{N} R_n.$$

Kelly's idea was to maximize the mean log return. Intuitively, this objective can be justified as maximizing the geometric mean return

$$\left(\prod_{n=1}^{N} R_n \right)^{1/N} = \exp \left(\frac{1}{N} \sum_{n=1}^{N} \log R_n \right),$$

which is equivalent to maximization of the expected logarithmic utility of wealth

$$\log W_N = \log W_0 \left(\sum_{n=1}^{N} \log R_n \right).$$

The logarithmic utility provides an element of risk aversion. If the returns R_i are independent and identically distributed random variables then, in the long run,

$$\mathbb{E}[\log W_N] = \log W_0 \, \mathbb{E}[\log R],$$

where R is the total gross return (ratio of final wealth to initial wealth), such that the optimal log growth return will almost surely result in higher wealth than any other policy.

The classic presentation of Kelly's criterion is to consider repeated wagers on a gambling game with binary outcomes. For each stage, a wager of one unit returns $1 + b$ with probability p if successful, otherwise, the wager returns nothing. The number b refers to the "odds" of the game. The problem is determining which fraction of the gambler's wealth should be wagered on each game instance. Let w be the fraction of wealth wagered in each instance. Since the rest of the wealth $1 - w$ is kept risk-free, the gross return at any stage is

$$R = \begin{cases} R_1 := (1 - w) + w(1 + b) = 1 + bw, & \text{with probability } p \\ R_2 := 1 - w, & \text{with probability } 1 - p. \end{cases}$$

In Kelly's setting, the objective is to maximize the expected log return

$$\max_{w \geq 0} \; p \log(1 + bw) + (1 - p) \log(1 - w). \tag{6.4}$$

The well-known analytical solution to this problem is

$$w^* = \begin{cases} p - \frac{1-p}{b}, & p(b+1) > 1 \\ 0, & p(b+1) \leq 1. \end{cases}$$

We can use this analytical solution to validate the solution of this problem that we will now obtain using conic optimization. To be able to use the conic optimization toolbox, we introduce two auxiliary variables q_1 and q_2 and taking the exponential of the terms in the objective function (6.4), we obtain

$$q_1 \leq \log(1 + bw) \iff \exp q_1 \leq 1 + bw \iff (1 + bw, 1, q_1) \in \boldsymbol{K}_{\exp}$$

and

$$q_2 \leq \log(1 - w) \iff \exp q_2 \leq 1 - w \iff (1 - w, 1, q_2) \in \boldsymbol{K}_{\exp}.$$

With these constraints, Kelly's problem becomes

$$
\begin{aligned}
\max \quad & pq_1 + (1 - p)q_2 \\
\text{s.t.} \quad & (1 + bw, 1, q_1) \in \boldsymbol{K}_{\exp} \\
& (1 - w, 1, q_2) \in \boldsymbol{K}_{\exp} \\
& w \geq 0 \\
& q_1, q_2 \in \mathbb{R}.
\end{aligned}
$$

The following code shows how to obtain the optimal solution using the conic optimization functions of the Pyomo kernel library and the Mosek solver.

```python
import pyomo.kernel as pmo

# parameter values
b = 1.25
p = 0.51


# conic optimization solution to Kelly's problem
def kelly(p, b):
    m = pmo.block()

    # decision variables
    m.q1 = pmo.variable()
    m.q2 = pmo.variable()
    m.w = pmo.variable(lb=0)

    # objective
    m.ElogR = pmo.objective(p * m.q1 + (1 - p) * m.q2, sense=pmo.maximize)

    # conic constraints
    m.t1 = pmo.conic.primal_exponential.as_domain(1 + b * m.w, 1, m.q1)
    m.t2 = pmo.conic.primal_exponential.as_domain(1 - m.w, 1, m.q2)

    return m


# conic optimization solution for w
m = kelly(p, b)
pmo.SolverFactory("mosek_direct").solve(m)
print(f"Conic optimization solution for w: {m.w(): 0.4f}")
```

```
# analytical solution to Kelly's problem
w_analytical = p - (1 - p) / b if p * (b + 1) > 1 else 0
print(f"Analytical solution for w: {w_analytical: 0.4f}")
```

For the given parameters $b = 1.25$ and $p = 0.51$, the two solutions agree and are equal to $w^{\text{opt}} = w^* \approx 0.1180$.

Example 6.9 (Risk-constrained Kelly's problem) Kelly's criterion of Example 6.8 suffers from a crucial downside: It optimizes for the long-term outcome but does not make sure that the investor does not go bankrupt in the first place.

Following Busseti, Ryu, and Boyd (2016), we therefore consider the risk-constrained version of Kelly's problem, in which we add the constraint $\mathbb{E}[R^{-\lambda}] \leq 1$, where $\lambda \geq 0$ is a risk-aversion parameter. For the case with two outcomes with known probabilities considered here, this constraint rewrites as

$$p_1 R_1^{-\lambda} + p_2 R_2^{-\lambda} \leq 1.$$

When $\lambda = 0$, the constraint is always satisfied and no risk-aversion is in effect. Choosing $\lambda > 0$ requires that low-return outcomes occur with low probability and this effect increases for larger values of λ. A feasible solution can always be found by setting the bet size $w = 0$ to give $R_1 = 1$ and $R_2 = 0$.

This constraint can be reformulated using exponential cones. To see this, let us perform the following equivalent rewriting:

$$e^{\log(p_1) - \lambda \log(R_1)} + e^{\log(p_2) - \lambda \log(R_2)} \leq 1.$$

Previously, we introduced two auxiliary real variables $q_1, q_2 \in \mathbb{R}$ such that $q_i \leq \log(R_i)$, using which we can reformulate the risk constraint as

$$e^{\log(p_1) - \lambda q_1} + e^{\log(p_2) - \lambda q_2} \leq 1.$$

Introducing two more nonnegative auxiliary variables $u_1, u_2 \geq 0$ such that $u_i \geq e^{\log(p_i) - \lambda q_i}$, the risk constraint is given by

$$u_1 + u_2 \leq 1$$
$$(u_1, 1, \log(p_1) - \lambda q_1) \in \boldsymbol{K}_{\exp}$$
$$(u_2, 1, \log(p_2) - \lambda q_2) \in \boldsymbol{K}_{\exp}.$$

For fixed probabilities $p_1 = p$ and $p_2 = 1 - p$, odds b, and risk-aversion parameter $\lambda \geq 0$, the risk-constrained Kelly bet is a solution to the conic problem, which can be rewritten as

$$\begin{aligned}
\max \quad & pq_1 + (1-p)q_2 \\
\text{s.t.} \quad & (1 + bw, 1, q_1) \in \boldsymbol{K}_{\exp} \\
& (1 - w, 1, q_2) \in \boldsymbol{K}_{\exp} \\
& u_1 + u_2 \leq 1 \\
& (u_1, 1, \log(p) - \lambda q_1) \in \boldsymbol{K}_{\exp}
\end{aligned}$$

$$(u_2, 1, \log(1 - p) - \lambda q_2) \in \boldsymbol{K}_{\exp}$$
$$u_1, u_2, w \geq 0$$
$$q_1, q_2 \in \mathbb{R}.$$

The following cell solves this problem with a Pyomo model using the Mosek solver.

```python
import pyomo.kernel as pmo

# parameter values
b = 1.25
p = 0.51
lambd = 3


def kelly_rc(p, b, lambd):
    m = pmo.block()

    # decision variables
    m.q1 = pmo.variable()
    m.q2 = pmo.variable()
    m.w = pmo.variable(lb=0)

    # objective
    m.ElogR = pmo.objective(p * m.q1 + (1 - p) * m.q2, sense=pmo.maximize)

    # conic constraints
    m.t1 = pmo.conic.primal_exponential.as_domain(1 + b * m.w, 1, m.q1)
    m.t2 = pmo.conic.primal_exponential.as_domain(1 - m.w, 1, m.q2)

    # risk constraints
    m.u1 = pmo.variable(lb=0)
    m.u2 = pmo.variable(lb=0)
    m.r0 = pmo.constraint(m.u1 + m.u2 <= 1)
    m.r1 = pmo.conic.primal_exponential.as_domain(m.u1, 1, np.log(p) -
    ↪    lambd * m.q1)
    m.r2 = pmo.conic.primal_exponential.as_domain(m.u2, 1, np.log(1 - p) -
    ↪    lambd * m.q2)

    return m


# conic optimization solution to the risk-constrained version of Kelly's
↪    problem
m = kelly_rc(p, b, lambd)
pmo.SolverFactory("mosek_direct").solve(m)
print(f"Risk-constrainend solution for w: {m.w(): 0.4f}")
```

For the same parameters $b = 1.25$ and $p = 0.51$ considered in Example 6.8, the optimal risk-adverse solution prescribes to invest a smaller fraction of the wealth, namely $w^{\text{opt}} = 0.0589 < w^* = 0.11800$.

Power Cone

Another powerful modeling tool is the power cone, which for a parameter $0 < \alpha < 1$, is defined as

$$K^n_{\text{pow},\alpha} = \left\{ x \in \mathbb{R}^n \,:\, x_1^\alpha x_2^{1-\alpha} \geq \sqrt{x_3^2 + \cdots + x_n^2}, \, x_1, x_2 \geq 0 \right\}.$$

Example 6.10 (p-power penalization) In some applications, one needs to penalize the deviations of a decision variable x from the zero vector by including its pth power $|x|^p$ in the objective, with $p > 1$. The meaning of such a term is that large deviations from zero are penalized much more than small deviations. The constraint $|x|^p \leq t$ can be rewritten equivalently as

$$(t, 1, x) \in K^3_{\text{pow}, 1/p}.$$

More examples of power cones can be found in [Mos23b].

6.2.2 Calculus of Conic-Representable Constraints

In the previous subsection, we saw examples of simple conic-representable constraints. The usual question for large-scale real-world problems is how to quickly determine whether a given constraint can be represented via (a system of) conic constraints. We need a set of "rules" that allow us to recognize such a constraint from its form and subcomponents.

To establish such a set of rules, let us define that a constraint $g(x) \leq t$ is **conic-representable** if it can be equivalently rewritten as one or more constraints of the form

$$A \begin{bmatrix} x \\ t \end{bmatrix} + b \in K \tag{6.5}$$

for some matrix A, some vector b, and some cone K.

While all the examples in the previous section have been defined for a *single* function $g(x)$ of a given vector of variable, there are operations that one can perform on several conic-representable functions $g_i(\cdot)$ that will lead to another conic-representable constraint:

- **linear substitution**: $g(x) = h(Cx + d)$ is conic-representable if h is;
- **summation**: $g(x) = \sum_{i=1}^m \alpha_i g_i(x)$ with $\alpha_i \geq 0$ is conic-representable if every g_i is;
- **maximum**: $g(x) = \max_{i=1,\ldots,m} g_i(x)$ is conic-representable if every g_i is;
- **partial minimization**: $g(x) = \min_y h(x, y)$ is conic-representable if h is.

Consider now a few examples of how the above rule set can help us recognize and model conic optimization problems.

Example 6.11 (General convex quadratic constraint) Consider a quadratic constraint of the form

$$x^\top A x + q^\top x + r \leq 0,$$

where the matrix $Q = H^\top H$ is positive semidefinite. This constraint can be equivalently rewritten as follows:

$$\left\| \begin{matrix} Hx \\ \frac{1+q^\top x + r}{2} \end{matrix} \right\|_2 \le \frac{1 - q^\top x - r}{2}.$$

This readily follows from the decomposition $Q = H^\top H$ of the positive semidefinite matrix Q (which exists by definition) by rewriting

$$x^\top Q x + q^\top x + r \le 0 \quad \Longleftrightarrow \quad x^\top H^\top H x \le -q^\top x - r,$$

and then by using the equivalence introduced in Example 6.2 plus the linear substitution property mentioned above.

Convex quadratic optimization is applied in all kinds of contexts in which, for example, you are solving a least-squares problem, minimizing the variance of a portfolio subject to a constraint on minimum mean return, signal processing computations in electrical engineering, etc.

Example 6.12 (Markowitz portfolio optimization part 2) Consider again the portfolio optimization in Example 5.10. Recall that the matrix Σ describes the covariance among uncertain return rates r_i, $i = 1, \ldots, n$. Since Σ is positive semidefinite by definition, it allows for a Cholesky factorization, that is, $\Sigma = BB^\top$. We can then rewrite the quadratic constraint as $\|B^\top x\|_2 \le \gamma$ and thus as $(\gamma, B^\top x) \in K_{\text{SOCO}}^{n+1}$ using the Lorentz cone. In this way, we realize that the original portfolio problem we formulated in (5.5) is in fact a conic quadratic optimization problem, which can thus be solved faster and more reliably. The optimal solution of that problem was the one with the maximum expected return while allowing for a specific level γ of risk.

However, an investor could aim for a different trade-off between return and risk and formulate a slightly different optimization problem, namely

$$\max \quad R\tilde{x} + \mu^\top x - \alpha x^\top \Sigma x \tag{6.6}$$
$$\text{s.t.} \quad \sum_{i=1}^{n} x_i + \tilde{x} = C$$
$$\tilde{x} \ge 0$$
$$x \ge 0,$$

where $\alpha \ge 0$ is a "risk tolerance" parameter that describes the relative importance of return vs. risk for the investor.

The risk, quantified by the variance of the investment return $x^\top \Sigma x = x^\top B^\top B x$, appears now in the objective function as a penalty term. Note that even in this new formulation we have a conic problem since we can rewrite it as

$$\max \quad R\tilde{x} + \mu^\top x - \alpha\gamma \tag{6.7}$$
$$\text{s.t.} \quad \sum_{i=1}^{n} x_i + \tilde{x} = C$$
$$\|B^\top x\|_2^2 \le s$$

$$\tilde{x} \geq 0$$
$$\boldsymbol{x} \geq 0$$
$$s \geq 0.$$

Solving for all values of $\alpha \geq 0$, one can obtain the so-called **efficient frontier**, which is shown in Figure 6.3

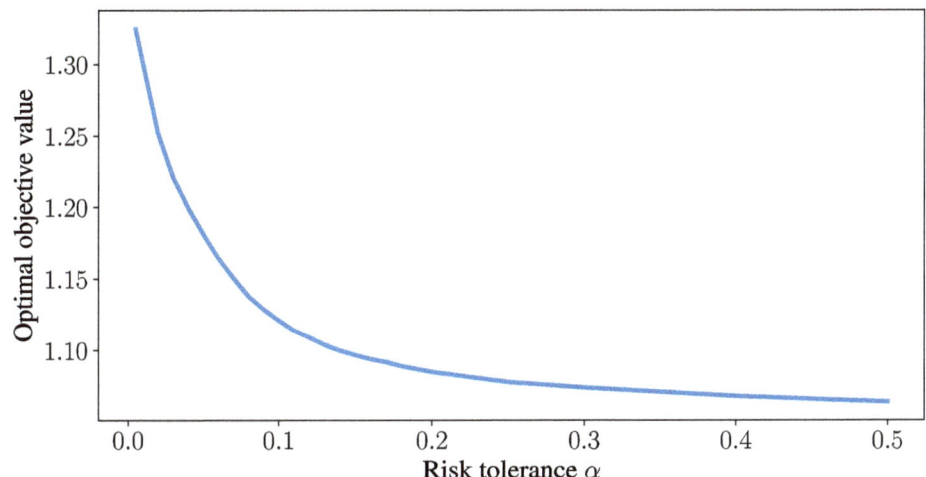

Figure 6.3 Trade-off between the optimal objective function value and the risk tolerance level.

Example 6.13 (Ordinary least squares regression part 2) SOCO can be used to formulate linear regression problems if the loss function used is quadratic, as for the OLS regression in Example 5.6. Let $A = [a_1 \ldots a_N] \in \mathbb{R}^{N \times n}$ be a dataset of N points, with n features/explanatory variables each, and $\boldsymbol{y} \in \mathbb{R}^N$ be the vector of the corresponding observables/dependent variables.

In a typical regression problem, we want to estimate the vector $\boldsymbol{x}$ of regressors that best predicts the observations, where "best" means that it is such that the "distance" between $A\boldsymbol{x}$ and $\boldsymbol{y}$ is the "smallest." For OLS regression, the goal is to minimize the 2-norm of this vector; compare Example 5.6. The problem can be rewritten as a SOCO problem as follows:

$$\min_{\boldsymbol{x}} \sqrt{\sum_{i=1}^{N} (a_i \boldsymbol{x} - y_i)^2} \quad \Longleftrightarrow \quad \min_{\boldsymbol{x}} \|A\boldsymbol{x} - \boldsymbol{y}\|_2$$

$$\Longleftrightarrow \quad \min_{\boldsymbol{x},t} \quad t$$
$$\text{s.t.} \quad \|A\boldsymbol{x} - \boldsymbol{y}\|_2 \leq t.$$

Example 6.14 (Logistic regression) **Logistic regression** is a classification tool used in machine learning where one needs to fit a parameter vector $x \in \mathbb{R}^n$ to the dataset $\{(a_1, y_1), \ldots, (a_N, y_N)\}$ where $a_i \in \mathbb{R}^n$ and with binary "labels" $y_i \in \{-1, 1\}$, by minimizing the loss function

$$\min \quad \sum_{i=1}^{N} \log\left(1 + \exp\left(-y_i\left(a_i^\top x\right)\right)\right)$$

$$\text{s.t.} \quad x \in \mathbb{R}^n.$$

Using the summation and linear substitution rules, we can reformulate it into a conic optimization problem:

$$\min \quad \sum_{i=1}^{N} t_i$$

$$\begin{aligned}
\text{s.t.} \quad & u_i + v_i \le 1, & \forall i = 1, \ldots, N \\
& (u_i, 1, s_i - t_i) \in K_{\exp}, & \forall i = 1, \ldots, N \\
& (v_i, 1, -t_i) \in K_{\exp}, & \forall i = 1, \ldots, N \\
& s_i = -y_i(a_i^\top x), & \forall i = 1, \ldots, N \\
& t, s, u, v \in \mathbb{R}^N \\
& x \in \mathbb{R}^n.
\end{aligned}$$

Example 6.15 (Variance-to-mean ratio) Consider the constraint

$$\frac{(Ax + b)^\top (Ax + b)}{c^\top x + d} \le t,$$

which is a typical example of a bound of a variance-to-mean ratio for, for example, random variables optimized in portfolio optimization. Again, we can use the linear substitution property and Example 6.4 to show that this constraint is SOCO-representable.

Example 6.16 (Partial minimization – Variance) Consider a portfolio optimization problem where you have a discrete uniform probability distribution for a vector $z \in \mathbb{R}^n$ of the returns on n assets such that $\mathbb{P}(z = z^i) = 1/N$ for some N. If we denote by $x \in \mathbb{R}^n$ the vector of decision variable that describes the portfolio weights for each of the assets, then the variance of the portfolio's return is

$$F(x) = \frac{1}{N} \sum_{i=1}^{N} \left(x^\top z^i - \left(\frac{1}{N} \sum_{j=1}^{N} x^\top z^j \right) \right)^2.$$

One can show that this quantity can be conveniently reformulated as the optimal value of an optimization problem involving SOCO constraints. More specifically, $F(x)$ can be rederived as

$$F(\boldsymbol{x}) = \inf_{y \in \mathbb{R}} f(\boldsymbol{x}, y),$$

where the function $f(\boldsymbol{x}, y) := \frac{1}{N} \sum_{i=1}^{N} \left(\boldsymbol{x}^\top \boldsymbol{z}^i - y \right)^2$ is very easily seen to be convex and SOCO-representable in $(\boldsymbol{x}, y)$ and, as a result, the minimizer – the variance $F(\boldsymbol{x})$ – is convex in $\boldsymbol{x}$ as well.

6.3 Duality in Conic Optimization

6.3.1 Dual Optimization Problem

We now introduce how simple the construction of a dual problem is for conic optimization problems. We do so by first making an analogy to how the derivation of the dual for LO worked via the Lagrange function (5.8):

Primal problem:
$\begin{aligned} \min \quad & \boldsymbol{c}^\top \boldsymbol{x} && \text{(P-conic)} \\ \text{s.t.} \quad & A_i \boldsymbol{x} + \boldsymbol{b}_i \in \boldsymbol{K}_i, \quad i = 1, \dots, m. \end{aligned}$

Dual problem:
$\begin{aligned} \max \quad & -\sum_{i=1}^{m} \boldsymbol{\lambda}_i^\top \boldsymbol{b}_i && \text{(D-conic)} \\ \text{s.t.} \quad & \boldsymbol{c} = \sum_{i=1}^{m} A_i^\top \boldsymbol{\lambda}_i \\ & \boldsymbol{\lambda}_i \in \boldsymbol{K}_i^*, \quad i = 1, \dots, m. \end{aligned}$

In this formulation, the sets $\boldsymbol{K}_i^*$ are the **dual cones** of the respective cones $\boldsymbol{K}_i$ used in the primal problem formulation. To motivate this formulation, we will first study the LO case.

> **Example 6.17 (LO duality as a special case of conic duality)** Consider deriving the dual of the LO problem
>
> $$\begin{aligned} \min \quad & \boldsymbol{c}^\top \boldsymbol{x} \\ \text{s.t.} \quad & A\boldsymbol{x} + \boldsymbol{b} \geq 0 \end{aligned}$$
>
> via Lagrange relaxation. To relax the constraints, we construct a vector of dual multipliers $\boldsymbol{\lambda} \in \mathbb{R}_+^m$:
>
> $$L(\boldsymbol{x}, \boldsymbol{\lambda}) = \boldsymbol{c}^\top \boldsymbol{x} - \boldsymbol{\lambda}^\top (A\boldsymbol{x} + \boldsymbol{b}),$$
>
> such that, because $\boldsymbol{\lambda} \in \mathbb{R}_+^m$, the product $-\boldsymbol{\lambda}^\top (A\boldsymbol{x} + \boldsymbol{b})$ is nonpositive (and thus, there is no positive penalty for violating any of constraints) whenever
>
> $$A\boldsymbol{x} + \boldsymbol{b} \leq 0 \quad \Longleftrightarrow \quad A\boldsymbol{x} + \boldsymbol{b} \in -\mathbb{R}^+.$$
>
> Thanks to the easiness of minimizing the expression
>
> $$\boldsymbol{c}^\top \boldsymbol{x} - \boldsymbol{\lambda}^\top (A\boldsymbol{x} + \boldsymbol{b}) = (\boldsymbol{c} + A^\top \boldsymbol{\lambda})^\top \boldsymbol{x} + \boldsymbol{\lambda}^\top \boldsymbol{b}$$
>
> over $\boldsymbol{x}$, we obtain that the dual function is

$$\ell(\boldsymbol{\lambda}) = \min_{\boldsymbol{x}} L(\boldsymbol{x}, \boldsymbol{\lambda}) = \begin{cases} \boldsymbol{\lambda}^\top \boldsymbol{b} & \text{if } \boldsymbol{c} - A^\top \boldsymbol{\lambda} = 0 \\ -\infty, & \text{otherwise,} \end{cases}$$

for which the weak duality $\ell(\boldsymbol{\lambda}) \leq \boldsymbol{c}^\top \boldsymbol{x}$ for any feasible $\boldsymbol{x}$, and the dual problem is formulated as the search for the best lower bound:

$$\max_{\boldsymbol{\lambda}} \quad \ell(\boldsymbol{\lambda})$$

$$\text{s.t.} \quad \boldsymbol{\lambda} \in \mathbb{R}_+^m.$$

As you can see, this problem formulation fits D-conic exactly if $\mathbb{R}_+^m$ is the dual cone of $\mathbb{R}_+^m$ which is indeed the case.

In the above derivation, the critical fact is the pairing of the sets to which $\boldsymbol{\lambda}$ belongs and to which $A\boldsymbol{x} + \boldsymbol{b}$ belongs when the problem's constraints are satisfied. In this case, both sets are linear cones $\mathbb{R}_+^m$, which are each other's **dual cones**.

Definition 6.1 For a convex cone $\boldsymbol{K} \subset \mathbb{R}^m$, its dual cone $\boldsymbol{K}^*$ is defined as the subset

$$\boldsymbol{K}^* := \left\{ \boldsymbol{x} \in \mathbb{R}^m \ : \ \boldsymbol{x}^\top \boldsymbol{y} \geq 0 \quad \forall \boldsymbol{y} \in \boldsymbol{K} \right\}.$$

For the general problem with a linear objective function and conic constraints formulated in (6.2), we define the Lagrangian as

$$L\left(\boldsymbol{x}, \boldsymbol{\lambda}^1, \ldots, \boldsymbol{\lambda}^m\right) := \boldsymbol{c}^\top \boldsymbol{x} - \sum_{i=1}^m \boldsymbol{\lambda}_i^\top \left(A_i \boldsymbol{x} + \boldsymbol{b}_i\right).$$

The last term "appropriately penalizes" the conic constraints because $\boldsymbol{x}$ is feasible, then thanks to $\boldsymbol{\lambda}_i \in \boldsymbol{K}^*$ we have that $-\boldsymbol{\lambda}_i^\top (A_i \boldsymbol{x} + \boldsymbol{b}_i) \leq 0$ for all $i = 1, \ldots, m$. Thanks to this, we have the weak duality result

$$\boldsymbol{c}^\top \boldsymbol{x} \geq \max_{\boldsymbol{\lambda}_i \in \boldsymbol{K}^*} \min_{\boldsymbol{x}} L\left(\boldsymbol{x}, \boldsymbol{\lambda}^1, \ldots, \boldsymbol{\lambda}^m\right) = v(D), \qquad \forall \boldsymbol{x} \text{ s.t. } A_i \boldsymbol{x} + \boldsymbol{b}_i \in \boldsymbol{K}_i.$$

The derivation of the dual function proceeds then in the same way as in the LO case by minimizing the Lagrangian over $\boldsymbol{x}$ as a strategy to obtain lower bounds on the primal problem:

$$\ell(\boldsymbol{\lambda}) := \min_{\boldsymbol{x}} L\left(\boldsymbol{x}, \boldsymbol{\lambda}^1, \ldots, \boldsymbol{\lambda}^m\right) = \min_{\boldsymbol{x}} \boldsymbol{c}^\top \boldsymbol{x} - \sum_{i=1}^m \boldsymbol{\lambda}_i^\top \left(A_i \boldsymbol{x} + \boldsymbol{b}_i\right)$$

$$= \min_{\boldsymbol{x}} -\sum_{i=1}^m \boldsymbol{\lambda}_i^\top \boldsymbol{b}_i + \boldsymbol{x}^\top \left(\boldsymbol{c} - \sum_{i=1}^m A_i^\top \boldsymbol{\lambda}_i\right)$$

$$= \begin{cases} -\sum_{i=1}^m \boldsymbol{\lambda}_i^\top \boldsymbol{b}_i, & \text{if } \boldsymbol{c} - \sum_{i=1}^m A_i^\top \boldsymbol{\lambda}_i = 0, \\ -\infty, & \text{otherwise.} \end{cases}$$

Searching for the best lower bound $\ell(\boldsymbol{\lambda})$, that is, maximizing this function over $\boldsymbol{\lambda}_i \in \boldsymbol{K}_i^*$, $i = 1, \ldots, m$ gives us D-conic.

The only missing step in this approach is being able to calculate the duals of the cones we introduced in this chapter.

Table 6.1 *Cones and their dual cones.*

Cone	Definition K	Dual cone K^*
K_{LO}	$\mathbb{R}^n_+$	$\mathbb{R}^n_+$
K_{SOCO}	$\{(x,t) \in \mathbb{R}^n : \|x\|_2 \le t\}$	$\{(x,t) \in \mathbb{R}^n : \|x\|_2 \le t\}$
K_{exp}	$\{(x_1,x_2,x_3) \in \mathbb{R}^3 : x_2 \exp(x_1/x_2) \le x_3,\, x_2 > 0\}$ $\cup$ $\{(x_1,x_2,x_3) \in \mathbb{R}^3 : x_1 \ge 0, x_3 \le 0\}$	$\text{cl}\left\{y \in \mathbb{R}^3 : y_1 \ge -y_3 e^{y_2/y_3 - 1},\, y_1 > 0, y_3 < 0\right\}$
K_{pow}	$\left\{x \in \mathbb{R}^n : x_1^\alpha x_2^{1-\alpha} \ge \sqrt{x_3^2 + \cdots + x_n^2},\, x_1,x_2 \ge 0\right\}$	$\left\{y \in \mathbb{R}^n : \left(\frac{y_1}{\alpha_1}, \frac{y_2}{1-\alpha}, y_3, \ldots, y_n\right) \in K_{\text{pow}}^{\alpha, 1-\alpha}\right\}$

Linear and SOCO Cone

The linear and SOCO cones are both self-dual, meaning the dual of any cone of these types is the cone itself, that is,

$$(\mathbb{R}^n_+)^* = \mathbb{R}^n_+ \qquad \text{and} \qquad (K^n_{\text{SOCO}})^* = K^n_{\text{SOCO}}.$$

Exponential Cone

For the exponential cone, we have the following formulation of its corresponding dual cone:

$$K^*_{\text{exp}} = \text{cl}\left\{y \in \mathbb{R}^3 \,:\, y_1 \ge -y_3 e^{y_2/y_3 - 1},\, y_1 > 0,\, y_3 < 0\right\},$$

where $\text{cl}(S)$ is the operation of taking the closure of a set S, which roughly means that we also include all points on the "boundary" of S.

Power Cone

The power cone has the following dual cone:

$$(K^{\alpha, 1-\alpha}_{\text{pow}})^* = \left\{y \in \mathbb{R}^n \,:\, \left(\frac{y_1}{\alpha_1}, \frac{y_2}{1-\alpha}, y_3, \ldots, y_n\right) \in K^{\alpha, 1-\alpha}_{\text{pow}}\right\}.$$

6.3.2 Strong Duality and KKT Conditions

Conic optimization is a special case of convex optimization, so the convex duality theory immediately applies here. Thus, the notion of strong duality is well-defined also for conic problems as the situation in which $v(P) = v(D)$. Also, in this case, Slater's condition is sufficient to prove strong duality as soon as we can exhibit a strictly feasible point for either the primal or the dual problem. More explicitly, any of the two conditions below is sufficient for strong duality:

$$\exists x \,:\, A_i x + b_i \in \text{int} K_i \qquad \text{or} \qquad \exists \lambda_i \in \text{int} K^*_i \,:\, c = \sum_{i=1}^m A_i^\top \lambda_i,$$

where $\text{int}(S)$ is the operation of taking the interior of a set S, which roughly means "including only points strictly inside it."

The KKT conditions for the conic optimization problem (6.2) are:

$$
\begin{cases}
\nabla_x L(x, \lambda) = c - \sum_{i=1}^{m} A_i^\top \lambda_i = 0 & \text{(stationarity)} \\
A_i x + b_i \in \text{int} K_i & \text{(primal feasibility)} \\
\lambda_i \in K_i^*, & \text{(dual feasibility)} \\
\lambda_i^\top (A_i x + b_i) = 0 & \text{(complementarity slackness)}.
\end{cases}
$$

6.4 *SDO Cone

One more cone is considered frequently in the literature – the positive semidefinite cone (SDO cone), which is somewhat less common in business analytics applications and requires a slightly more general notation. For sake of completeness, we discuss it here.

Let $S^n = \{Q \in \mathbb{R}^{n \times n} : Q = Q^\top\}$ be the set of all real symmetric $n \times n$ matrices. The **positive semidefinite cone** or **SDO cone** is then defined as the set of symmetric positive semidefinite $n \times n$ matrices, that is,

$$
K_{\text{SDO}}^n := \{Q \in S^n : Q \succeq 0\} \subset S^n,
$$

where "SDO" is an abbreviation for "semidefinite optimization." Since this is a cone in the space of matrices and not vectors, a note of care is needed to handle the corresponding notation. In that case, a mapping $\mathcal{A}$ of a variable vector x to S^n takes the form

$$
\mathcal{A}x + B := \sum_{i=1}^{n} A_i x_i + B.
$$

Example 6.18 (An SDO constraint) The following is an example of an SDO constraint:

$$
\underbrace{\begin{bmatrix} 1 & 0 \\ 0 & 2 \end{bmatrix}}_{B} + x_1 \underbrace{\begin{bmatrix} -2 & 1 \\ 1 & 3 \end{bmatrix}}_{A_1} + x_2 \underbrace{\begin{bmatrix} 1 & 4 \\ 4 & -5 \end{bmatrix}}_{A_2} \in K_{\text{SDO}}^2,
$$

since B, A_1, and A_2 are all symmetric matrices.

Every constraint expressible with the SOCO cone (and thus also with the nonnegative orthant cone) can be expressed using the SDO cone.

Because the SDO cone consists of matrices instead of vectors, a slightly more involved notation is needed to formulate the dual problem involving SDO constraints, where in the original case we used transposes of relevant matrices. Let us redefine the following two operations:

- the "dot product" of two symmetric matrices, that is, the analog of $\lambda_i^\top b_i$, is

$$
\langle A, B \rangle_{S^n} = \sum_{i,j=1}^{n} A_{ij} B_{ij},
$$

which is a matrix analog of the vector inner product;

- the "transpose" of $\mathcal{A}\colon \mathbb{R}^n \to \mathbf{S}^n$ is taken to be the mapping $\mathcal{A}^\top \colon \mathbf{S}^n \to \mathbb{R}^n$ such that

$$\langle B, \mathcal{A}z \rangle_{\mathbf{S}^n} = \langle \mathcal{A}^\top B, z \rangle_{\mathbb{R}^n}, \quad \forall B \in \mathbf{S}^n, z \in \mathbb{R}^n.$$

An example best illustrates this notion.

Example 6.19 Consider the map $\mathcal{A}\colon \mathbb{R}^3 \to \mathbf{S}^2$ defined as

$$\mathcal{A}\colon z \to z_1 \begin{bmatrix} 1 & 1 \\ 1 & 1 \end{bmatrix} + z_3 \begin{bmatrix} 2 & -2 \\ -2 & 1 \end{bmatrix}.$$

Its transpose is $\mathcal{A}^\top \colon \mathbf{S}^n \to \mathbb{R}^3$ such that for every $B \in \mathbf{S}^2$ and every $z \in \mathbb{R}^3$ it holds that:

$$\left\langle B, z_1 \begin{bmatrix} 1 & 1 \\ 1 & 1 \end{bmatrix} + z_3 \begin{bmatrix} 2 & -2 \\ -2 & 1 \end{bmatrix} \right\rangle_{\mathbf{S}^n} = \langle \mathcal{A}^\top B, z \rangle_{\mathbb{R}^3}.$$

More explicitly, we can describe the transpose map $\mathcal{A}^\top$ as

$$\mathcal{A}^\top \colon B \to \begin{bmatrix} \left\langle B, \begin{bmatrix} 1 & 1 \\ 1 & 1 \end{bmatrix} \right\rangle_{\mathbf{S}^2} \\ 0 \\ \left\langle B, \begin{bmatrix} 2 & -2 \\ -2 & 1 \end{bmatrix} \right\rangle_{\mathbf{S}^2} \end{bmatrix}.$$

In words, the mapping $\mathcal{A}^\top$ returns a vector whose elements are the coefficients of the z variables when writing out in full the following dot product:

$$\left\langle B, z_1 \begin{bmatrix} 1 & 1 \\ 1 & 1 \end{bmatrix} + z_3 \begin{bmatrix} 2 & -2 \\ -2 & 1 \end{bmatrix} \right\rangle_{\mathbf{S}^n}$$
$$= z_1 \left\langle B, \begin{bmatrix} 1 & 1 \\ 1 & 1 \end{bmatrix} \right\rangle_{\mathbf{S}^n} + 0 \cdot z_2 + z_3 \left\langle B, \begin{bmatrix} 2 & -2 \\ -2 & 1 \end{bmatrix} \right\rangle_{\mathbf{S}^n}.$$

The SDO cone is useful for many phenomena related to combinatorics or geometry, such as fitting the smallest-volume ellipsoid around a given dataset.

Example 6.20 (Minimum-volume ellipsoid using a SDO conic problem) We are provided with N data points $y_1, \ldots, y_N$ in $\mathbb{R}^n$ and we are to construct an optimization tool that will work on these data and all data that is sufficiently "similar" to it, that is, does not fall exactly into one of the observations, but is close to them. For this purpose, we might want to estimate the set of "similar" outcomes so that it can be useful to construct a small convex set that will include all data points.

One specific case of such an idea is to fit a smallest-possible ellipsoid to include all $y_1, \ldots, y_N$. A general ellipsoid can be described as $\{y : \|Xy - b\|_2 \leq 1\}$, where $X \succeq 0$ is a positive semidefinite matrix that "scales" the ellipsoid and $b \in \mathbb{R}^n$ is the center of the ellipsoid.

We aim to keep the ellipsoid as small as possible while including all the data points. The volume of an ellipsoid is inversely proportional to the determinant of X that defines it. Therefore, the optimization problem that minimizes the volume of an ellipsoid while including all the points is

$$
\begin{aligned}
\max \quad & \det(X) \\
\text{s.t.} \quad & \|X\boldsymbol{y}_k - \boldsymbol{b}\|_2 \leq 1, \qquad k = 1, \ldots, N \\
& X \in \boldsymbol{K}^n_{\mathrm{SDO}} \\
& \boldsymbol{b} \in \mathbb{R}^n.
\end{aligned}
$$

6.5 Numerical Solution Methods

There are many solvers for conic optimization problems. The best solver for SDO and SOCO is MOSEK. Other commercial solvers like CPLEX and Gurobi support SOCO but not SDO. ECOS is a free SOCO solver and many free SDO solvers are also available: SeDuMi, SDPT3, SDPA, CSDP, PENLAB, SDPNAL.

6.6 A Complete Example: Multilayered Building Insulation

Thermal insulation is installed in buildings to reduce annual energy costs. However, installation costs money, so the decision of how much insulation to install is a trade-off between the annualized capital costs of insulation and the annual operating costs for heating and air conditioning. This notebook shows the formulation and solution of an optimization problem using conic optimization.

Consider a wall or surface that separates the conditioned interior space in a building at temperature T_i from the external environment at temperature T_o. At a given point in time, the conduction of heat through the wall (measured as the amount of heat transferred per unit of time) is given by

$$
\frac{dQ}{dt} = UA(T_i - T_o),
$$

where U is the overall heat transfer coefficient and A is the heat transfer area. For a wall constructed from N layers of different insulating materials, the inverse of the overall heat transfer coefficient U is given by a sum of serial thermal "resistances"

$$
\frac{1}{U} = R_0 + \sum_{n=1}^{N} R_n,
$$

where R_0 is the thermal resistance of the structural elements. The thermal resistance of the nth insulating layer is equal to $R_n = \frac{x_n}{k_n}$ for a material with thickness x_n and thermal conductivity k_n, so we can rewrite

$$
\frac{1}{U} = R_0 + \sum_{n=1}^{N} \frac{x_n}{k_n}.
$$

The economic objective is to minimize the cost C, obtained as the combined annual energy operating expenses and capital cost of insulation.

We assume that annual energy costs are proportional to the overall heat transfer coefficient U and let $\alpha \geq 0$ be the coefficient for the proportional relationship of the overall heat transfer coefficient U to annual energy costs. Furthermore, we assume that the cost of installing a unit area of insulation in the nth layer is given by the affine expression $a_n + b_n x_n$. The combined annualized costs are then

$$C = \alpha U + \beta \sum_{n=1}^{N} (a_n y_n + b_n x_n),$$

where β is a discount factor for the equivalent annualized insulation cost and y_n is a binary variable indicating whether layer n is included in the installation. The feasible values for x_n are subject to constraints

$$\begin{cases} x_n \leq T y_n \\ \sum_{n=1}^{N} x_n \leq T, \end{cases}$$

where T is an upper bound on insulation thickness.

Analytical Solution for $N = 1$ Layer

In the case of a single layer, $N = 1$, Figure 6.4 illustrates the trade-off between operating and capital costs. Since $N = 1$, we have a one-dimensional cost optimization problem of which we can directly obtain a closed-form analytical solution. Indeed, the expression for the cost $C(x)$ as a function of the thickness x reads

$$C(x) = \frac{\alpha k}{k R_0 + x} + \beta(a + bx).$$

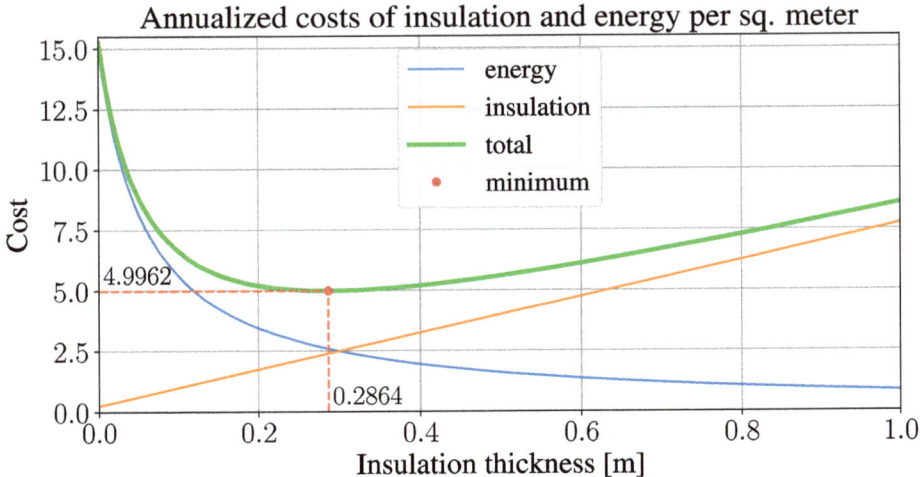

Figure 6.4 The objective function $C(x)$ broken down into its two components (energy operating costs and capital insulation costs) and the corresponding optimal solution x^*.

For fixed parameters k, R_0, β, b, we can calculate the optimal thickness x^* as

$$x^* = -kR_0 + \sqrt{\frac{\alpha k}{\beta b}}.$$

Multilayer Solutions as a Mixed-Integer Quadratic Constraint Optimization (MIQCO) Problem

We cannot easily find an analytical optimal layer composition for multiple layers and must resort to conic optimization.

In the general case with N layers, the objective function is given by

$$\frac{\alpha}{R} + \beta \sum_{n=1}^{N} (a_n y_n + b_n x_n y_n),$$

where the first term is nonlinear in the variables $x_1, \ldots, x_N$ since the denominator of the first term is equal to

$$R = R_0 + \sum_{n=1}^{N} \frac{x_n}{k_n}.$$

To overcome this issue, we can include U as a decision variable and include a constraint

$$\frac{1}{R} \leq U.$$

Since we minimize the objective and U has no other constraint, the problem will guarantee that U is equal to $1/R$. The extra constraint $RU \leq 1$ can be reformulated using an additional decision variable z as

$$1 \leq RU \quad \Longleftrightarrow \quad \begin{cases} z^2 \leq 2RU \\ z = 2 \end{cases} \quad \Longleftrightarrow \quad \begin{cases} \left\| \begin{matrix} \sqrt{2}z \\ R - U \end{matrix} \right\|_2 \leq R + U \\ z = 2, \end{cases}$$

from which we see that the whole problem can be reformulated as a conic optimization problem. The middle formulation above can be implemented in Pyomo as a **conic rotated constraint** [San23], of the form

$$\sum_{n=0}^{N-1} z_n^2 \leq 2 r_1 r_2.$$

In our case, we choose $n = 1$, $z_0 = z$, $r_1 = R$, and $r_2 = U$. Adopting this formulation, the full multilayer optimization problem then reads

$$\min \quad \alpha U + \beta \sum_{n=1}^{N} (a_n y_n + b_n x_n)$$

$$\text{s.t.} \quad R = R_0 + \sum_{n=1}^{N} \frac{x_n}{k_n}$$

$$x_n \leq T y_n, \qquad\qquad\qquad\qquad\qquad \forall n = 1, \ldots, N$$

$$\sum_{n=1}^{N} x_n \leq T$$

$$z^2 \leq 2RU$$

$$z = \sqrt{2}$$

$$R, U > 0$$

$$x_n \geq 0, \qquad\qquad\qquad \forall\, n = 1, \ldots, N$$

$$y_n \in \mathbb{B}, \qquad\qquad\qquad \forall\, n = 1, \ldots, N.$$

The following is the Pyomo code that implements this model to obtain an optimal solution.

```python
import pyomo.kernel as pmo


def insulate(df, alpha, beta, R0, T):
    m = pmo.block()

    # index set
    m.N = df.index

    a = df["a"]
    b = df["b"]
    k = df["k"]

    # decision variables
    m.R = pmo.variable(lb=0)
    m.U = pmo.variable(lb=0)
    m.x = pmo.variable_dict({n: pmo.variable(lb=0) for n in m.N})
    m.y = pmo.variable_dict({n: pmo.variable(domain=pmo.Binary) for n in m.N})

    # objective
    m.cost = pmo.objective(
        alpha * m.U + beta * sum(a[n] * m.y[n] + b[n] * m.x[n] for n in m.N)
    )

    # insulation model
    m.insulation = pmo.constraint(m.R == R0 + sum(m.x[n] / k[n] for n in m.N))

    # total thickness limit
    m.thickness = pmo.constraint(sum(m.x[n] for n in m.N) <= T)

    # layer model
    m.layers = pmo.constraint_dict(
        {n: pmo.constraint(m.x[n] <= T * m.y[n]) for n in m.N}
    )

    # conic constraint
    m.q = pmo.conic.rotated_quadratic.as_domain(m.R, m.U, [np.sqrt(2)])
```

```
    return m


# application parameters
alpha = 30   # $ K / W annualized cost per sq meter per W/sq m/K
beta = 0.05  # equivalent annual cost factor
R0 = 2.0   # Watts/K/m**2
T = 0.15   # maximum insulation thickness

df = pd.DataFrame(
    {
        "Foam": {"k": 0.015, "a": 0.0, "b": 110.0},
        "Wool": {"k": 0.010, "a": 0.0, "b": 200.0},
    }
).T


m = insulate(df, alpha, beta, R0, T)
pmo.SolverFactory("mosek_direct").solve(m)
```

The optimal solution for the two types of material specified in the pandas dataframe above is $x_0^* = x_{\text{foam}}^* = 0.0628$ and $x_1^* = x_{\text{wool}}^* = 0.0872$, and is illustrated in Figure 6.5.

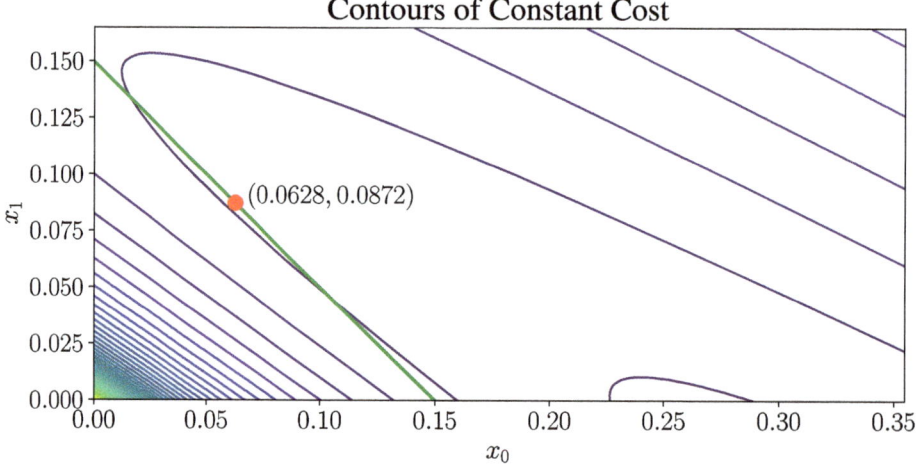

Figure 6.5 Graphical representation of the two-layer problem solved above. The green line represents the thickness constraint $x_0 + x_1 \leq T$, the curves are the isolines of the objective function, and the optimal solution $x^* = (x_0^*, x_1^*)$ is highlighted in red.

Exercises

Exercise 6.1 Consider a company with two divisions A and B that together need to produce one product for a period of n weeks. Each week there is a demand of d_i that has to be met. The company can produce goods now and store them to use them for later demand, but since the size of the warehouse is limited it is only possible to store C items (you may assume that it is possible to store a fractional product).

Each week, divisions A and B have a minimum and a maximum number of hours they can work, because of labor agreements. These minimum and maximum hours may differ per week. The amount of produced goods is equal to the square root of the hours worked by division A times the square root of the hours worked by division B. (Importantly, it is possible that the minimum number of hours that A and B need to work may result in more products than that can be stored. In that scenario, the workers at the two divisions still have to work the minimum, but do not produce anything more than the maximum number of goods that can be stored.)

You may assume that all goods produced in a week are completed at the end of the week and that all demand is concentrated at the end of the week (you do not have to worry about the capacity or demand *during* a week).

The goal is to minimize the total hours worked by divisions A and B. Define a second-order conic optimization problem to solve this problem.

Exercise 6.2 Formulate the following optimization problem as a conic optimization problem:

$$\min \quad \|Ax + b\|_1^2$$

$$\text{s.t.} \quad e^{a_1^\top x + b_1} + e^{a_2^\top x + b_2} + a_5^\top x - \sum_{i=1}^{n} \ln(x_i) + b_5 \leq 0$$

$$\|Dx + d\|_\infty \leq 1$$

$$\sqrt{\left(a_3^\top x + b_3\right)^2 + \left(a_4^\top x + b_4\right)^2 + \sum_{i=1}^{n} x_i^2 + a_3^\top x} \leq 1$$

$$x \in \mathbb{R}$$

in which $x, a_1, a_2, a_3, a_4, a_5 \in \mathbb{R}^n$, $D \in \mathbb{R}^{m \times n}$, $A \in \mathbb{R}^{m \times n}$, $b \in \mathbb{R}^m$, $b_1, b_2, b_3, b_4, b_5 \in \mathbb{R}$, and $d \in \mathbb{R}^m$.

Exercise 6.3 Determine for each of the following sets whether they are convex and whether they are cones.

(a) $S_1 = \{x \in \mathbb{R}^2 \mid 2x_2 \leq x_1 \text{ and } 2x_1 \geq x_2\}$
(b) $S_2 = \{x \in \mathbb{R}^2 \mid 2x_2 \leq x_1 + 1 \text{ and } 2x_1 \geq x_2 + 1\}$
(c) $S_3 = \{x \in \mathbb{R}^2 \mid x_1 \geq -2x_2\}$
(d) $S_4 = \{x \in \mathbb{R}^2 \mid x_1^2 \leq x_2\}$
(e) $S_5 = \{x \in \mathbb{R}^2 \mid x_1^2 \geq x_2\}$

Hint: All these sets are two-dimensional, so if it helps you could try to draw/visualize the set.

Exercise 6.4 Show that every second-order (or Lorentz) cone $K_{\text{SOCO}}^{n+1} = \{(x,t) : \|x\| \leq t\} \subseteq \mathbb{R}^{n+1}$ is a convex cone.

Exercise 6.5 Prove or disprove the following statements:

(a) The union of two cones is itself a cone.
(b) The union of two convex cones is itself a convex cone.
(c) The intersection of two convex cones is itself a convex cone.

Exercise 6.6 Show that $\mathbb{R}_n^+$ is a self-dual cone, that is, $(\mathbb{R}_n^+)^* = \mathbb{R}_n^+$.

Exercise 6.7

(a) Prove that for $x_1, x_2 \geq 0$ the following equivalence holds:

$$x_3^2 \leq 2x_1 x_2 \iff \sqrt{2x_3^2 + (x_1 - x_2)^2} \leq x_1 + x_2.$$

(b) Use (a) and the fact that any second-order cone is a convex cone to show that the set

$$\mathcal{Q}_3^r = \{x \in \mathbb{R}^3 \mid x_3^2 \leq 2x_1 x_2, \ x_1 \geq 0, \ x_2 \geq 0\}$$

is a convex cone.

Exercise 6.8 Write each of the following constraints in the form $A_i x - b_i \in K$.

(a) $x(x+1) \leq t$.

(b) $\dfrac{(x^2 + 1)}{(2x + 5)} \leq t$ for $x \geq 0$.

(c) $(2x+1)^4 + x \leq t$.

(You may use more than one conic inequality, and the x in the conic inequality does not need to be the scalar appearing in the inequalities in (a), (b), and (c).)

Exercise 6.9 Consider the following constraint:

$$\mathbb{P}\left(\sum_{i=1}^{10} x_i(a_i + b_i z) < 10\right) > 0.95, \tag{6.8}$$

where a_i, b_i, $i = 1, \ldots, 10$ are known constants, and z is a random variable with probability distribution

$$\mathbb{P}(z = y_j) = p_j, \quad j = 1, \ldots, N.$$

One way of conservatively approximating constraints like this is through the use of the so-called Markov inequality which says that for a random variable X and a constant α it holds that:

$$\mathbb{P}(X \geq \alpha) \leq \frac{\mathbb{E} \exp(X)}{\exp(\alpha)}.$$

Use this inequality and the exponential cone to obtain a conic, conservative approximation of (6.8).

Exercise 6.10 Consider again the situation of Exercise 6.1 in which a warehouse has to be built. The number of stores is now equal to m and the coordinates of these stores are given by $v_1, \ldots, v_m \in \mathbb{R}^2$. We want to minimize *the sum of the distances* from the warehouse to all stores. However, we have the additional constraint that none of the stores should be distant more than R from the warehouse. Formulate this problem as a second-order conic optimization problem.

7

Accounting for Uncertainty: Optimization Meets Reality

7.1 Introduction

In all the optimization problems discussed so far, we treated the quantities in the problem description as exact, but, in reality, they cannot always be trusted or assumed to be what we think. Uncertainty might negatively affect solutions to an optimization problem in the following forms:

- Estimation/forecast errors (increasingly important in an ML-driven world):
  - in a production planning problem, future customer demand is a forecast;
  - in a vehicle routing problem, travel times along various roads are real-time updated forecasts;
  - in a wind farm layout problem, power production levels are based on wind forecasts.

- Measurement errors:
  - a warehouse manager might have errors in the data records regarding current stock levels;
  - the concentration level of a given chemical substance is different from expected.

- Implementation errors:
  - a given quantity of an ingredient is sent to production in a chemical company, but due to device errors, a slightly smaller amount is actually received;
  - electrical power sent to an antenna is subject to the generator's errors.

Since there is so much uncertainty, natural questions are:

- How to check how bad "not knowing the numbers exactly in advance" can be for the performance of solutions?
- If a given solution is very sensitive to changes in the problem parameters, can a better solution be found by taking uncertainty into account in advance, and if so, how?

In this brief chapter, we will first show three examples of how solutions can be inspected for their sensitivity to data changes. Then, we will set the direction for the remainder of the book where two principal approaches – robust optimization and stochastic optimization – will be "added" on top of everything we learned so far.

7.2 Three Illustrative Examples

The three following examples are presented according to an increasing level of complication in which the uncertainty affects the optimal solution:

1. in the easiest case, uncertainty does not affect the feasibility of the solution but might imply that the real-life execution of the solution is more costly/takes more time;
2. the next level is when the uncertainty might affect the feasibility of the solution but only within the context of a single time period;
3. in the most difficult case, uncertainty affects the feasibility of the solution in a given decision time period, and the consequences of it carry over to subsequent time periods of the decision process.

We begin with a well-known fleet assignment problem that airlines solve daily.

Example 7.1 (Fleet assignment problem) Given a set of flights, an airline company needs to determine the specific route operated by each airplane in the most cost-effective way. The airline company should try to use as few airplanes as possible, but the same airplane can operate two subsequent flights only if the time interval between the arrival of the first flight and the departure of the next flight is longer than or equal to one hour.

The task of the airline operations team is to determine the minimum number of airplanes needed to operate the given list of flights. This problem, known as the **fleet assignment problem** or **aircraft rotation problem**, can be formulated and solved as a MILO problem.

The idea of the MILO formulation is to construct a "feasible path" for each aircraft in a graph where the flights are nodes with indices $1, \ldots, F$, and there is one source node 0 and a sink node $F+1$ such that only arcs can be used between nodes that can be operated after each other. Denote the set of flight indices by $\mathcal{F} = \{1, \ldots, F\}$, and the set of aircraft indices by $\mathcal{M} = \{1, \ldots, M\}$.

The set of arcs $\mathcal{A} \subseteq \mathcal{F} \cup \{0\} \times \mathcal{F} \cup \{F+1\}$ that can be used by each aircraft is then:

$$\mathcal{A} = \bigcup_{f_1, f_2 \in \mathcal{F}} \{(f_1, f_2) : f_1 \text{ arrives at least 1h before the departure of } f_2\}$$

$$\cup \bigcup_{f \in \mathcal{F}} \{(0, f), (f, F+1)\},$$

where 0 and $F+1$ play the role of dummy flights/nodes.

In this context, for $a \in \mathcal{A}$ and $i \in \mathcal{M}$, $x_{a,i} \in \mathbb{B}$ is the binary decision variable indicating whether the pair of flights $a = (f_1, f_2)$ is operated by an aircraft i ($x_{a,i} = 1$) or not ($x_{a,i} = 0$). Furthermore, we use binary variables $y_i \in \mathbb{B}$, for $i \in \mathcal{M}$, to track whether or not the aircraft i is used. The corresponding MILO is then:

$$\min \sum_{i \in \mathcal{M}} y_i \tag{7.1a}$$

$$\text{s.t.} \sum_{\substack{g \in \mathcal{F} \cup \{0, F+1\}: \\ a = (g, f) \in \mathcal{A}}} x_{a,i} = \sum_{\substack{h \in \mathcal{F} \cup \{0, F+1\}: \\ a' = (f, h) \in \mathcal{A}}} x_{a',i}, \quad \forall f \in \mathcal{F}, \forall i \in \mathcal{M} \tag{7.1b}$$

$$\sum_{\substack{i\in\mathcal{M}}}\sum_{\substack{g\in\mathcal{F}\cup\{0,F+1\}:\\a=(g,f)\in\mathcal{A}}} x_{a,i} = 1, \qquad\qquad \forall f\in\mathcal{F} \qquad (7.1\mathrm{c})$$

$$\sum_{\substack{f\in\mathcal{F}\cup\{0,F+1\}:\\a=(0,f)\in\mathcal{A}}} x_{a,i} \le 1, \qquad\qquad \forall i\in\mathcal{M} \qquad (7.1\mathrm{d})$$

$$x_{(f,0),i} \le y_i, \qquad\qquad \forall f\in\mathcal{F}, \forall i\in\mathcal{M} \qquad (7.1\mathrm{e})$$

$$y_i \le y_{i-1}, \qquad\qquad \forall i\in\mathcal{M}: i>0 \qquad (7.1\mathrm{f})$$

$$x_{a,i} \in \mathbb{B}, \qquad\qquad \forall a\in\mathcal{A}, i\in\mathcal{M} \qquad (7.1\mathrm{g})$$

$$y_i \in \mathbb{B}, \qquad\qquad \forall i\in\mathcal{M}, \qquad (7.1\mathrm{h})$$

where:

- the objective function (7.1a) aims to minimize the number of airplanes used;
- constraint (7.1b) enforces that for a given "real" flight the number of used "incoming arcs" with airplane i must be equal to the number of "outgoing arcs" with this airplane;
- constraint (7.1c) ensures that each flight is operated by exactly one aircraft;
- constraint (7.1d) enforces that each airplane serves at most one "path";
- constraint (7.1e) ensures that if at least one arc $(f,0)$ is used using a given airplane, then this airplane is used;
- (7.1f) is a symmetry-breaking constraint (that does not change the problem but makes feasible only one out of many equivalent solutions), which ensures we utilize airplanes in the order they appear in the aircraft set.

Let us try to solve an instance of this problem using the following data:

Table 7.1 *The collection of $F = 30$ flights that a small airline needs to operate.*

Flight	Dep. time	Arr. time	Flight	Dep. time	Arr. time	Flight	Dep. time	Arr. time
0	13	22	10	6	18	20	15	21
1	14	23	11	10	19	21	11	14
2	10	19	12	0	9	22	3	6
3	10	22	13	14	23	23	15	18
4	1	4	14	8	14	24	4	10
5	0	12	15	16	19	25	19	22
6	11	23	16	5	8	26	20	23
7	2	11	17	7	13	27	17	20
8	3	15	18	13	19	28	6	9
9	12	18	19	5	8	29	7	10

The minimum number of airplanes to operate such that the flight list in Table 7.1 meets all time constraints is 14, see Figure 7.1 for an optimal schedule (it is not unique!).

When two subsequent flights are scheduled with the same aircraft with only one hour between them, the timely departure of the second flight is not always possible. Indeed, in practice, the second flight can only be operated without delay only if "everything goes smoothly," that is, the first flight arrives on time and the aircraft service time does not exceed one hour. This might not always be the case, and hence, the delay of the earlier flight may propagate into the later flight as well, making two flights delayed.

How to get a quick idea of how "risky" a schedule is? A possible proxy for this is the number of subsequent flights separated by a one-hour interval in the optimal schedule. As visible in Figure 7.1, there are five such pairs; see the red arcs. What is the likelihood that all these transitions will go smoothly? In some cases, it might be low.

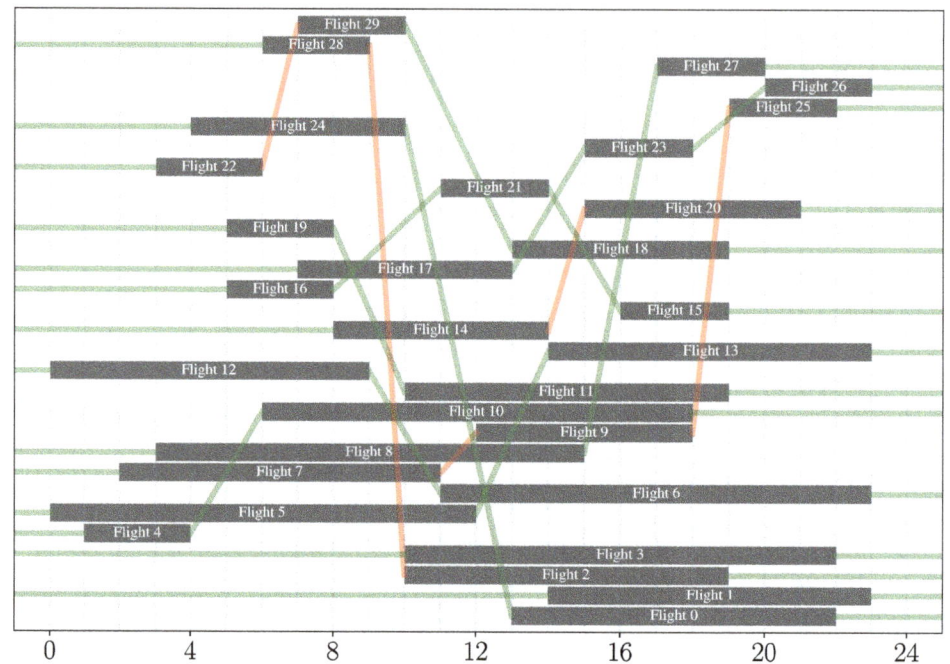

Figure 7.1 An optimal fleet assignment using the minimum number of airplanes for the instance given in Table 7.1. These arcs draw feasible paths through the graph corresponding to the assignment of one aircraft to service one or more flights. The arcs are green if the time between the two flights is strictly larger than the minimum layover time (one hour) and red if it is equal.

The above example was a fairly simple one: if something goes wrong, say a flight gets delayed, its delay may propagate to the subsequent flight to which the same aircraft was assigned, but this latter could probably still be operated, albeit later. In other words, a parameter/data perturbation might cause some annoyance/discomfort for passengers (unless they are transferring and might miss their next flight), but does not affect the feasibility of the problem itself (of course, it affects the feasibility of the underlying MILO formulation, but the real-life implementation can be adapted easily).

There are, however, optimization problems that become infeasible or whose original optimal solution becomes "impossible to fix" when some of the parameters/data deviate from their original values. Consider the following, which is a slightly modified version of problem 12.3 from [Wil13].

Example 7.2 (Production planning using shared resources) A production planner must decide on the mix of products to be manufactured in a factory for the upcoming month. His goal is to maximize the combined profit from all products. A unit of each product requires a certain amount of time (in hours) spent on various processing machines. Data per product are given in Table 7.2. The number of machines available of each type is as follows: four grinders, two vertical drills, three horizontal drills, one borer and one planer.

Table 7.2 *Product data for the production planning problem.*

Product	Grinding	Vertical drilling	Horizontal drilling	Boring	Planing	Profit
0	0.5	0.1	0.2	0.05	—	7.0
1	0.7	0.2	—	0.03	—	6.0
2	—	—	0.8	—	0.01	8.0
3	—	0.3	—	0.07	—	4.0
4	0.3	—	—	0.10	0.05	8.0
5	0.2	0.6	—	—	—	9.0
6	0.5	—	0.6	0.08	0.05	3.0

It is assumed that throughout the month there are 24 working days of 8 working hours each, hence 192 working hours to use in total. The entire optimization problem can be formulated as an integer linear optimization problem:

$$\max \sum_{p \in P} c_p x_p$$

$$\text{s.t.} \sum_{p \in P} a_{p,r} x_p \le b_r, \qquad \forall r \in R$$

$$x_p \in \mathbb{Z}_+, \qquad \forall p \in P,$$

where $p \in P$ are the products, $r \in R$ are the machines (resources) with availability b_r each, $a_{p,r}$ is the resource requirement for product p on machine r, c_p is the unit profit for product p, and the decision variables x_p are the production quantities.

The optimal production plan is given by $x_0 = 0$, $x_1 = 116$, $x_2 = 720$, $x_3 = 0$, $x_4 = 1885$, $x_5 = 601$, and $x_6 = 0$, with a profit of 26,945.

Assume now that each of the machine requirement data per product is not known exactly but that it is expected to vary by up to 5% in both directions. This means that we might be in a situation where we do not have enough resources to produce the desired quantities. How do we investigate the potential impact of such a resource shortage?

One of the possible methods is a simulation study in which we decide to produce the nonzero-quantity products in a random order and, for each of them, we check how much of each resource is left. If the total amount of resources is insufficient for the full quantity, we cut the production down to what is possible. For each such random permutation of product order, a specific realization of the processing times $a_{p,r}$ becomes $(1 + z_{p,r})a_{p,r}$ where $z_{p,r}$ is a real number randomly sampled from the interval $[-0.05, 0.05]$. The simulated profit, in this case, is equal to 26,474.55, and the comparison of the desired and average simulated amounts of production is given in Table 7.3.

Table 7.3 *Desired and average realized production levels under simulation in the production problem of Example 7.2.*

Product	Desired	Realized
0	0.0	0.00
1	116.0	74.63
2	720.0	708.00
3	0.0	0.00
4	1885.0	1839.65
5	601.0	585.59
6	0.0	0.00

Therefore, the average realized profit is about 3.13% less than assumed, and the produced quantities also differ from the expected ones. In case contracts have been signed for the product amounts delivered, the company could be in trouble because of the unmet delivery targets.

In multi-stage problems, things become even more complicated, since perturbations even of a single parameter/data can propagate through multiple time periods. For instance, in Example 7.1, if a single flight is delayed, it affects the arrival times of all subsequent flights of the same aircraft. When evaluating the impact of uncertainty in such settings, it is not enough to "try to make the constraints violated by manipulating the parameter values with the solution being fixed," but we also need to take into account the fact that the decision-maker has a chance to adapt to the change through later decisions, as in the example to follow.

Example 7.3 (Robustness analysis of BIM production plan against uncertain demand)
Consider the raw material acquisition planning problem of the BIM company described in Section 2.5. Assume further that the purchase plan is "frozen in time," that is, in each month the amount of raw materials purchased is exactly the one from the optimal solution to this example but that now the demand for chips can differ between the months.

Specifically, we assume that the demand for both types of chips can increase or decrease by $\rho = 10\%$ relative to the forecast. How can we check the viability of the acquisition plan in this case?

We run a small simulation to observe how the system behaves under various demand scenarios. There are several quantities we can track in such a simulation: the state of the inventory, as well as possible missed demand if we do not produce enough in a given month. Because demand can differ from the assumed one, we need to assume how the decision-maker reacts to this because the following can happen:

- in the case of smaller-than-expected demand, it is possible that we produce the initially assumed number of chips, which would mean that there would be excess chips left – this creates a need for including an "inventory of chips" quantity or an assumption that the excess chips are lost;

- in the case of excess demand, we can either scale up the production to meet the demand as long as the inventory allows, or we can stick to the original production plan – in that case, it would be helpful to keep track of how much of the excess demand is missed.

For sake of illustration, let us assume that in each month, given the observed demand, we adjust the chip production to minimize the amount of missed demand, remaining within the capacity constraints.

Figure 7.2 depicts the average number of missed chip demands of each type over time. It indicates that if meeting the demand is a priority, simply optimizing for the forecast demand is not the best solution in the case where the raw material orders remain fixed.

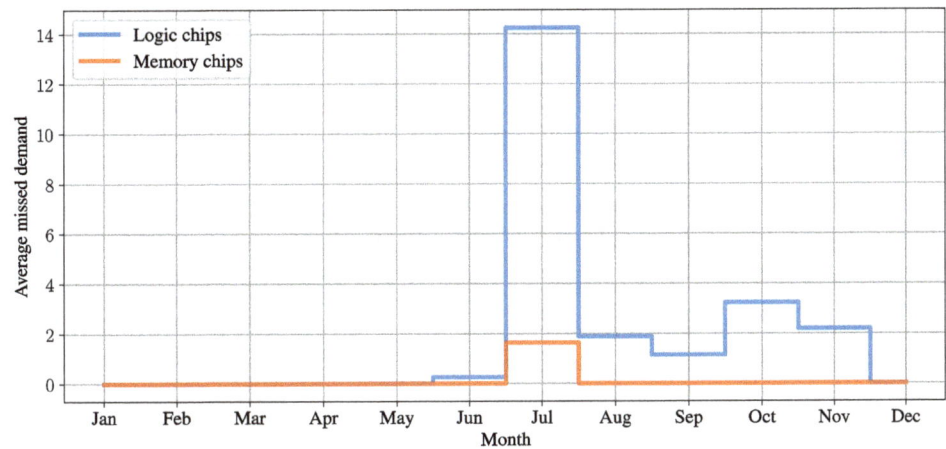

Figure 7.2 Average missed demand per chip type in a simulation consisting of 50 samples with uncertainty $\rho = 10\%$.

This indicates that if meeting the demand is a priority, then simply optimizing for the forecast chip demand is not the best solution in the case where the raw material orders remain fixed.

7.3 Uncertainty-Including Optimization

Optimizing *and* taking uncertainty into account means that we seek a solution that accounts for "different outcomes of the data." Logically, this sounds like solving "multiple versions of the same problem at the same time" so we can expect that it will increase the difficulty level of the problem compared to the deterministic setting.

Additionally, the problem is very vague because there is no "single outcome of the objective function" anymore – everything depends on the data realization. For example, if the value of the objective function depends on the uncertain parameter, what should be the value to minimize? The average across all of them, or maybe the worst-possible value? The same applies to the constraints – if the constraint is satisfied or not depends on the unknown values, do we want them to be satisfied across 90% of the possible outcomes, or for *all* possible outcomes?

The key decisions are the following:

- What are the most important uncertainties?
- Can some decisions be modified after the uncertainties become known, or do we need to fix all of them beforehand?
- What do we know about the uncertainties? Do we have past data or forecasts that can be used as a probability distribution, or do we only know that a given parameter cannot differ from its estimated value by, say, 5%?
- What mindset do we adopt while solving the problem? Do we optimize for the average or the worst-case performance? Do we want the constraints to be satisfied at all times or with sufficiently high probability?

There are many optimization-using research and applied communities, each of which likes to answer the above questions differently, which leads to different (meta)algorithms. To make things even more complex, including uncertainty in general problems such as those given in Chapters 5 and 6 is significantly more difficult than in the MILO-based problems of Chapters 2 to 4.

To remain concise and be aware that the material of Chapters 2 to 4 is sufficient to cover the vast majority of real-world problems, we adopt the mindset "we have a MILO-formulated problem and need to deal with uncertainty inside." To this end, we propose three approaches:

- *Simple hacks* added to deterministic approaches – sometimes, a simple and common-sense-based modification of the deterministic problem will "do the job" of accounting for uncertainty.
- *Robust optimization*: a general paradigm in which we define a set of outcomes of the uncertainty and require the solution to work best assuming the worst-case outcome for your specific solution will always be realized.
- *Stochastic optimization*: a general paradigm where we define (based on data) the probability distribution of the unknown parameters, and want the solution to perform best on average and/or to meet certain probabilistic guarantees.

These are best illustrated in the examples we have already studied, and Table 7.4 gives possible implementations of the three approaches (which are not the only possible ones).

From the table, it might seem like the recipes proposed in the "robust" and "stochastic" optimization columns are overly complicated. While true that their implementation is more complex than the "hacks," it might pay off in the solutions being less conservative, that is, they yield a better objective function without becoming "robust to events that never happen."

7.4 Setup and Outlook

As already mentioned, from now on we will stick to possible-to-state-as-MILO problems affected by uncertainty. Mathematically, this means problems such as

$$
\begin{aligned}
\min \quad & c^\top x & (7.2)\\
\text{s.t.} \quad & A(z)x \leq b(z),\\
& x \in \mathbb{R}^n,
\end{aligned}
$$

Table 7.4 *Outline of key uncertainty-handling approaches on the example of the three problems considered.*

Example	Uncertainty-minded hack	Robust optimization	Stochastic optimization
Fleet assignment Example 7.1	Keep the number of airplanes fixed and minimize the number of "risky 1 h connections"	Assume that at most five flights are late by at most 1 h. Minimize the number of airplanes making sure no flight's delay propagates into another flight	Assume each flight is delayed by 1 h with 5% probability, minimize the number of airplanes such that the expected sum of delays is less than 4 h
Production planning Example 7.2	Assume an extra 5% duration of every task	Assume at most two tasks can take at most 5% longer but you do not know which, find a profit-maximizing production plan making sure that constraints hold under each such scenario	Assume each task duration can be longer by 10% with probability 0.3. Maximize profit with 80% guarantee of plan feasibility
Inventory control Example 7.3	Keep 5% more inventory of every material	Assume ±5% demand change per period and at most 10% total absolute demand perturbation over 12 months	Assume demand per period distributed uniformly over [95%, 105%] of nominal value, minimize inventory cost with 95% probability of not missing customer demand

where we assume that the coefficient matrix $A(z) \in \mathbb{R}^{m \times n}$ and vector $b(z) \in \mathbb{R}^m$ depend on the uncertain parameters z linearly. More specifically, for every $i = 1, \ldots, m$ the ith constraint of (7.2) is of the form

$$a_i(z)^\top x \leq b_i(z),$$

or, even more explicitly,

$$(a_i + A_i z)^\top x \leq b_i + b_i^\top z,$$

provided that $a_i(z) = a_i + A_i z$ and $b_i(z) = b_i + b_i^\top z$.

For simplicity, we assumed that the objective function of (7.2) does not depend on uncertainty, but in subsequent chapters we will show how to deal with the situations where it actually does.

Problem (7.2) is an example of a single-stage optimization problem under uncertainty where x needs to be decided upon before z is observed. In general, it is possible that, after observing (some of) the uncertainties, some of the decisions can be adapted. For that reason, we will also discuss solving optimization problems in a multi-stage setting, allowing the uncertainty to take a stochastic or worst-case nature. The following scheme outlines the structure of the remainder of the book:

		Decision stages	
		Single-stage	Multi-stage
Uncertainty mindset	Worst-case	Chapter 8	Chapter 10
	Stochastic	Chapter 9	

Because there are no widely adopted packages for solving optimization problems under uncertainty, we will explain more details of the implementation of the particular algorithms.

Exercises

Exercise 7.1 The purpose of this exercise is to reflect on the impact of uncertainty in the data on optimization solutions used by several systems and the choice of the best solution method for them.

(a) Name at least three types of uncertainties that are certainly part of map navigation systems. How do you assume that these are dealt with?
(b) Suppose that your data fit very well some statistical distribution. For instance, you deal with travel times and these are very well modeled using a (shifted) exponential distribution. Which of the presented approaches (stochastic, robust, simple hacks) would seem most plausible to determine the shortest path?
(c) Suppose that your optimization model deals with life-critical decisions, such as determining the radiation intensity to apply to a given body tissue (the optimized quantities are strengths of multiple beams used at the same time), and there is uncertainty about the exact positioning of the patient during therapy. Which approach would you choose?
(d) Suppose that your model is very complex, but you know very well the adversarial effects of uncertainty. For example, in a complex supply chain, you can very reliably translate changes in customer demand into the average cost of inventory or lost customer trust. What approach would you be inclined to use?

8

Robust Optimization

8.1 Introduction

This chapter introduces robust optimization (RO), where we aim to solve a MILO in which some of the parameters/data can take multiple (possibly infinitely many) values and we want the optimal solution to perform "the best possible," assuming that the unknown problem parameters can always turn out to be "the worst possible."

Example 8.1 Consider the production planning problem in Example 7.2 where uncertainty affects the duration of tasks. Recall that P denotes the collection of products and R is the collection of machines, which are the shared resource. To approach this problem in a robust way, let us introduce a set Z of values that the uncertain parameters, that is, the possible task duration, can take. More specifically, let us assume that all tasks can exceed their nominal duration $a_{p,r}$ by an "extra margin" $\rho > 0$ and thus take

$$Z = Z(\rho) := \{\tilde{a}_{p,r} \ : \ \tilde{a}_{p,r} \in [a_{p,r}, (1+\rho)a_{p,r}], \ p \in P, \ r \in R\}.$$

With this definition, the robust formulation of the same problem becomes:

$$\min \quad \sum_{p \in P} c_p x_p \tag{8.1}$$

$$\text{s.t.} \quad \sum_{p \in P} \tilde{a}_{p,r} x_p \le b_r, \qquad \forall r \in R, \forall \tilde{a}_{p,r} \in Z$$

$$x_p \in \mathbb{Z}_+, \qquad \forall p \in P.$$

One can verify that this problem is equivalent to one in which we simply assume every task duration to attain its maximum duration and, thus, solving it just entails substituting the coefficients $\tilde{a}_{p,r}$ with their maximum possible value $(1+\rho)a_{p,r}$.

However, assuming that every single task takes ρ percent longer is a bit too conservative, since we do not expect all tasks to attain their maximum duration. An alternative could be to assume that the total sum of the absolute differences of the task durations from their nominal values (hence our "prediction errors") across all tasks is bounded by a certain scalar $\Gamma > 0$. This would lead to a different set Z for the uncertain parameters, namely

$$Z(\Gamma) := \left\{\tilde{a}_{p,r} : \tilde{a}_{p,r} \in [a_{p,r}, (1+\rho)a_{p,r}], \ p \in P, \ r \in R, \ \sum_{p,r} |\tilde{a}_{p,r} - a_{p,r}| \le \Gamma\right\}.$$

$$\tag{8.2}$$

> Construction of a set of plausible outcomes such as the one in (8.2) is not unreasonable, but how can we solve the problem (8.1) assuming that $Z(\Gamma)$ is defined as in (8.2) and that given our decisions, the worst-possible realization of the duration materializes? The answer is not immediate because it is not clear what is the worst-case scenario in $Z(\Gamma)$.

This chapter aims to develop methods to deal with uncertainty sets and robust problems, such as the one in the above example. To develop methods, however, some rather fundamental questions need to be addressed first for the optimization problems to be even well defined:

- How should the optimization problem be formulated mathematically to reflect the idea of "best possible solution under worst-possible scenario"?
- How to construct "reasonable" **uncertainty sets** Z of scenarios/uncertain parameter realizations?

We address the first of these questions in Section 8.2 and the second in Section 8.3, only afterward moving to specific solution approaches in Section 8.4.

8.2 Modeling for RO

8.2.1 Best Solution in Worst Circumstances

To correctly formulate an RO version of a (MI)LO problem, we first introduce some standard notation. Assume that we are solving a general MILO problem, whose coefficients depend on a parameter vector $z \in \mathbb{R}^l$, namely

$$
\begin{aligned}
\min \quad & (c + Cz)^\top x \\
\text{s.t.} \quad & (a_i + A_i z)^\top x \leq b_i + b_i^\top z, \qquad i = 1, \ldots, m \\
& x \in \mathbb{R}^n,
\end{aligned}
$$

with $x \in \mathbb{R}^n$ being the vector of decision variables (we are free to add integrality constraints if needed) and $c, a_i \in \mathbb{R}^n$, and $b_i \in \mathbb{R}$. The matrices $C, A_i \in \mathbb{R}^{n \times l}$ and the vectors $b_i \in \mathbb{R}^l$ describe how z impacts the objective function and the constraints, respectively. Because C, A_i, and b_i become irrelevant when $z = 0$ (the "no data perturbation case"), we call c, a_i, and b_i the **nominal values**. Therefore, we refer to Z as the uncertainty set and to any single element $z \in Z$ as **scenario**.

Our goal is to optimize and find the *best* decisions x *before* knowing the realized scenario z, which is assumed to be able to take its worst possible value given our decisions. Thus, to formulate a **robust MILO**:

- we minimize the worst-possible value of the objective over all $z \in Z$, that is,

$$
\sup_{z \in Z} (c + Cz)^\top x,
$$

- we rewrite each constraint as

$$
\sup_{z \in Z} \{ (a_i + A_i z)^\top x - b_i - b_i^\top z \} \leq 0
$$

$$
\iff (a_i + A_i z)^\top x - b_i - b_i^\top z \leq 0, \quad \forall z \in Z,
$$

since we want each of them to hold for all $z \in Z$.

If we combine the two elements, we obtain the following formulation:

$$\min_{z \in Z} \sup \left\{ (c + Cz)^\top x \right\} \tag{8.3}$$

$$\text{s.t. } (a_i + A_i z)^\top x \le b_i + b_i^\top z, \qquad i = 1, \ldots, m, \quad \forall z \in Z$$

$$x \in \mathbb{R}^n.$$

This is the problem formulation to which we need to convert any MILO problem that we want to solve using RO. For that reason, in the next sections we will focus on modeling, explaining how to equivalently reformulate the problem at hand as (8.3).

8.2.2 Removing Equality Constraints Affected by Uncertainty

The first important RO modeling aspect is that we cannot directly deal with problems that have equality constraints with uncertain parameters. To understand this, consider the following example.

Example 8.2 (Car manufacturing part 1) Suppose a car factory needs to plan its production over n months. The production cost of a single car can vary over different months and is equal to p_t for month t. There is an inventory cost h (constant over the entire horizon) for any unsold car at the end of each month. The customer demand (which has to be satisfied) in period t is given by z_t. The optimization problem can be written as

$$\min \quad \sum_{t=1}^{n} p_t x_t + h \sum_{t=1}^{n} v_t$$

$$\text{s.t.} \quad v_1 = x_1 - z_1$$

$$v_t = v_{t-1} + x_t - z_t, \qquad t = 2, \ldots, n$$

$$x_t, v_t \ge 0, \qquad t = 1, \ldots, n,$$

where the decision variable x_t describes many cars to produce in month t and v_t is the inventory after month t. The constraints $v_t \ge 0$ guarantee that the customer demand is satisfied. Suppose now that we add an uncertainty set Z for the customer demand, obtaining the following problem:

$$\min \quad \sum_{t=1}^{n} p_t x_t + h \sum_{t=1}^{n} v_t$$

$$\text{s.t.} \quad v_1 = x_1 - z_1, \qquad\qquad\qquad \forall z \in Z$$

$$v_t = v_{t-1} + x_t - z_t, \qquad t = 2, \ldots, n, \quad \forall z \in Z$$

$$x_i, v_i \ge 0, \qquad i = 1, \ldots, n.$$

This formulation does not make sense because the constraint

$$v_i = v_{i-1} + x_i - z_i, \quad z \in Z,$$

cannot hold for a fixed v and x if z_i can take multiple distinct values.

The above example does not mean that problems involving equality constraints with uncertain parameters cannot be "robustified." We can still treat these problems by eliminating some variables. The intuitive idea is that if there is an equality constraint, then one variable can be rewritten in terms of the other variables.

Example 8.3 (Car manufacturing part 2) In the car manufacturing problem described in Example 8.2, we can eliminate all variables v_i by writing each of them as $v_i = \sum_{s=1}^{i}(x_s - z_s)$. After substituting these expressions, the problem can be rewritten as

$$\min \quad \sum_{i=1}^{n} p_i x_i + h \sum_{i=1}^{n} \sum_{s=1}^{i}(x_s - z_s)$$

$$\text{s.t.} \quad \sum_{s=1}^{i}(x_s - z_s) \geq 0, \qquad\qquad i = 1, \ldots, n$$

$$x_i \geq 0, \qquad\qquad i = 1, \ldots, n,$$

which has only inequality constraints and thus makes sense when z can take multiple values.

8.2.3 Nonequivalence of "Equivalent" Problems

In earlier chapters, it was clear that the same problem can be written in different ways that lead to the same optimal solution. However, this may not be the case anymore once we introduce robustness to two "equivalent" optimization problems.

Example 8.4 Consider the constraint $z_1 x_1 + z_2 x_2 \leq 4$ and the following system of constraints:

$$\begin{cases} z_1 x_1 \leq t_1 \\ z_2 x_2 \leq t_2 \\ t_1 + t_2 \leq 4. \end{cases}$$

For any fixed $z = (z_1, z_2)$ the two systems are equivalent, since the two corresponding feasible sets

$$X_1 = \{x \in \mathbb{R}^2 : z_1 x_1 + z_2 x_2 \leq 4\} \text{ and}$$
$$X_2 = \{x \in \mathbb{R}^2 : \exists t_1, t_2 : t_1 + t_2 \leq 4, z_1 x_1 \leq t_1, z_2 x_2 \leq t_2\}$$

are equal. However, one may consider the uncertainty set $Z = \{z : -1 \leq z_1, z_2 \leq 1, z_1 + z_2 = 0\}$, which results in the feasible sets

$$\widetilde{X}_1 = \{x : z_1 x_1 + z_2 x_2 \leq 4 \; \forall z \in Z\} \text{ and}$$
$$\widetilde{X}_2 = \{x : \exists t_1, t_2 : t_1 + t_2 \leq 4, z_1 x_1 \leq t_1 \; \forall z \in Z, z_2 x_2 \leq t_2 \; \forall z \in Z\}.$$

It can be verified that

$$\widetilde{X}_1 = \{x : \max\{x_1 - x_2, x_2 - x_1\} \leq 4\} \text{ and } \widetilde{X}_2 = \{x : |x_1| + |x_2| \leq 4\},$$

and that, in particular, $\widetilde{X}_1 \neq \widetilde{X}_2$.

In the above example, the two uncertain parameters are clearly strongly related (we have $z_1 + z_2 = 0$, so z_1 and z_2 cannot freely attain their extreme values) and in the original constraint, they appear together. However, once we decouple them one from another into different constraints and impose robustness, different scenarios of $\boldsymbol{z}$ will make the two different constraints "worst-possible." Preparing for this, an event that cannot happen, makes the resulting formulation overly conservative.

8.2.4 Removing Uncertainty From the Objective

The last aspect of our modeling approach is a rather intuitive trick, which turns out to help a lot in developing solution methods. Without loss of generality, we can consider RO problems with uncertainty only in the constraints. Indeed, if there is uncertainty in the objective, say $(\boldsymbol{c} + C\boldsymbol{z})^\top \boldsymbol{x}$, we can "remove" it by introducing an auxiliary variable $t \in \mathbb{R}$ and using the following standard trick:

$$\begin{aligned} \min \quad & t & & (8.4) \\ \text{s.t.} \quad & (\boldsymbol{c} + C\boldsymbol{z})^\top \boldsymbol{x} \leq t, & & \forall \boldsymbol{z} \in Z \\ & (\boldsymbol{a}_i + A_i \boldsymbol{z})^\top \boldsymbol{x} - b_i - \boldsymbol{b}_i^\top \boldsymbol{z} \leq 0, & i = 1, \dots, m, \quad & \forall \boldsymbol{z} \in Z \\ & \boldsymbol{x} \in \mathbb{R}^n, t \in \mathbb{R}. \end{aligned}$$

Problems (8.3) and (8.4) are *equivalent*, which means that from an optimal solution to any of the two problems, one can construct an optimal solution to the other one with the same value of the objective. What is good about this fact is that it enables us to reduce the entire discussion about RO to a discussion about constraints only (similar to the role of a "canonical form" of an LO problem).

8.2.5 Modeling – Summary

We can summarize this section in the following remarks:

- if there are uncertainty-affected equality constraints, they should be eliminated by substitution;
- separating related uncertainties by splitting constraints into smaller ones can lead to overly conservative solutions;
- uncertainty can be removed from the objective function.

In view of this, from now on we will consider the following standardized problem:

$$\begin{aligned} \min \quad & \boldsymbol{c}^\top \boldsymbol{x} & & (8.5) \\ \text{s.t.} \quad & (\boldsymbol{a}_i + A_i \boldsymbol{z})^\top \boldsymbol{x} \leq b_i + \boldsymbol{b}_i^\top \boldsymbol{z}, & i = 1, \dots, m, \quad & \forall \boldsymbol{z} \in Z \\ & \boldsymbol{x} \in \mathbb{R}^n. \end{aligned}$$

8.3 Uncertainty Set Design

8.3.1 Basic Modeling Choices

The construction of Z scenarios is central in the formulation and solution of RO problems. For the RO methodology to be mathematically feasible, we assume that the set Z is compact and convex. Apart from that ground assumption, we will always broadly fall into one of the two following cases:

- *No data or statistical information about z.* In this case, there is relatively little to be done on the technical side, and the best strategy is to construct a "reasonable" set of outcomes/scenarios. For instance, in the case of Example 7.2 one could assume, based on expert knowledge, that at most five tasks simultaneously can take longer by at most 10% compared to their nominal duration.
- *Statistical information about z is available.* In this case, which can imply historical data, probability distribution, or forecasts with certain uncertainty estimates (Bayesian forecasts, confidence regions), we can leverage this to construct an uncertainty set that includes the outcome of z with *high probability*.

The reader is invited to come up with "common sense" versions of uncertainty sets for many of the optimization problems discussed in the book so far (e.g., what number of flights can be expected to be delayed typically in Example 7.1). In the following, we present two examples that illustrate the possible sources of data-driven uncertainty sets, and the modeling choices to be made there.

Example 8.5 (Building uncertainty sets from data) When data are available on possible perturbation values, one can specify the uncertainty set Z so that it includes $\alpha \in (0,1)\%$ of the realizations. In doing so, it is important to keep the uncertainty set as small as possible because the larger the volume it has, the safer the solutions become, but at the expense of the objective function value.

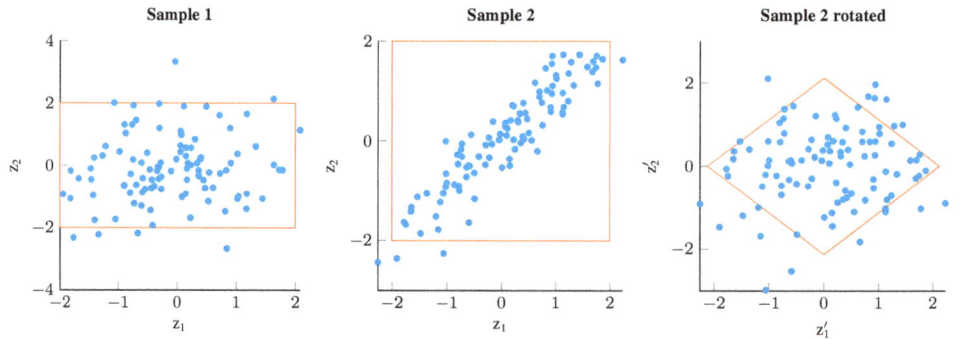

Figure 8.1 Examples of data samples with fit uncertainty sets.

Figure 8.1 illustrates several examples of sets fitted to two-dimensional data samples, where the data sample could correspond, for example, to actual raw material usage in chip production (Example 2.1) – we will elaborate on this example in full at the end of this chapter.

In the left-hand panel of Figure 8.1 we see a data sample to which a rectangular uncertainty set has been fit that includes 87 out of 100 points (*high probability* of covering the actual realization). In the middle panel, we can observe another sample, this time a highly correlated one. Fitting a rectangle "as is" to this sample gives a set with a lot of "empty space," which can lead to over-conservative solutions. In such a situation, one can first rotate the data and then fit a shape to the data, as in the right-hand panel.

Example 8.6 (Building uncertainty sets from prediction errors) In forecasting-based contexts, one can use the specific forecasting engine to estimate, for example, the 90% confidence interval for the uncertain parameters. In Figure 8.2 we illustrate a point forecast for customer demand in 10 subsequent weeks (the black curve) and the default 5% and 95% percentile curves for the demand for ("confidence set" – the blue region). The uncertainty set (red-shaded area) has been defined for each value of the x-coordinate, as "half of the confidence interval."

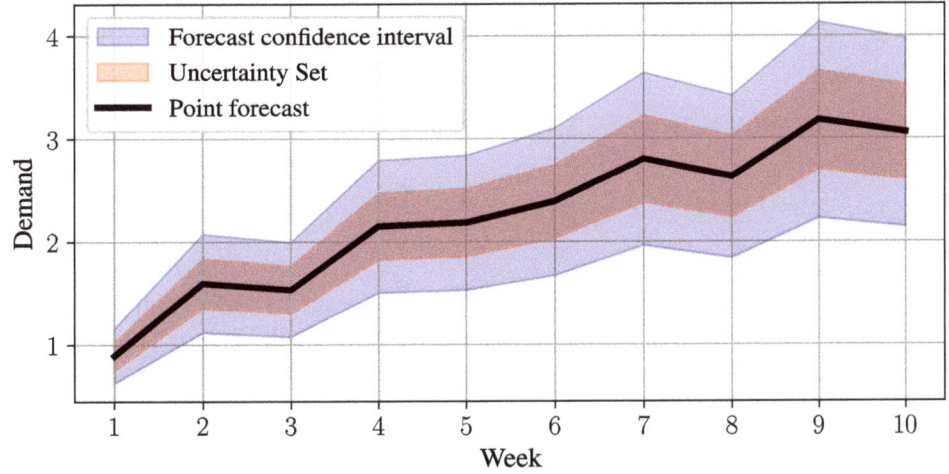

Figure 8.2 Illustration of uncertainty sets for customer demand obtained from the confidence intervals of a data-driven forecast method.

This form of uncertainty set could be constructed, for example, in the context of demand forecasting for chip production in Example 7.3.

8.3.2 Shape of the Uncertainty Set

We now outline several "shapes" an uncertainty set can take. As we shall see later in this chapter, the specific shape an uncertainty set can take has a substantial impact on the complexity of the final problem to solve. For this reason, it is important to have a shortlist of "shapes for every occasion" that keeps problem (8.5) easy to solve.

Box Uncertainty

The simplest idea of modeling uncertainty consists of simply letting each of the unknown quantities vary on its own, independently of the other. That is, we create an interval of possible values for each of the uncertain parameters, leading to the simplest possible **box uncertainty set**:

$$Z = \left\{ \boldsymbol{z} \in \mathbb{R}^l \ : \ l_j \leq z_j \leq u_j, j = 1, \ldots, l \right\} = \prod_{j=1}^{l} [l_j, u_j]. \tag{8.6}$$

This is the type of uncertainty set that was used in Example 7.2 to test the robustness of the initial solution.

Budgeted Uncertainty

The box uncertainty set of the previous section has simplicity as its advantage. However, in real-life situations, it can be quite uncommon for all parameters z_j to *simultaneously* attain their most *extreme* (upper or lower) values. For this reason, the box uncertainty set can be "cut" while still including a statistical majority of scenarios of z. A very popular idea for implementing this approach is the **cardinality-constrained uncertainty set** or **budgeted uncertainty set** [BS04]. The high-level idea of this set is the following:

> *For each entry z_i of the uncertain parameter z, we know its nominal value $\bar{z}_i$ and that it lies in the interval $[\bar{z}_i - \hat{z}_i, \bar{z}_i + \hat{z}_i]$, but at most Γ entries of z_i can simultaneously deviate from their nominal values.*

Mathematically, this idea is formulated as follows:

$$Z = \left\{ z \in \mathbb{R}^l : \bar{z}_j - \hat{z}_j \leq z_j \leq \bar{z}_j + \hat{z}_j, \, j = 1, \ldots, l, \quad \sum_{j=1}^{l} \frac{|z_j - \bar{z}_j|}{\hat{z}_j} \leq \Gamma \right\}.$$

The benefit is that it "cuts the corners" of a box uncertainty set while remaining a set that can be formulated using a finite number of LO constraints (which will lead to the robust problem being MILO-representable).

Note that in the special case in which $\bar{z}_j = 0$ and $\hat{z}_j = 1$ for every $j = 1, \ldots, l$, the budget uncertainty set allows the more compact description, namely

$$Z = \left\{ z \in \mathbb{R}^l : \|z\|_\infty \leq 1, \|z\|_1 \leq \Gamma \right\}.$$

Ellipsoidal Uncertainty

As already mentioned, in the data-driven context, statistical information can be used to construct high-probability uncertainty sets. In statistics, a typical confidence region for a multi-dimensional parameter vector is an ellipsoid, which arises naturally when we assume that the uncertain parameters follow a multivariate normal distribution.

This gives rise to a natural idea to take the uncertainty set to be an ellipsoid, that is,

$$Z = \left\{ z \in \mathbb{R}^l : \|Pz - c\|_2 \leq r \right\}.$$

Similarly to the budgeted uncertainty set, an ellipsoid has the benefit of "not including the corners" of scenarios and can be built in a statistically sound way starting from real data. For instance, one can construct the smallest possible ellipsoid that includes most of the available data points similar to Example 6.20.

General Conic Uncertainty Set

All the considered uncertainty sets so far can be encompassed in the following general formulation:

$$Z = \left\{ z \in \mathbb{R}^l : F_j z + f_j \in K_j, \, j = 1, \ldots, K \right\}, \tag{8.7}$$

where each K_i is one of the LO, SOCO, or SDO cones we introduced in Chapter 6. Due to its generality, we shall use this formulation in Section 8.4.2 to provide a one-fits-all derivation of the so-called **robust counterparts** for all types of uncertainty set.

8.3.3 Sizing the Uncertainty Set

Regardless of the uncertainty set "shape," the obtained solution needs to be tested (typically using a simulation like we did in Chapter 7) to see if it meets the desired goals. The reason why this is necessary is the following. If an uncertainty set is constructed to make the solution feasible with at least $(1 - \alpha)\%$ guarantee, the solution will typically be more robust than that.

To better understand this, we visualize the situation in the space of z in Figure 8.3 where the uncertainty set taken was an ellipsoid. For a fixed feasible solution $\bar{x}$, the set of z's for which a single constraint holds is a half-space defined as

$$\left\{ z : \left(A_i^\top x + b_i \right)^\top z \le b_i - a_i^\top x \right\}.$$

By the very construction of our solution, we know that this half-space includes the uncertainty set (because the solution is constructed to work for all realizations in Z). However, if we intersect all such half-spaces across all constraints, we obtain the triangular shaded area in Figure 8.3. This "robustness zone" of a solution includes the uncertainty set Z, but also more than that.

Building an uncertainty set is more an art rather than an exact science. It should be based on data or expert knowledge as much as possible and preferably use the simplest constraints possible (linear or SOCO constraints). It is unlikely that the "first guess" uncertainty set will lead to the perfect solution for which reason the process of calibration of the set size, coupled with a simulation of the performance of a solution, is a must.

8.4 Solution Methods

Having discussed the necessary details that make problems such as (8.5) meaningful, we can now give two approaches to solving a RO problem. This boils down to a choice between the following options:

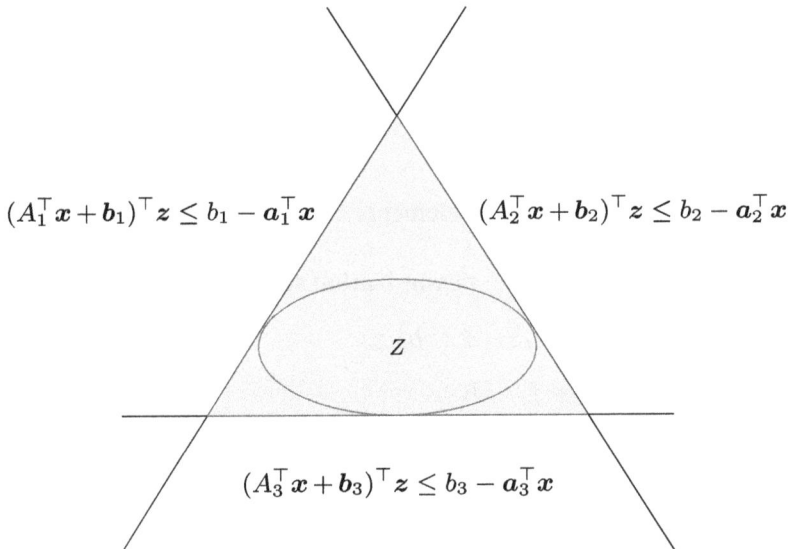

Figure 8.3 Illustration of "robustness zone" of an RO problem solved with an uncertainty set Z.

- iteratively building a finite list of scenarios from Z to which our solution is robust;
- using mathematics to transform the problem to an equivalent form for which every feasible solution is immediately feasible for all $z \in Z$.

The first approach is simple but might take many iterations, while the second approach is solved only once, but requires more mathematics.

8.4.1 Adversarial Approach

Robust problems have a two-layer structure: There is the optimizer that tries to pick the best xs and there is the adversary that, given a fixed decision x, will select the worst possible scenario z. It is therefore logical that this problem could be solved sequentially: Find some x, then the corresponding worst-case z, correct the x to be robust to this scenario as well, and keep iterating in this way.

This intuition is partly correct: Although Z contains infinitely many elements, in practice, only a relatively small number of elements of Z *really matters*. In other words, if instead of considering the entire set Z one would only keep a finite set of scenarios $\overline{Z} \subset Z$, we could expect the optimal solution to be the same. This is the core idea of the iterative approach, which gradually expands a finite set of scenarios using only the new scenarios that matter until a robust solution is found.

Is there a guarantee that such an iterative scheme ever ends? If the set Z is a polytope (i.e., it is formulated using linear constraints only), the answer is yes: Since there are finitely many vertices in a bounded polyhedron, this iterative procedure of adding scenarios must end. However, the number of scenarios that need to be added can be quite large.

Taking into account a general uncertainty set defined in (8.7), this algorithmic procedure can be described as follows:

1. Initialize a separate scenario set as $\overline{Z}^i = \{0\}$ for each constraint i, with $i = 1, \dots, m$.
2. Solve the master problem

$$\min \quad c^\top x \tag{8.8}$$
$$\text{s.t.} \quad (a_i + A_i z)^\top x \le b_i + b_i^\top z, \qquad i = 1, \dots, m, \quad \forall z \in \overline{Z}^i$$
$$x \in \mathbb{R}^n,$$

   by enumerating constraint i over all elements of the finite set $\overline{Z}^i$ and denote the optimal solution as $\bar{x}$.
3. For each i solve the corresponding **pessimization subproblem**:

$$\max_z \quad (a_i + A_i z)^\top \bar{x} - b_i^\top z \tag{8.9}$$
$$\text{s.t.} \quad F_j z + f_j \in K_j, \qquad j = 1, \dots, K,$$

   finding its optimal solution z^*. If $(a_i + A_i z^*)^\top \bar{x} - b_i^\top z^* > b_i$, then $\overline{Z}^i := \overline{Z}^i \cup \{z^*\}$, otherwise, leave the set $\overline{Z}^i$ unchanged.
4. If in Step 3 no new scenario has been added to any of the sets $\overline{Z}^i$, the algorithm stops and returns $\bar{x}$ as the optimal solution. Otherwise, go to Step 2.

This approach has many advantages. First, its simplicity: All one needs is (i) a formulation of the original problem and (ii) the ability to solve the pessimization subproblems for each constraint i. This simplicity comes at a cost: At each iteration, the size of the problem solved in Step 2 increases as the sets of scenarios grow larger and, at some point, it might no longer be possible to solve it. The good news is that, even if this happens, the solution $\bar{x}$ obtained in the current stage is already robust to all scenarios in the sets $\overline{Z}^i$ and, even if these do not fully describe the entire uncertainty set, the current solution $\bar{x}$ typically provides some reasonable partial robustness.

8.4.2 Robust Counterparts of Constraints

In the iterative approach of the previous section, for each x we need to solve for the worst-case z explicitly, and, in addition, the collection $\overline{Z}$ increases at every step. One can ask: Can we invent a trick such that given a decision x, the right worst-case z "chooses itself" automatically? In this section, we will show that something like this is indeed possible.

The goal will be to eliminate the quantifiers "$\forall z \in Z$" without losing robustness. The main idea to tackle this issue is to reformulate each constraint with $\forall z \in Z$ expression to an equivalent set of closed-form constraints.

We focus on each constraint separately because x needs to satisfy *all* constraints, and thus *each of them separately*.

We first illustrate our idea using an example.

Example 8.7 Consider the following constraint:

$$(2 + z_1)x_1 + (3 - z_1 + z_2)x_2 + (4 - z_2)x_3 \le 4, \quad \forall z \in Z = \{(z_1, z_2) : |z_1| + |z_2| \le 1\}.$$

Let us derive the worst-case value of the left-hand side:

$$\sup_{z \in Z}\{(2 + z_1)x_1 + (3 - z_1 + z_2)x_2 + (4 - z_2)x_3\}$$

$$= 2x_1 + 3x_2 + 4x_3 + \sup_{z \in Z}\{(x_1 - x_2)z_1 + (x_2 - x_3)z_2\}$$

$$= 2x_1 + 3x_2 + 4x_3 + \max\{|x_1 - x_2|, |x_2 - x_3|\}.$$

From this, we obtain the following equivalent closed form of the constraint:

$$2x_1 + 3x_2 + 4x_3 + \max\{|x_1 - x_2|, |x_2 - x_3|\} \le 4,$$

which in turn is equivalent to the following system of linear inequalities:

$$\begin{cases} 2x_1 + 3x_2 + 4x_3 + t \le 4 \\ x_1 - x_2 \le t \\ -x_1 + x_2 \le t \\ x_2 - x_3 \le t \\ -x_2 + x_3 \le t. \end{cases}$$

The original robust constraint has been translated into a system of deterministic and linear ones. To reformulate the constraint as an equivalent system of linear inequalities, we had to add some extra variable(s), which is to be expected in RO.

We will now proceed to deriving a robust counterpart of a linear constraint with a general conic uncertainty set

$$Z = \{z \in \mathbb{R}^l \ : \ F_i z + f_i \in K_i\}.$$

We will assume that the interior of Z is nonempty and that Z is a bounded set. With an uncertainty set like this, the pessimization problem of maximizing the left-hand side of a constraint is an easy conic problem:

$$\max_{z} \quad a_i^\top x + \left(A_i^\top x - b_i\right)^\top z \tag{8.10}$$

$$\text{s.t.} \quad F_i z + f_i \in K_i. \tag{8.11}$$

We will treat (8.10) as the primal (see Chapters 2 and 6) and see what we can infer about its dual that could be useful to devise a "self-choosing worst scenario." The optimal solution of (8.10) exists and is bounded because we assumed Z to be a compact set. By conic duality (see Section 6.3), we know that if the primal problem is feasible, then strong duality holds. Because the primal is not only feasible but also bounded, we know that the corresponding dual problem is also feasible and that the optimal values of the two optimization problems coincide.

The dual of (8.10) is

$$\min \quad a_i^\top x + \sum_{i=1}^m \lambda_i^\top f_i$$

$$\text{s.t.} \quad \left(A_i^\top x - b_i\right) + \sum_{i=1}^m F_i^\top \lambda_i = 0$$

$$\lambda_i \in K_i^*, \qquad\qquad\qquad i = 1, \ldots, m.$$

In words, for any x, if we have vectors $\lambda_i \in K_i^*$ such that $\left(A_i^\top x - b_i\right) + \sum_{i}^m F_i^\top \lambda_i = 0$, then

$$a_i^\top x + \sum_{i=1}^m \lambda_i^\top f_i \geq \sup_{z \in Z} \left(\left(a_i + A_i z\right)^\top x - b_i^\top z \right).$$

So if $a_i^\top x + \sum_{i=1}^m \lambda_i^\top f_i \leq b_i$, then we are guaranteed that

$$\sup_{z \in Z} (a_i + A_i z)^\top x \leq b_i + b_i^\top z.$$

We have thus the equivalence

$$\sup_{z \in Z} (a_i + A_i z)^\top x \leq b_i + b_i^\top z$$

$$\Longleftrightarrow \quad \inf_{\lambda_i \in K_i^*, \ (A_i^\top x - b_i) + \sum_{i=1}^m F_i^\top \lambda_i = 0} \quad a_i^\top x + \sum_{i=1}^m \lambda_i^\top f_i \leq b_i.$$

Looking at the right-hand-side inequality, we see an optimization problem. The only remaining trick to notice is that if we have a set of feasible $\lambda_i \in K_i^*$ that makes the objective function at most b_i, then it surely holds for the best $\lambda_i \in K_i^*$. At the same time, λ is not a variable controlled by the adversary – it is "our own" variable that we use to construct valid upper bounds on the adversary's behavior. Therefore, we can drop the infimum sign and conclude that our original constraint is equivalent to the following system:

$$
\begin{cases}
a_i^\top x + \sum_{i=1}^m \lambda_i^\top f_i \le b_i \\
\left(A_i^\top x - b_i \right) + \sum_{i=1}^m F_i^\top \lambda_i = 0 \\
\lambda_i \in K_i^*, \quad i = 1, \dots, m.
\end{cases}
$$

We shall see now how this general formulation translates into different final forms, depending on the "shape" of the uncertainty set. If the uncertainty set Z is polyhedral, the robust counterpart of any constraint can be obtained by considering the standard LO dual of the adversary's problem.

Example 8.8 (Polyhedral and box uncertainty sets) Consider the constraint

$$
(a + Az)^\top x \le b, \qquad \forall z \in Z,
$$

where the uncertainty set Z is polyhedral, say $Z = \{z \in \mathbb{R}^l : Dz \le d\}$. Then, the following logical equivalence holds:

$$
(a + Az)^\top x \le b, \qquad \forall z \in Z = \{z \in \mathbb{R}^l : Dz \le d\}
$$

$$
\iff \max_{z \in \mathbb{R}^l : Dz \le d} (a + Az)^\top x \le b.
$$

Consider the final problem, which is a standard LO problem with z as variables and x as parameters, that is,

$$
\begin{aligned}
a^\top x + \max \quad & \left(A^\top x \right)^\top z && \text{(SP)}\\
\text{s.t.} \quad & Dz \le d \\
& z \in \mathbb{R}^l.
\end{aligned}
$$

The corresponding dual subproblem with dual variables $\lambda \ge 0$ is

$$
\begin{aligned}
\min \quad & d^\top \lambda && \text{(DSP)}\\
\text{s.t.} \quad & D^\top \lambda = A^\top x \\
& \lambda \ge 0.
\end{aligned}
$$

By strong LO duality, the subproblem (SP) and its dual (DSP) have the same optimal value and, hence, for any x the following identity holds:

$$
a^\top x + \min_{\lambda \ge 0:\, D^\top \lambda = A^\top x} d^\top \lambda = a^\top x + \max_{z \in \mathbb{R}^l :\, Dz \le d} \left(A^\top x \right)^\top z \le b.
$$

Note that the existence of λ that satisfies $D^\top \lambda = A^\top x$ and $d^\top \lambda \le b$ is sufficient for the above constraint to hold for x. Therefore, we can conclude that the original constraint is equivalent to the set of constraints

$$\begin{cases} \boldsymbol{a}^\top \boldsymbol{x} + \boldsymbol{d}^\top \boldsymbol{\lambda} \le b \\ D^\top \boldsymbol{\lambda} = A^\top \boldsymbol{x} \\ \boldsymbol{\lambda} \ge 0. \end{cases} \tag{8.12}$$

The box uncertainty set is a special case of a polyhedral uncertainty set, in which each uncertain parameter z_j can vary in a given interval $[l_j, u_j]$, that is,

$$Z_{\text{box}} = \prod_{j=1}^{l} [l_j, u_j] \subset \mathbb{R}^l$$

is the Cartesian product of finite intervals. The box uncertainty set Z_{box} can be equivalently rewritten as $Z_{\text{box}} = \{\boldsymbol{z} \in \mathbb{R}^n : D\boldsymbol{z} \le \boldsymbol{d}\}$ with

$$D = \begin{pmatrix} I \\ -I \end{pmatrix} \quad \text{and} \quad \boldsymbol{d} = \begin{pmatrix} \boldsymbol{u} \\ -\boldsymbol{l} \end{pmatrix}.$$

The tractable robust counterpart of the constraint $(\boldsymbol{a} + A\boldsymbol{z})^\top \boldsymbol{x} \le b$ for every $\boldsymbol{z} \in Z_{\text{box}}$ can be obtained from the more general case above and reads as follows:

$$\begin{cases} \boldsymbol{a}^\top \boldsymbol{x} + \boldsymbol{u}^\top \boldsymbol{\nu} - \boldsymbol{l}^\top \boldsymbol{\mu} \le b \\ \boldsymbol{\nu} - \boldsymbol{\mu} = A^\top \boldsymbol{x} \\ \boldsymbol{\nu}, \boldsymbol{\mu} \ge 0, \end{cases}$$

where $\boldsymbol{\nu}, \boldsymbol{\mu} \in \mathbb{R}^l$ originate as (dual) variables of the Lagrangian dual subproblem and they correspond to the upper bounds (the $\boldsymbol{u}$-vector) and to the lower bounds (the $\boldsymbol{l}$-vector), respectively.

Since the final form of the constraint "stores" the information about the shape of the uncertainty set, it should not come as a surprise that in the following example a robust LO constraint translates into a SOCO constraint under an ellipsoidal uncertainty set.

Example 8.9 (Ellipsoidal uncertainty set) Consider a single robust constraint of the form $(\boldsymbol{a} + A\boldsymbol{z})^\top \boldsymbol{x} \le b$ for every $\boldsymbol{z} \in Z$ in the case in which the uncertainty set is the ellipsoid

$$Z = \{\boldsymbol{z} \in \mathbb{R}^l : \|P\boldsymbol{z}\|_2 \le r\} = \left\{ \boldsymbol{z} \in \mathbb{R}^l : \begin{bmatrix} P \\ \boldsymbol{0}^\top \end{bmatrix} \boldsymbol{z} + \begin{bmatrix} \boldsymbol{0} \\ r \end{bmatrix} \in \boldsymbol{K}_{\text{SOCO}}^{l+1} \right\},$$

where P is a positive definite $l \times l$ matrix and $r > 0$ a positive scalar. According to our general approach, the robust counterpart can then be derived by taking a dual vector $\boldsymbol{\lambda} = (\boldsymbol{\kappa}, \eta) \in \boldsymbol{K}_{\text{SOCO}}^{l+1}$, where $\boldsymbol{\kappa} \in \mathbb{R}^l$, $\eta \in \mathbb{R}$, and writing the following system:

$$\begin{cases} \boldsymbol{a}^\top \boldsymbol{x} + \begin{bmatrix} \boldsymbol{0} \\ r \end{bmatrix}^\top \boldsymbol{\lambda} \le b \\ (A^\top \boldsymbol{x} - b) + \begin{bmatrix} P \\ \boldsymbol{0}^\top \end{bmatrix}^\top \boldsymbol{\lambda} = 0 \\ \boldsymbol{\lambda} = (\boldsymbol{\kappa}, \eta) \in \boldsymbol{K}_{\text{SOCO}}^{l+1}, \end{cases}$$

which can be further simplified into

$$\begin{cases} \boldsymbol{a}^\top \boldsymbol{x} + r\eta \leq b \\ (A^\top \boldsymbol{x} - \boldsymbol{b}) + \boldsymbol{P}^\top \boldsymbol{\kappa} = 0 \\ \boldsymbol{\lambda} = (\boldsymbol{\kappa}, \eta) \in \boldsymbol{K}_{\text{SOCO}}^{l+1}. \end{cases}$$

Because the matrix $\boldsymbol{P}$ is positive definite, it is also invertible and we can convert the inequality to a formula for $\boldsymbol{\kappa}$ that reads $\boldsymbol{\kappa} = -\boldsymbol{P}^{-1}(A^\top \boldsymbol{x} - \boldsymbol{b})$. Next, we can observe that η appears only in the first inequality and, therefore, we can assume that it takes the lowest value that is possible for this term. From the constraint $(\boldsymbol{\kappa}, \eta) \in \boldsymbol{K}_{\text{SOCO}}^{l+1}$ we have that $\eta \geq \|\boldsymbol{P}^{-1}(A^\top \boldsymbol{x} - \boldsymbol{b})\|_2$. There are no other constraints on η and, apart from that, it makes sense for it to take the lowest value possible so we can assume $\eta = \|\boldsymbol{P}^{-1}(A^\top \boldsymbol{x} - \boldsymbol{b})\|_2$. This simplifies our system of inequalities to a single inequality:

$$\boldsymbol{a}^\top \boldsymbol{x} + r \|\boldsymbol{P}^{-1}(A^\top \boldsymbol{x} - \boldsymbol{b})\|_2 \leq b.$$

The same result could have been obtained in a much simpler way by rewriting the original constraint and the uncertainty set formulation in terms of a different variable $\boldsymbol{u} = \boldsymbol{P}\boldsymbol{z}$ (and equivalently $\boldsymbol{z} = \boldsymbol{P}^{-1}\boldsymbol{u}$):

$$\boldsymbol{a}^\top \boldsymbol{x} + (\boldsymbol{P}^{-1}(A^\top \boldsymbol{x} - \boldsymbol{b}))^\top \boldsymbol{u} \leq b, \quad \forall \boldsymbol{u} \in \mathbb{R}^l : \|\boldsymbol{u}\|_2 \leq r.$$

Cauchy's inequality tells us that the expression $\boldsymbol{y}^\top \boldsymbol{u}$ attains its maximum over $\|\boldsymbol{u}\|_2 \leq r$ when $\boldsymbol{u} = r\boldsymbol{y}/\|\boldsymbol{y}\|_2$, that is, when the two vectors $\boldsymbol{u}$ and $\boldsymbol{y}$ point in the same direction. This immediately yields the worst-case value of the left-hand side, that is, $\boldsymbol{a}^\top \boldsymbol{x} + r\|\boldsymbol{P}^{-1}(A^\top \boldsymbol{x} - \boldsymbol{b})\|_2$.

Although the original constraint was linear, the robust constraint became a SOCO constraint. This is often because the robust constraint implicitly incorporates the description of a nonlinear uncertainty set.

Table 8.1 gives an overview of the robust counterparts for several uncertainty sets, after certain simplifications – most of the results can be derived by considering the standard LO dual of the adversary's problem.

The entire discussion above applies to a single constraint, which means that we need another set of dual vectors $\boldsymbol{\lambda}_i$ for each constraint. Adding the robust counterparts of all the constraints together plus the objective function, we can formulate the following equivalent form of problem (8.5):

$$\min \quad \boldsymbol{c}^\top \boldsymbol{x} \tag{8.13}$$

$$\text{s.t.} \quad \boldsymbol{a}_i^\top \boldsymbol{x} + \sum_{j=1}^{K} \boldsymbol{\lambda}_{i,j}^\top \boldsymbol{f}_i \leq b_i, \qquad i = 1, \ldots, m$$

$$A_i^\top \boldsymbol{x} + \sum_{j=1}^{K} F_i^\top \boldsymbol{\lambda}_i = 0, \qquad i = 1, \ldots, m$$

$$\boldsymbol{\lambda}_{i,j} \in \boldsymbol{K}_{i,j}^*, \qquad i = 1, \ldots, m$$

$$\boldsymbol{x} \in \mathbb{R}^n.$$

Table 8.1 *Robust counterparts of the constraint* $(a + Az)^\top x \le b, \; \forall z \in Z$ *for several types of uncertainty sets* Z.

Original constraint: $(a + Az)^\top x \le b, \quad \forall z \in Z$		
Uncertainty set Z	Robust counterpart	Tractability
Symmetric box $\;\|z\|_\infty \le d_{\max}$	$a^\top x + d_{\max}\|A^\top x\|_1 \le b$	LO
General box $\;l_j \le z_j \le u_j$	$\begin{cases} a^\top x + u^\top \nu - l^\top \mu \le b, \\ \nu - \mu = A^\top x, \\ \nu, \mu \ge 0, \end{cases}$	LO
Polyhedral $\;Dz \le d$	$\begin{cases} a^\top x + d^\top \lambda \le b \\ D^\top \lambda = A^\top x \\ \lambda \ge 0 \end{cases}$	LO
Budgeted $\;\|z\|_\infty \le 1, \|z\|_1 \le \Gamma$	$\begin{cases} a^\top x + \|(A^\top x) - \alpha + \beta\|_1 + \lambda\Gamma \le b \\ \alpha + \beta \le \lambda \\ \alpha, \beta, \lambda \ge 0 \end{cases}$	LO
Ellipsoidal $\;\|Pz\|_2 \le r$	$a^\top x + r\|P^{-1}A^\top x\|_2 \le b$	SOCO

8.5 A Complete Example: Robust Microchip Production Planning

Suppose now that there is uncertainty affecting microchip production at BIM. Specifically, the company notices that the amount of copper needed for the two types of microchips is not *exactly* 4 and 2 grams, but varies due to some external factors affecting the production process. How does this uncertainty affect the optimal production plan?

Box Uncertainty

Assume that we have access to $n = 2000$ observed copper consumption pairs for the production of logic and memory chips. As illustrated in Figure 8.4, the amounts vary around the original values, 4 and 2 grams, respectively. A very simple and somehow naive uncertainty set can be the minimal box that contains all the data provided.

Using the data provided, we can calculate the vectors containing the lower and upper bounds, respectively l and u. Using the box uncertainty set defined by them, we can consider the following robust variant of their optimization model:

$$\begin{aligned}
\max \quad & 12x_1 + 9x_2 \\
\text{s.t.} \quad & x_1 \le 1000 \\
& x_2 \le 1500 \\
& x_1 + x_2 \le 1750 \\
& z_1 x_1 + z_2 x_2 \le 4800, \qquad\qquad \forall l \le z \le u \\
& x_1, x_2 \ge 0.
\end{aligned}$$

Using LO duality, we can take the constraint affected by uncertainty, that is,

$$z_1 x_1 + z_2 x_2 \le 4800, \qquad \forall l \le z \le u$$

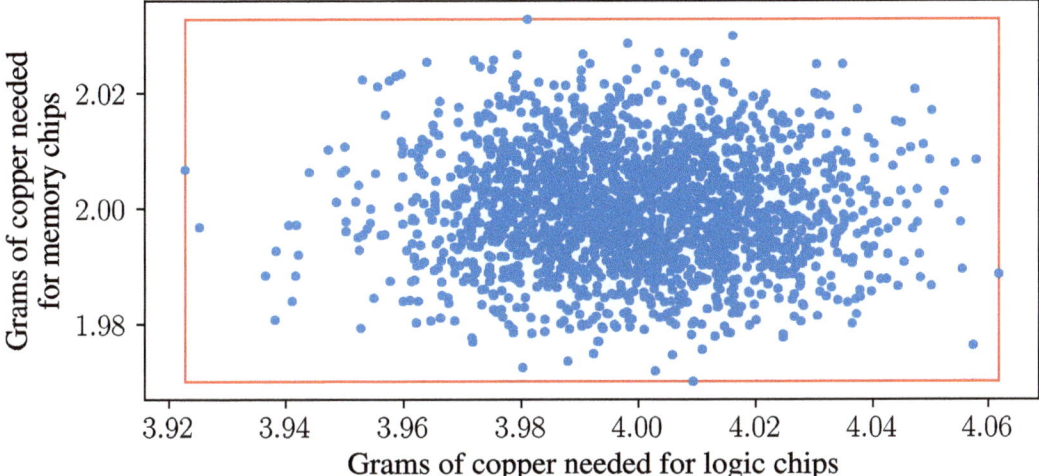

Figure 8.4 Box uncertainty set defined to include all copper consumption pairs provided in the dataset.

and derive the following equivalent set of constraints:

$$\begin{cases} \boldsymbol{u}^\top \boldsymbol{y} - \boldsymbol{l}^\top \boldsymbol{w} \leq 4800 \\ \boldsymbol{y} - \boldsymbol{w} = \boldsymbol{x} \\ \boldsymbol{y}, \boldsymbol{w} \geq 0, \end{cases}$$

which will allow us to obtain a robustified LO problem that we can solve.

```python
def BIMWithBoxUncertainty(lower, upper, domain=pyo.NonNegativeReals):
    m = pyo.ConcreteModel("BIM with Box Uncertainty")

    m.chips = pyo.Set(initialize=chips)
    m.x = pyo.Var(m.chips, within=domain)

    m.profit = pyo.Objective(
        expr=sum([profits[c] * m.x[c] for c in m.chips]), sense=pyo.maximize
    )

    m.silicon = pyo.Constraint(expr=m.x["logic"] <= 1000)
    m.germanium = pyo.Constraint(expr=m.x["memory"] <= 1500)
    m.plastic = pyo.Constraint(expr=sum([m.x[c] for c in m.chips]) <= 1750)

    m.y = pyo.Var(m.chips, domain=pyo.NonNegativeReals)
    m.w = pyo.Var(m.chips, domain=pyo.NonNegativeReals)

    m.robustcopper = pyo.Constraint(
        expr=sum([upper[c] * m.y[c] - lower[c] * m.w[c] for c in m.chips]) <=
        ↪    4800
    )

    @m.Constraint(m.chips)
```

```
    def PerVariable(m, c):
        return m.x[c] == m.y[c] - m.w[c]

    return m


# continuous variables
m = BIMWithBoxUncertainty(lower, upper)
SOLVER.solve(m)

# integer variables
m = BIMWithBoxUncertainty(lower, upper, domain=pyo.NonNegativeIntegers)
SOLVER.solve(m)
```

The optimal solution is $x = (612.4, 1137.6)$ and yields a profit of 17,587.20, while if we impose the integrality constraint, we get $x = (612, 1138)$ and a profit of 17,586.

Budget Uncertainty Set

Assume now that each uncertain coefficient z_j may deviate by at most $\pm\delta$ from the nominal value $\bar{z}_j$ but no more than Γ will actually deviate, so that

$$
\begin{aligned}
\max \quad & 12x_1 + 9x_2 \\
\text{s.t.} \quad & x_1 \leq 1000 \\
& x_2 \leq 1500 \\
& x_1 + x_2 \leq 1750 \\
& z_1 x_1 + z_2 x_2 \leq 4800, \qquad \forall y \in \mathbb{R}^2 : z_j = \bar{z}_j + \delta y_j, \|y\|_\infty \leq 1, \|y\|_1 \leq \Gamma \\
& x_1, x_2 \geq 0,
\end{aligned}
$$

with $(\bar{z}_1, \bar{z}_2) = (4, 2)$. Using Lagrangian duality, we can derive the robustified equivalent LO problem:

$$
\begin{aligned}
\max \quad & 12x_1 + 9x_2 \\
\text{s.t.} \quad & x_1 \leq 1000 \\
& x_2 \leq 1500 \\
& x_1 + x_2 \leq 1750 \\
& \bar{z}_1 x_1 + \bar{z}_2 x_2 + \lambda\Gamma + t_1 + t_2 \leq 4800 \\
& -\delta x_1 + \lambda + t_1 \geq 0 \\
& -\delta x_2 + \lambda + t_2 \geq 0 \\
& \delta x_1 + \lambda + t_1 \geq 0 \\
& \delta x_2 + \lambda + t_2 \geq 0 \\
& x_1, x_2, \lambda, t_1, t_2 \geq 0.
\end{aligned}
$$

The Pyomo implementation of this robust problem is the following:

```python
def BIMWithBudgetUncertainty(delta, gamma, domain=pyo.NonNegativeReals):
    m = pyo.ConcreteModel("BIM with Budget Uncertainty")

    m.chips = pyo.Set(initialize=chips)
    m.x = pyo.Var(m.chips, domain=domain)

    m.profit = pyo.Objective(
        expr=sum([profits[c] * m.x[c] for c in m.chips]), sense=pyo.maximize
    )

    m.silicon = pyo.Constraint(expr=m.x["logic"] <= 1000)
    m.germanium = pyo.Constraint(expr=m.x["memory"] <= 1500)
    m.plastic = pyo.Constraint(expr=sum([m.x[c] for c in m.chips]) <= 1750)

    m.t = pyo.Var(m.chips, domain=pyo.NonNegativeReals)
    m.lam = pyo.Var(domain=pyo.NonNegativeReals)

    m.robustcopper = pyo.Constraint(
        expr=sum([copper[c] * m.x[c] for c in m.chips])
        + gamma * m.lam
        + sum(m.t[c] for c in m.chips)
        <= 4800
    )

    @m.Constraint(m.chips)
    def up_rule(m, c):
        return m.t[c] >= delta * m.x[c] - m.lam

    @m.Constraint(m.chips)
    def down_rule(m, c):
        return m.t[c] >= -delta * m.x[c] - m.lam

    return m


m = BIMWithBudgetUncertainty(0.01, 2, domain=pyo.NonNegativeIntegers)
SOLVER.solve(m)
```

The optimal integer solution that we obtain with $\delta = 0.01$ and $\Gamma = 2$ is $x = (641, 1109)$ and yields a profit of $17,673$.

Adversarial Approach for the Budgeted Uncertainty Set

Instead of adopting the approach of robust counterparts, we could also use the adversarial approach, where we initially solve the problem for the nominal value of the data. Then, we iteratively search for scenarios that make the current solution violate the copper constraint and presolve the problem to take this scenario into account.

To do so, we need (i) to slightly modify our problem formulation function to allow many scenarios for the parameter z and (ii) a function that for a given solution finds the worst-possible realization of the uncertainty restricted by the parameters Γ and δ.

The role of the function in (ii) is to solve the following maximization problem for a given solution $(\bar{x}_1, \bar{x}_2)$:

$$\max (\bar{z}_1 + \delta y_1)\bar{x}_1 + (\bar{z}_2 + \delta y_2)\bar{x}_2 - 4800$$
$$\text{s.t. } |y_1| + |y_2| \leq \Gamma$$
$$-1 \leq y_i \leq 1, \qquad\qquad i = 1,2.$$

We wrap the two functions above into a loop of the adversarial approach, which begins with a nonperturbation assumption and gradually generates violating scenarios, reoptimizing until the maximum constraint violation is below a tolerable threshold. The full Pyomo code is provided below.

```python
def BIMWithSetOfScenarios(
    delta, Z=[{"logic": 0, "memory": 0}], domain=pyo.NonNegativeReals
):
    chips = ["logic", "memory"]
    profits = {"logic": 12, "memory": 9}
    copper = {"logic": 4, "memory": 2}

    m = pyo.ConcreteModel("BIM basic problem")

    m.chips = pyo.Set(initialize=chips)
    m.scenarios = pyo.Set(initialize=range(len(Z)))
    m.x = pyo.Var(m.chips, within=pyo.NonNegativeReals)

    m.profit = pyo.Objective(
        expr=pyo.quicksum([profits[c] * m.x[c] for c in m.chips]),
        sense=pyo.maximize,
    )

    m.silicon = pyo.Constraint(expr=m.x["logic"] <= 1000)
    m.gemanium = pyo.Constraint(expr=m.x["memory"] <= 1500)
    m.plastic = pyo.Constraint(expr=pyo.quicksum([m.x[c] for c in m.chips]) <=
    ↪    1750)

    @m.Constraint(m.scenarios)
    def balance(m, i):
        z = Z[i]
        return pyo.quicksum(copper[c] * m.x[c] * (1 + z[c]) for c in m.chips)
        ↪    <= 4800

    return m


def BIMPessimization(x, delta, gamma):
    chips = ["logic", "memory"]
    copper = {"logic": 4, "memory": 2}

    m = pyo.ConcreteModel("BIM pessimization problem")
    m.chips = pyo.Set(initialize=chips)
```

```python
    m.z = pyo.Var(m.chips, within=pyo.Reals)
    m.u = pyo.Var(m.chips, within=pyo.NonNegativeReals)

    @m.Constraint(m.chips)
    def absolute_value_1(m, i):
        return m.z[i] <= m.u[i]

    @m.Constraint(m.chips)
    def absolute_value_2(m, i):
        return -m.z[i] <= m.u[i]

    @m.Constraint(m.chips)
    def absolute_value_less_than_one(m, i):
        return m.u[i] <= 1.0

    m.budget_constraint = pyo.Constraint(
        expr=pyo.quicksum([m.u[i] for i in m.chips]) <= gamma
    )
    m.violation = pyo.Objective(
        expr=-4800
        + pyo.quicksum([copper[c] * x[c] * (1 + delta * m.z[c]) for c in
        ↪ m.chips]),
        sense=pyo.maximize,
    )

    return m


# Parameters
adversarial_converged = False
stopping_precision = 0.1
max_iterations = 5
adversarial_iterations = 0
delta = 0.2
gamma = 1.5
chips = ["logic", "memory"]

# Initialize the null scenario - no perturbation
Z = [{"logic": 0, "memory": 0}]

while (not adversarial_converged) and (adversarial_iterations <
↪ max_iterations):
    # Building and solving the master problem
    model = BIMWithSetOfScenarios(delta, Z, domain=pyo.NonNegativeIntegers)
    SOLVER.solve(model)

    # Saving the current solution
    x = {i: model.x[i]() for i in model.chips}

    print(f"\nIteration #{adversarial_iterations}")
```

```
print("Current solution: ")
for c in chips:
    print(f"x['{c}']= {x[c]:.2f}")

# Pessimization
m = BIMPessimization(x, delta, gamma)
SOLVER.solve(m)

worst_z = {i: m.z[i]() for i in chips}
constraint_violation = m.violation()

# If pessimization yields no violation, stop the procedure,
# otherwise add a scenario and repeat
if constraint_violation < stopping_precision:
    print("No violation found. Stopping the procedure.")
    adversarial_converged = True
else:
    print(
        f"Violation found: z['logic'] = {worst_z['logic']}",
        f"z['memory'] = {worst_z['memory']}".
        f"constraint violation: {constraint_violation:6.2f}"
    )
    Z.append(worst_z)

adversarial_iterations += 1
```

As outlined in Table 8.2, it takes only two scenarios to be added to the baseline scenario to arrive at the solution, which is essentially robust! For that reason, in many settings, the adversarial approach is very viable. In fact, because the budgeted uncertainty set for two uncertain parameters has at most eight vertices, we are guaranteed that it would never take more than eight iterations to reach a fully robust solution (constraint violation of exactly 0).

This is not true, however, if the uncertainty set is not a polytope but is, for example, an ellipsoid (ball) with infinitely many extreme points – it is expected that there will always be some minuscule constraint violation remaining after a certain number of iterations.

Ball Uncertainty Set

Let us now make yet another different assumption regarding the uncertainty related to copper consumption. More specifically, we assume that the two uncertain coefficients z_1 and z_2 can vary in a two-dimensional ball centered around the point $(\bar{z}_1, \bar{z}_2) = (4, 2)$ and with radius r. A straightforward reformulation leads to the equivalent constraint:

$$\bar{z}_1 x_1 + \bar{z}_2 x_2 + r\|x\|_2 \leq 4800.$$

Table 8.2 *Summary of the iterations.*

Iteration #	Current solution	Violation found	Constraint violation
0	$x_1 = 650.00, x_2 = 1100.00$	$z_1 = 1.0, z_2 = 0.5$	740.00
1	$x_1 = 37.50, x_2 = 1500.00$	—	no violation

By defining auxiliary variables $y = 4800 - \bar{z}_1 x_1 - \bar{z}_2 x_2 \geq 0$ and $\boldsymbol{w} = r\boldsymbol{x} \geq 0$, we can rewrite:

$$\|\boldsymbol{w}\|_2^2 \leq y^2.$$

The resulting optimization problem is nonlinear, but dedicated solvers can take advantage of the fact that it is conic and solve it efficiently. Specifically, the commercial solvers `cplex`, `gurobi`, and `xpress` support convex quadratic inequalities, which we can use to model this variant in `pyomo.environ` as follows:

```python
def BIMWithBallUncertaintyAsSquaredSecondOrderCone(r,
↪    domain=pyo.NonNegativeReals):
    m = pyo.ConcreteModel("BIM with Ball Uncertainty as SOC")

    m.chips = pyo.Set(initialize=chips)
    m.x = pyo.Var(m.chips, within=domain)

    # the nonnegativity of this variable is essential!
    m.y = pyo.Var(within=pyo.NonNegativeReals)

    m.profit = pyo.Objective(
        expr=sum([profits[c] * m.x[c] for c in m.chips]), sense=pyo.maximize
    )

    m.silicon = pyo.Constraint(expr=m.x["logic"] <= 1000)
    m.germanium = pyo.Constraint(expr=m.x["memory"] <= 1500)
    m.plastic = pyo.Constraint(expr=sum([m.x[c] for c in m.chips]) <= 1750)
    m.copper = pyo.Constraint(
        expr=m.y == 4800 - sum(copper[c] * m.x[c] for c in m.chips)
    )
    m.robust = pyo.Constraint(expr=sum((r * m.x[c]) ** 2 for c in m.chips) <=
↪    m.y**2)

    return m


m = BIMWithBallUncertaintyAsSquaredSecondOrderCone(r,
↪    domain=pyo.NonNegativeIntegers)
pyo.SolverFactory("gurobi_direct").solve(m)
```

This Pyomo implementation solves the instance corresponding to $r = 0.05$ using `gurobi`, obtaining $\boldsymbol{x} = (617, 1133)$ as the optimal solution and a profit of 17,601. You can try implementing the adversarial approach in the ellipsoidal uncertainty case to see how many iterations it might take to converge to an optimal solution.

Exercises

Exercise 8.1 Consider the shortest path problem defined on a graph $G = (V, E)$, whose goal is to traverse from node s to node t, and each edge has weight w_{ij}. Assume that the length of each edge can be perturbed through the mechanism $(1 + z_{ij})w_{ij}$ where $0 \leq z_{ij} \leq 1$ for all

$(i,j) \in E$, and where the sum of all perturbations cannot be greater than $\Gamma \geq 1$. Your goal is to determine the shortest worst-case path through the graph. Formulate this problem as a MILO problem. Is it true that you can relax the integrality constraints on the variables and still arrive at an integer optimal solution?

Exercise 8.2 Derive a linear optimization reformulation for the following robust counterpart:

$$(4 + a_1)x_1 + (5 - a_2)x_2 + a_3x_3 \leq 5, \quad \forall\, (a_1, a_2, a_3) : \; -1 \leq a_1, a_2, a_3 \leq 1.$$

Exercise 8.3 Derive a second-order cone reformulation for the following robust counterpart:

$$(4 + a_1)x_1 + (5 - a_2)x_2 + a_3x_3 \leq 5, \quad \forall\, (a_1, a_2, a_3) : \; \|a\|_2 \leq 1.$$

Exercise 8.4 Consider an optimization problem with an n-dimensional variable x and M linear inequalities of the form

$$\hat{a}_k^\top x \leq b_k, \qquad k = 1, \ldots, M.$$

Assume that for the optimal solution x^* there are m binding constraints, that is, there are $0 < m \leq M$ constraints for which the equality $\hat{a}_k^\top x^* = b_k$ holds. Consider the situation in which the coefficients are not known exactly and, more specifically, they are equal to

$$a_{ik} = \hat{a}_{ik} + z_k, \qquad \forall\, i = 1, \ldots, n$$

where $\{z_k\}_{k=1,\ldots,M}$ are i.i.d. (independent and identically distributed) one-dimensional uniform random variables on the interval $[-1, 1]$. Note that the same random quantity is added to each entry of the vector $\hat{a}$; in other words, in vector form we have $a_k = \hat{a}_k + z_k\mathbf{1}$. What is the probability that for all m binding constraints it still holds that $a_k^\top x^* \leq b_k$?

Exercise 8.5 Rewrite the following robust constraint into a second-order conic constraint:

$$z_1 x_1 + z_2 x_2 \leq 4 \qquad \forall\, z \in Z = \{z \in \mathbb{R}^2 \; : \; \sqrt{(z_1 - 1)^2 + (z_2 - 2)^2} \leq 6\}.$$

Exercise 8.6 Consider the robust constraint: $-z_1 \log(x_1) + z_2 x_2 \leq 2, \quad \forall\, z \in \mathbb{R}^2 \; : \; z \geq 0, z_1 + z_2 \leq 1\}$.

(a) Using duality, rewrite the robust constraint as a finite set of constraints.
(b) In solving (a), where did you use strong duality? Why does it hold?

Exercise 8.7 In (re)formulating a robust constraint, you have to be careful that the constraints are indeed equivalent, and this exercise illustrates how things can go wrong. Consider the constraint

$$|x_1 - z| + |x_2 - z| \leq 2, \qquad \forall\, z \in Z = \{z \in \mathbb{R} \; : \; |z| \leq 1\}. \tag{8.14}$$

Using the standard trick for absolute values, we set $y_i = |x_i - z| = \max\{x_i - z, z - x_i\}$ for $i = 1, 2$. Then (8.14) can be rewritten as

$$
\begin{aligned}
y_1 + y_2 &\leq 2 \\
y_1 &\geq x_1 - z \quad \forall z : |z| \leq 1 \\
y_1 &\geq z - x_1 \quad \forall z : |z| \leq 1 \\
y_2 &\geq x_2 - z \quad \forall z : |z| \leq 1 \\
y_2 &\geq z - x_2 \quad \forall z : |z| \leq 1.
\end{aligned} \tag{8.15}
$$

(a) Show that the robust constraint in (8.14) and the set of robust constraints in (8.15) are not equivalent by giving a value for x that is feasible for (8.14), but not feasible for (8.15).

(b) In the set of constraints in (8.15) the original constraint (8.14) was split in two parts using the y-variables. Give a correct linear reformulation of (8.14) in which you do *not* split up the constraint. You do not need to derive a robust counterpart, that is, you could stop once you have obtained linear constraints with $\forall z : |z| \leq 1$. *Hint*: You need four constraints.

Exercise 8.8 Consider the knapsack problem:

$$
\begin{aligned}
\max \quad & \sum_{i=1}^{n} p_i x_i \\
\text{s.t.} \quad & \sum_{i=1}^{n} z_i x_i \leq V \\
& 0 \leq x_i \leq 1, \qquad \forall i = 1, \ldots, n,
\end{aligned}
$$

where the weights parameters $z = (z_1, \ldots, z_n)$ appearing in the capacity constraint are uncertain and vary in a specific uncertainty set $Z \subset \mathbb{R}^n$.

(a) Assume box uncertainty, that is, $Z_{\text{box}} = \prod_{i=1}^{n} [l_i, u_i]$, which means $l_i \leq z_i \leq u_i$ for every $i = 1, \ldots, n$. Formulate the robust counterpart for the capacity constraint $\sum_{i=1}^{n} z_i x_i \leq V$ for all $z \in Z_{\text{box}}$.

(b) Assume a budget uncertainty set, that is

$$
Z_{\text{budget}} = \left\{ z \in \mathbb{R}^n : \bar{w}_j - \hat{w}_j \leq z_j \leq \bar{w}_j + \hat{w}_j, \forall j, \quad \sum_{j=1}^{n} \frac{|z_j - \bar{w}_j|}{\hat{w}_j} \leq \Gamma \right\}.
$$

Formulate the robust counterpart for the capacity constraint $\sum_{i=1}^{n} z_i x_i \leq V$ for all $z \in Z_{\text{budget}}$.

(c) Assume ball uncertainty, that is, $Z = Z_{\text{ball}}(0, R)$, which means $\sum_{i=1}^{n} z_i x_i \leq V$ for all $\|z\|_2 \leq R$. Formulate the robust counterpart for the capacity constraint.

(d) Assume a different ball uncertainty set, that is, $Z = Z_{\text{ball}}(c, R)$, which means $\sum_{i=1}^{n} z_i x_i \leq V$ for all $\|z - c\|_2 \leq R$, where $c \in \mathbb{R}^n$ is a given constant vector. Formulate the robust counterpart for the capacity constraint.

(e) For each of the three uncertainty sets (box, budget, and ball), find an advantage and a disadvantage.

Exercise 8.9 You want to bring three suitcases on your next vacation. The airline you are traveling with does not restrict the number of suitcases as long as their total weight is at most 15 kg. The suitcases weight respectively $w_1 = 3$, $w_2 = 10$ and $w_3 = 15$ kg. Moreover, the three suitcases have different "values" (in view of the different items you put in each), namely $p_1 = 1$, $p_2 = 2$, and $p_3 = 5$.

(a) Formulate an ILO to solve this problem. What is the optimal solution?
(b) Assume that the scale you used has a 10% measurement error, which means that the actual weight can be either 10% higher or lower than the measured one. What is the robust solution of this problem?
(c) Now assume that the scale has an *absolute* error of 2 kg. What is the robust solution now?

Exercise 8.10 Suppose that a budget of B needs to be invested in n assets. Each asset has an expected return of $\hat{p}_i$. However, these returns are stochastic in nature and can be modeled as follows: The return of asset i is equal to $p_i = \hat{p}_i + z_i$, where $z \in \mathbb{R}^n$ is the vector of fluctuations that satisfies the constraint $\|z\|_2 \leq R$.

(a) Formulate this problem as a robust optimization problem.
(b) Which method do you think has the highest expected total return, the one in which investments are decided based on *expected* returns or the one obtained from robust optimization? In which real-life situation would you use the robust optimization method, and in which other would you use it instead of the expected returns?

9

Stochastic Optimization

9.1 Introduction

In this chapter, compared to Chapter 8 we assume that data or expert knowledge can tell us not only something about the possible values of the problem's parameters but also about their relative likelihood, that is, the probability distribution.

Distributional information about z can come in very different ways, for example:

1. raw data is available in the form of data points – this can be interpreted as a discrete probability distribution assuming each data point is equally likely;
2. based on the data, a specific probability distribution is estimated, for example, a normal distribution;
3. a predictive tool is used to forecast the value of parameter z but instead of a point forecast, a probability distribution is provided.

All this opens up new possibilities to handle uncertainty because in such a setup we can care about different quantities than the worst-case value: for example, the average outcome, or the probability that the outcome reaches a certain value.

This makes particular sense when one repeatedly solves an optimization problem many times (say, every day or every 15 seconds). In such cases, one can expect that in the long run, the objective value "averages out," and that there is a reasonable amount of past data to infer the underlying probability distribution. It thus makes sense then to take that information into account within the optimization stage.

Example 9.1 (Markowitz portfolio optimization part 3) Consider again the portfolio optimization problem of Examples 5.10 and 6.12. Specifically, assume that there is an initial capital C that needs to be invested in a selection of n possible assets, each of them with an unknown return rate r_i, $i = 1, \ldots, n$ or in another risk-free asset with a guaranteed return rate R. Let r be the random vector that collects the unknown return rates. Let x be the vector whose ith component x_i describes the amount invested in asset i and $\tilde{x}$ the amount invested in the risk-free asset. We want the weights x_i to be nonnegative (we cannot short-sell assets) and fully invest our capital, that is, $\sum_{i=1}^{n} x_i + \tilde{x} = C$.

In this stochastic setting, it is not obvious what it means to *optimize* the portfolio. If we know the expected return rates $\mu = \mathbb{E}r$, the simplest possible answer is to determine the portfolio that maximizes the *expected return*:

$$\max \mathbb{E}\left(R\tilde{x} + \boldsymbol{r}^\top \boldsymbol{x}\right)$$

$$\text{s.t.} \quad \sum_{i=1}^{n} x_i + \tilde{x} = C$$

$$\tilde{x} \geq 0$$

$$\boldsymbol{x} \geq 0.$$

In this setting, the objective function is rewritten as $R\tilde{x} + \boldsymbol{\mu}^\top \boldsymbol{x}$ and we obtain a deterministic LO problem.

However, a portfolio that maximizes the expected return can also be risky. For that reason, additional constraints can be imposed that limit the risk level of a portfolio. One possible way to do so is enforcing an upper bound γ^2 on the variance of the portfolio, which is precisely the one in (5.5):

$$\max \quad R\tilde{x} + \boldsymbol{\mu}^\top \boldsymbol{x}$$

$$\text{s.t.} \quad \text{Var}\left(\boldsymbol{r}^\top \boldsymbol{x}\right) \leq \gamma^2$$

$$\sum_{i=1}^{n} x_i + \tilde{x} = C$$

$$\tilde{x} \geq 0$$

$$\boldsymbol{x} \geq 0.$$

Another way is by imposing that the probability that the capital invested in risky assets goes below α should not be greater than a fixed probability $\beta \in [0,1]$. The corresponding problem formulation is

$$\max \quad R\tilde{x} + \boldsymbol{\mu}^\top \boldsymbol{x}$$

$$\text{s.t.} \quad \mathbb{P}\left(\boldsymbol{r}^\top \boldsymbol{x} < \alpha\right) \leq \beta$$

$$\sum_{i=1}^{n} x_i + \tilde{x} = C$$

$$\tilde{x} \geq 0$$

$$\boldsymbol{x} \geq 0.$$

This example illustrates that, depending on the preference, there are many ways to "compress" the distribution of $\boldsymbol{z}$ into a single value to be bounded or optimized. In the terminology of this chapter, the three canonical ways are:

- expectations;
- probabilities; and
- other composite measures of "risk" such as the variance.

These different forms are "friendly" to different types of distributional knowledge, and thus, you always need to make your own selection as to what setup will suit your application best. We will go through these possibilities, indicating the key techniques used.

9.2 Expectations

The most basic way of accounting for randomness is to consider the average or expected value of a given uncertain quantity. For constraints, expectations are not as popular because protection against the average outcome implies a low level of protection. For that reason, we consider problems in which the expectation appears only in the objective, such as

$$\min_{x} \quad \mathbb{E} f(x, z). \tag{9.1}$$

Example 9.2 (Stock optimization for seafood distribution center part 1) Each day a seafood distribution center buys x tons of tuna at a unit cost of c. Next, a certain demand z is observed from the retailers, to whom the fish is sold at a unit price per ton $p > c$. The leftover fish must be stored in a cold warehouse for a unit holding cost of h. The seafood distribution center cannot sell more fish than it has in stock, therefore at most $\min\{z, x\}$ tons will be sold and there may be $\max\{0, x - z\}$ tons left. Therefore, the net profit is $p \min\{z, x\} - cx - h(x - z)^+$. Provided that a reasonable estimate of the probability distribution $\mathbb{P}$ of the tuna demand z is available, if we want to maximize the long-term net profit then we can formulate the following optimization problem:

$$\max \quad \mathbb{E}\left[p \min\{z, x\} - cx - h(x - z)^+ \right]$$
$$\text{s.t.} \quad x \geq 0.$$

Note that in the simple setting of this problem the feasible set for the decision variable x is not affected by uncertainty, since we have $x \geq 0$ regardless of the demand z.

This is an instance of the so-called **newsvendor problem**, which is the cornerstone of many optimization problems related to inventory and sales management.

We will now consider two basic methods to solve the problems: In Section 9.2.1 we look at the case in which $\mathbb{E} f(x, z)$ can be rewritten explicitly as a function of x. Afterward, in Section 9.2.2, we will look at the situation where this is not possible, but the expectation can still be approximated by using the data.

9.2.1 Easy-to-Calculate Expectations

It is sometimes possible to explicitly compute the expectation appearing in the objective function of (9.1). Although rare, it is an important case as it effectively transforms a stochastic problem into a deterministic one. Consider the following example.

Example 9.3 (Quadratic function and normal distribution) Assume that z follows a multivariate normal distribution $z \sim \mathcal{N}(\mu, \Sigma)$ and that $f(x, z)$ is a quadratic function of the form

$$f(x, z) = \left((a + Az)^\top x \right)^\top \left((a + Az)^\top x \right) + x^\top C z.$$

Then, the expectation $\mathbb{E}f(x, z)$ can be explicitly rewritten as

$$\mathbb{E}\left[\left((a + Az)^\top x\right)\left((a + Az)^\top x\right) + x^\top Cz\right]$$
$$= x^\top \left((a + A\mu)(a + A\mu)^\top + \Sigma\right)x + x^\top C\mu,$$

which is a convex quadratic function of x. This is not surprising because $f(x, z)$ was convex in x, and taking expectations preserves convexity. □

Similarly, one can derive explicit expectations for some other (simple) distributions and sufficiently simple functions $f(x, z)$. Even if no closed-form expressions are available, as long as one is able to compute and pass to the solver the value and the gradient (and possibly the Hessian) of $\mathbb{E}f(x, z)$, a problem like this can be solved effectively.

Example 9.4 (Maximizing the profit of a pop-up shop) There is an opportunity to operate a pop-up shop to sell a unique commemorative item for each event held at a famous location. The items cost $12 each when bought from the supplier and will sell for $40. Unsold items can be returned to the supplier at a value of only $2 (see Table 9.1).

Table 9.1 *Financial data for the pop-up shop problem.*

Parameter	Symbol	Value
sales price	r	$40
unit cost	c	$12
salvage value	w	$2

Demand for these items, however, will be high only if the weather is good. Historical data suggests three typical scenarios, as detailed in Table 9.2.

Table 9.2 *Weather and demand scenarios in the pop-up shop problem.*

Scenario (s)	Probability (p_s)	Demand (d_s)
sunny skies	0.1	650
good weather	0.6	400
poor weather	0.3	200

The problem is to determine how many items to order for the pop-up shop. The dilemma, of course, is that the weather will not be known until after the order is placed. Ordering enough items to meet the demand for a good weather day results in a financial penalty on returned goods if the weather is poor. On the other hand, ordering just enough to satisfy the demand on a poor weather day leaves "money on the table" for good weather days. How many items should be ordered for sale?

A naive solution to this problem is to place an order equal to the expected demand, which can be calculated as

$$\mathbb{E}(D) = \sum_{s \in S} p_s d_s = 0.1 \cdot 650 + 0.6 \cdot 400 + 0.3 \cdot 200 = 365.$$

Choosing an order size $\hat{x} = \mathbb{E}[D] = 365$ results in an expected profit, which we call the **expected value of the mean scenario (EVM)**. By ordering exactly this amount of items, we can calculate the net profit $f_s(\hat{x})$ for each scenario s as detailed in Table 9.3.

Table 9.3 *Performance of the mean scenario solution in the pop-up shop problem.*

Scenario (s)	Demand (d_s)	Probability (p_s)	Sold (y_s)	Salvage ($x - y_s$)	Profit (f_s)
sunny skies	650	0.30	365	0	\$10,220
good weather	400	0.60	365	0	\$10,220
poor weather	200	0.10	200	165	\$3,950
Average profit					\$8,339

The expected value of the mean scenario is equal to

$$\text{EVM} = \mathbb{E}(f) \sum_{s \in S} p_s f_s(\hat{x}) = \$8339.$$

As shown in the table above, no scenario shows a loss, which appears to be a satisfactory outcome. However, can we find an order that results in a higher expected profit?

To answer this question, let us formulate the problem in mathematical terms. Let x be a nonnegative number representing the number of items that will be ordered, and y_s be the nonnegative variable describing the number of items sold in scenario s in the set S comprising all scenarios under consideration. The number y_s of sold items is the lesser of the demand d_s and the order size x, that is,

$$y_s = \min(d_s, x), \qquad \forall s \in S.$$

Any unsold inventory $x - y_s$ remaining after the event will be sold at the salvage price w. Taking into account the revenue from sales ry_s, the salvage value of the unsold inventory $w(x - y_s)$, and the cost of the order cx, the profit f_s in scenario s is given by

$$f_s = ry_s + w(x - y_s) - cx, \qquad \forall s \in S.$$

Using the constants introduced earlier, the profit f_s in scenario $s \in S$ can then be written as

$$f_s = \underbrace{ry_s}_{\text{sales revenue}} + \underbrace{w(x - y_s)}_{\text{salvage value}} - \underbrace{cx}_{\text{order cost}}.$$

The expected profit is given by $\mathbb{E}(F) = \sum_s p_s f_s$. The optimization problem is to find the order size x that maximizes the expected profit subject to operational constraints on the decision variables. The variables x and y_s are nonnegative integers, while f_s is a real number that can take either positive or negative values. Putting these facts together, the optimization problem to be solved is

$$\max \quad \mathbb{E}(F) = \sum_{s \in S} p_s f_s$$

$$
\begin{aligned}
\text{s.t.} \quad & f_s = r y_s + w(x - y_s) - cx, & & \forall s \in S \\
& y_s \le x, & & \forall s \in S \\
& y_s \le d_s, & & \forall s \in S \\
& y_s \in \mathbb{Z}_+, & & \forall s \in S \\
& x \in \mathbb{Z}_+,
\end{aligned}
$$

where S is the set of all scenarios under consideration. Let us see how we can implement this problem in Pyomo.

```python
# price and scenario information
r = 40
c = 12
w = 2

scenarios = {
    "sunny skies": {"demand": 650, "p": 0.1},
    "good weather": {"demand": 400, "p": 0.6},
    "poor weather": {"demand": 200, "p": 0.3},
}


def PopUpShop(r, c, w, scenarios):
    m = pyo.ConcreteModel("Pop-up shop")

    # set of scenarios
    m.S = pyo.Set(initialize=scenarios.keys())

    # decision variables
    m.x = pyo.Var(domain=pyo.NonNegativeIntegers)
    m.y = pyo.Var(m.S, domain=pyo.NonNegativeIntegers)
    m.f = pyo.Var(m.S, domain=pyo.Reals)

    # parameters
    m.r = pyo.Param(initialize=r)
    m.c = pyo.Param(initialize=c)
    m.w = pyo.Param(initialize=w)

    # objective
    @m.Objective(sense=pyo.maximize)
    def EV(m):
        return sum([scenarios[s]["probability"] * m.f[s] for s in m.S])

    # constraints
    @m.Constraint(m.S)
    def profit(m, s):
        return m.f[s] == m.r * m.y[s] + m.w * (m.x - m.y[s]) - m.c * m.x
```

```
    @m.Constraint(m.S)
    def sales_less_than_order(m, s):
        return m.y[s] <= m.x

    @m.Constraint(m.S)
    def sales_less_than_demand(m, s):
        return m.y[s] <= scenarios[s]["demand"]

    return m


m = PopUpShop(r, c, w, scenarios)
SOLVER.solve(m)
```

The optimal quantity to order turns out to be $x = 400$ units, and detailed per-scenario results of the optimization are shown in Table 9.4.

Table 9.4 *Performance of the stochastic solution in the pop-up shop problem.*

Scenario (s)	Demand (d_s)	Probability (p_s)	Sold (y_s)	Salvage ($x - y_s$)	Profit (f_s)
sunny skies	650	0.30	400	0	$11,200
good weather	400	0.60	400	0	$11,200
poor weather	200	0.10	200	200	$3,600
Average profit					$8,920

Optimizing over all scenarios provides an expected profit of $8920, an increase of $581 (or 7%) over the naive strategy of simply ordering the expected number of items sold. The new optimal solution places a larger order, that is, $x = 400$. In poor weather conditions, there will be more returns and lower profits that are more than compensated for by increased profits in good weather conditions.

The additional value that results from solving this planning problem is called the **value of the stochastic solution (VSS)**. The VSS is the additional profit compared to ordering to simply meet the expected demand. In this case,

$$\text{VSS} = \text{EV} - \text{EVM} = 8920 - 8339 = \$581.$$

9.2.2 Sample Average Approximation

If $\mathbb{E}f(\boldsymbol{x}, \boldsymbol{z})$ cannot be computed explicitly, but we have an expression for the function $f(\boldsymbol{x}, \boldsymbol{z})$ for any fixed $\boldsymbol{z}$, then a strategy is to construct a finite random i.i.d. sample $\boldsymbol{z}^1, \ldots, \boldsymbol{z}^N$ of size N and use it to approximate the objective function as

$$\mathbb{E}f(\boldsymbol{x}, \boldsymbol{z}) \approx \bar{f}_N(\boldsymbol{x}) := \frac{1}{N} \sum_{j=1}^{N} f(\boldsymbol{x}, \boldsymbol{z}^j).$$

This means we approximate the optimal solution of the original problem $\min\limits_{x\in X} \mathbb{E}f(x,z)$ with that of the following problem:

$$\min_{x\in X} \mathbb{E}f(x,z) \approx \min_{x\in X} \bar{f}_N(x) = \frac{1}{N}\sum_{j=1}^{N} f(x,z^j).$$

This approach is known as the **sample average approximation (SAA)**. SAA usually works very well whenever reasonably large samples (compared to the dimensionality of z) are available. On small samples, however, this approach runs the risk over **overfitting** the decision to the specific sample at hand, instead of the entire distribution $\mathbb{P}$. In other words, the solution might perform well on-sample but not so well out-of-sample.

Example 9.5 (Stock optimization for seafood distribution center part 2) Consider again the problem illustrated in Example 9.2 and assume further that the demand for tuna in tons can be modeled as a continuous random variable z with cumulative distribution function $F(\cdot)$. We consider the following three distributions, visualized in Figure 9.1:

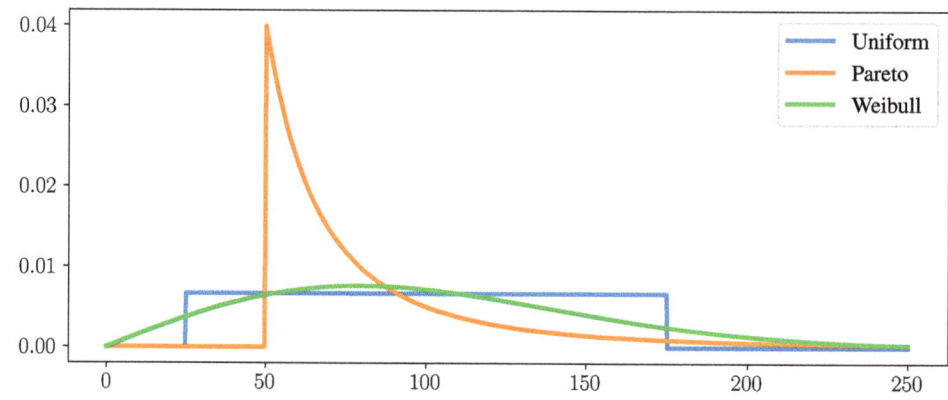

Figure 9.1 The PDFs for the three different demand distributions.

1. a uniform distribution in the interval $[25, 175]$;
2. a Pareto distribution on the interval $[50, +\infty)$ with $x_m = 50$ and exponent $\alpha = 2$;
3. a Weibull distribution on the interval $[0, +\infty)$ with shape parameter $k = 2$ and scale parameter $\lambda = 112.84$.

Note that all three distributions above have the same expected value, that is, $\mathbb{E}z = 100$ tons. We find the optimal solution of the seafood inventory problem using the explicit formula that features the inverse cumulative distribution functions (CDFs)/quantile functions F for the three distributions using the closed-form formula

$$x^* = F^{-1}\left(\frac{p-c}{p+h}\right).$$

If we take $c = 10$, $p = 25$, $h = 3$ as values for the model parameters, the quantile of interest is then $q = \frac{p-c}{p+h} \approx 0.5357$. Then we can deduce that the optimal order size is $x^* = 105.36$ for the uniform distribution, $x^* = 73.38$ for the Pareto distribution, and $x^* = 98.84$ for the Weibull distribution. The relationship between the quantile q and the three optimal solutions is shown in Figure 9.2.

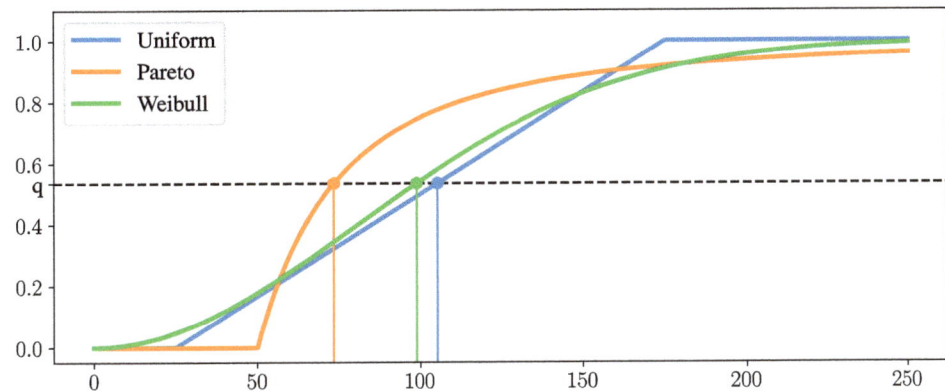

Figure 9.2 The CDFs for the three different demand distributions and the corresponding optimal solutions for $q \approx 0.5357$.

Ignoring the closed-form solution, we can approximate the optimal solution by sampling $N = 5000$ scenarios for each of the three distributions, see the following Python code.

```python
def SeafoodStockSAA(N, sample):
    model = pyo.ConcreteModel("Seafood Stock using SAA method")

    def indices_rule(model):
        return range(N)

    model.indices = pyo.Set(initialize=indices_rule)
    model.xi = pyo.Param(model.indices, initialize=dict(enumerate(sample)))

    # first stage variable: x (amount of fish bought)
    model.x = pyo.Var(domain=pyo.NonNegativeReals)

    def first_stage_profit(model):
        return -c * model.x

    model.first_stage_profit = pyo.Expression(rule=first_stage_profit)

    # second stage variables: y (sold) and z (unsold)
    model.y = pyo.Var(model.indices, domain=pyo.NonNegativeReals)   # sold
    model.z = pyo.Var(
        model.indices, domain=pyo.NonNegativeReals
    )   # unsold to be returned

    # second stage constraints
    model.cantsoldfishidonthave = pyo.ConstraintList()
    model.fishdonotdisappear = pyo.ConstraintList()
```

```
    for i in model.indices:
        model.cantsoldfishidonthave.add(expr=model.y[i] <= model.xi[i])
        model.fishdonotdisappear.add(expr=model.y[i] + model.z[i] ==
        ↪    model.x)

    def second_stage_profit(model):
        return sum([p * model.y[i] - h * model.z[i] for i in model.indices])
        ↪    / float(N)

    model.second_stage_profit = pyo.Expression(rule=second_stage_profit)

    def total_profit(model):
        return model.first_stage_profit + model.second_stage_profit

    model.total_expected_profit = pyo.Objective(rule=total_profit,
    ↪    sense=pyo.maximize)

    return model


np.random.seed(1)
N = 5000

samples = np.random.uniform(low=25.0, high=175.0, size=N)
model = SeafoodStockSAA(N, samples)
SOLVER.solve(model)

shape = 2
xm = 50
samples = (np.random.pareto(a=shape, size=N) + 1) * xm
model = SeafoodStockSAA(N, samples)
SOLVER.solve(model)

shape = 2
scale = 113
samples = scale * np.random.weibull(a=shape, size=N)
model = SeafoodStockSAA(N, samples)
SOLVER.solve(model)
```

We obtain the following approximate solutions: $x^* = 105.82$ for the uniform distribution, $x^* = 73.33$ for the Pareto distribution, and $x^* = 100.06$ for the Weibull distribution.

We can also be interested in how fast the optimal solution converges to the "true optimum" as the sample size increases in the SAA approach (and therefore, to what extent should we not trust SAA solutions based on too small samples). In Figure 9.3, we show that the solution tends to stabilize with a sample size $N < 500$.

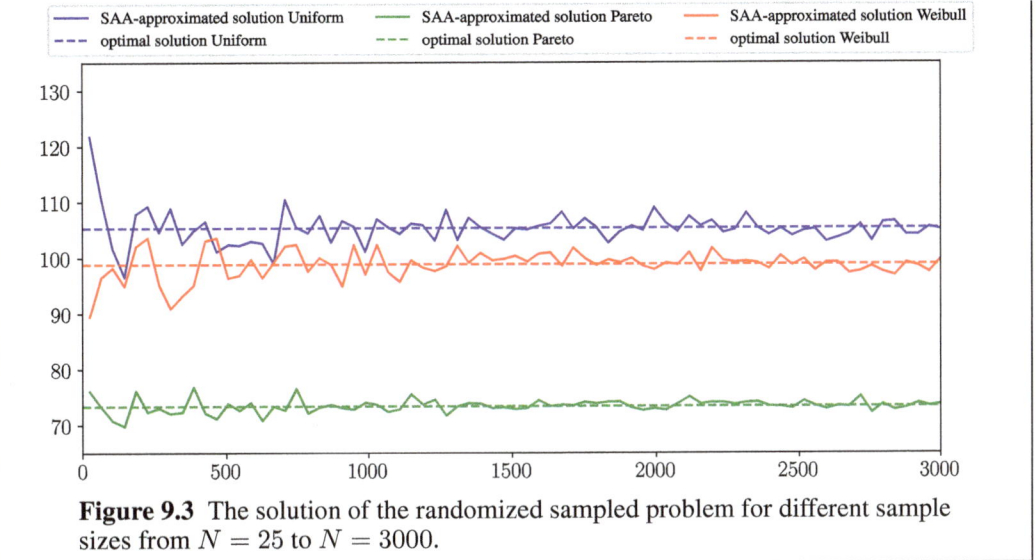

Figure 9.3 The solution of the randomized sampled problem for different sample sizes from $N = 25$ to $N = 3000$.

9.3 Probabilities

9.3.1 Basics and Modeling Choices

We now move on to the case where, instead of the objective function, we focus on the constraints. In particular, the goal is to ensure that the constraints hold with a certain probability. Consider the following example.

Example 9.6 (Blending problem part 1) We need to combine two ingredients (in amounts x_1 and x_2) in a cost-efficient way to obtain a sufficiently nutritious mix: In particular, the mix should contain 7 kg calcium and 4 kg protein. The amounts of calcium and protein in 1 kg of the second ingredient are deterministic and both equal to 1. However, we do not exactly know the nutritional value of the first ingredient, so we introduce two parameters z_1 and z_2, which we should treat as random. The corresponding LO problem is

$$
\begin{aligned}
\min \quad & x_1 + x_2 \\
\text{s.t.} \quad & z_1 x_1 + x_2 \geq 7 \\
& z_2 x_1 + x_2 \geq 4 \\
& x_1, x_2 \geq 0.
\end{aligned}
$$

An approach to solving this problem would be to specify the maximum acceptable level of "risk" that we are willing to take using **chance constraints**. In other words, we require that the constraints are satisfied *with a high probability*, greater than a threshold that we can choose. Introducing three threshold values $\varepsilon, \varepsilon_1, \varepsilon_2 \in (0,1)$ (but to be taken small), we could account for this requirement either *individually*, obtaining the problem

$$\min \quad x_1 + x_2 \tag{9.2}$$
$$\text{s.t.} \quad \mathbb{P}(z_1 x_1 + x_2 \geq 7) \geq 1 - \varepsilon_1$$
$$\mathbb{P}(z_2 x_1 + x_2 \geq 4) \geq 1 - \varepsilon_2$$
$$x_1, x_2 \geq 0.$$

or, *jointly*,

$$\min \quad x_1 + x_2 \tag{9.3}$$
$$\text{s.t.} \quad \mathbb{P}(z_1 x_1 + x_2 \geq 7, z_2 x_1 + x_2 \geq 4) \geq 1 - \varepsilon$$
$$x_1, x_2 \geq 0.$$

The two design choices lead to a different-looking feasible region for the problem, as illustrated in Figure 9.4.

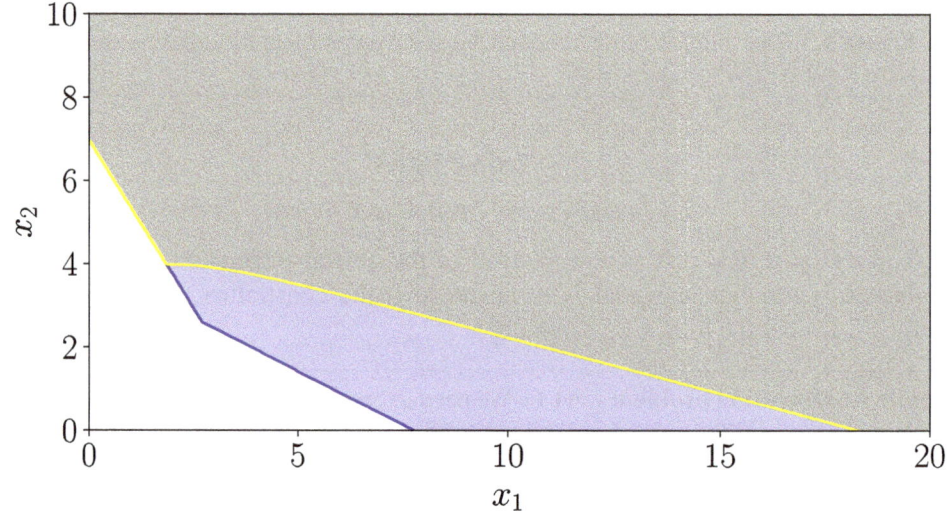

Figure 9.4 The feasible regions of problem (9.3) (in yellow) and that of problem (9.2) (in blue) corresponding to the same threshold value $\varepsilon = \varepsilon_1 = \varepsilon_2 = 0.7$ and assuming two independent χ^2 distributions for the uncertain coefficients z_1, z_2 with parameters $\nu_1 = 1.4$ and $\nu_2 = 0.6$, respectively.

In particular, the individual chance constraints could result in a larger (i.e., more optimistic) feasible region than that of a joint chance constraint; see Figure 9.4 for an example.

Example 9.6 highlights the key issue when aiming for probabilistic guarantees on the constraints $A(z)x - b(z) \leq 0$ the choice between going for:

- a collection of **individual chance constraints**

$$\mathbb{P}\left(a_1(z)^\top x - b_1(z) \leq 0\right) \geq 1 - \varepsilon_1$$
$$\vdots$$
$$\mathbb{P}\left(a_m(z)^\top x - b_m(z) \leq 0\right) \geq 1 - \varepsilon_m;$$

- a **joint chance constraint** of the form

$$\mathbb{P}\big(a_i(\boldsymbol{z})^\top \boldsymbol{x} - b_i(\boldsymbol{z}) \leq 0, \quad \forall\, i = 1, \ldots, m\big) \geq 1 - \varepsilon,$$

for some $\varepsilon \in (0,1)$.

There are two aspects to this choice: (i) the suitability to a given application context and (ii) the computational complexity of the resulting constraints.

If the rows of $A(\boldsymbol{z})\boldsymbol{x} - \boldsymbol{b}(\boldsymbol{z})$ *together* describe a precise reliability target, the joint option seems to be appropriate. If the constraints represent different and uncorrelated goals, it is more appropriate to consider them separately. In the latter case, we have more flexibility to differ the threshold values $\varepsilon_1, \ldots, \varepsilon_m$, possibly ranking the priority level of various constraints.

Computationally, it is easier to deal with individual chance constraints than with joint chance constraints (although none is really simple apart from a few specific cases). For that reason, even in situations where joint chance constraints would be more appropriate, it is often preferred to use individual chance constraints.

In what follows, we will introduce various ways of approaching chance constraints depending on the structure of the optimization problem at hand:

- Exact approaches for special cases in which the joint/individual constraints can be reformulated to their exact equivalents that can be solved using off-the-shelf software:

  - a MILO reformulation of individual and joint chance constraints in case of a discrete distribution of $\boldsymbol{z}$;
  - a list of situations when individual and joint chance constraints with continuous $\boldsymbol{z}$ distributions correspond to a convex optimization problem.

- An SAA-like approach to enforce the original constraints to hold for a sufficiently large number of sampled scenarios $\boldsymbol{z}$.

9.3.2 A MILO Reformulation for Discrete Distributions

Suppose that the distribution of $\boldsymbol{z}$ is known to be discrete, with probability distribution $\mathbb{P}(\boldsymbol{z} = \boldsymbol{z}^j) = p_j$, for $j = 1, \ldots, N$. In this case, if we want a generic individual chance constraint $a(\boldsymbol{z})^\top \boldsymbol{x} - b(\boldsymbol{z}) \leq 0$ to be satisfied with probability at least $1 - \varepsilon$, we can construct a MILO reformulation using the big-M technique as follows:

$$\begin{cases} a(\boldsymbol{z}^j)^\top \boldsymbol{x} - b(\boldsymbol{z}^j) \leq M u_j, & j = 1, \ldots, N \\ \sum_{j=1}^{N} p_j u_j \leq \varepsilon & \\ u_j \in \mathbb{B}, & j = 1, \ldots, N, \end{cases} \tag{9.4}$$

where M is a "sufficiently big" number. The idea is that each binary variable u_j determines whether the constraint related to sample $\boldsymbol{z}^j$ is allowed to be violated, by increasing the right-hand side by a number M. The second constraint in (9.4) imposes that the total probability mass of the violated constraints is not larger than ε.

The benefits of this approach are clear – it is simple. There are obviously also drawbacks – N new binary variables are added to the problem, which then becomes a MILO and possibly hard to solve.

The complete example at the end of this chapter will make use of this method to solve an energy dispatch problem that uses a finite sample of wind energy realizations.

9.3.3 Log-Concavity and Convexity for Continuous Distributions

Apart from the case of discrete probabilities leading to MILO reformulations, there can be general continuous probability distributions for which the chance-constrained problem is a convex optimization problem, or, more specifically, a linear or second-order conic optimization problem. The road to such special cases leads to considering a special class of probability distributions with **log-concave** density or cumulative density functions.

Definition 9.1 A function $f \colon \mathbb{R}^m \to \mathbb{R}$ is said to be **log-concave** if for every $x, y \in \mathbb{R}^m$ and every $0 < \theta < 1$ it holds that

$$f(\theta x + (1 - \theta)y) \geq f(x)^\theta f(y)^{1-\theta},$$

or, equivalently, if $\log f(x)$ is a concave function of x, that is for every $x, y \in \mathbb{R}^m$ and every $0 < \theta < 1$

$$\log f(\theta x + (1 - \theta)y) \geq \theta \log f(x) + (1 - \theta) \log f(y).$$

Although the definition might feel unfamiliar, many common probability distributions have log-concave densities, including:

- the standard normal and multivariate normal distributions;
- the exponential distribution;
- the uniform distribution over any convex set;
- the logistic distribution;
- the Weibull distribution with shape parameter ≥ 1.

It can be proven that the CDF $F(\cdot)$ of distributions with log-concave density $f(\cdot)$ is also log-concave. However, there exist some distributions with nonlog-concave densities that nonetheless have log-concave CDFs, for instance:

- the log-normal distribution;
- the Pareto distribution;
- the Weibull distribution with shape parameter < 1.

A possible strategy to prove the log-concavity of a CDF F of an m-dimensional distribution is to consider the corresponding probability density $h(\cdot)$ and check whether either of the following conditions holds:

- $\log f(\cdot)$ is concave (with $\log 0 = -\infty$), or
- $f(\cdot)^{-1/m}$ is convex (with $0^{-1/m} = \infty$).

We shall now distinguish a few important special cases of settings in which the corresponding feasible set

$$C(\varepsilon) := \{x \in \mathbb{R}^n \; : \; \mathbb{P}(A(z)x - b(z) \le 0) \ge 1 - \varepsilon\}$$

is convex (a minimum requirement) and/or can be expressed using convenient convex constraints (which would be better). These situations will depend, in particular, on the number of rows of $b(z), A(z)$ (individual or joint chance constraints), the form of $\mathbb{P}$, and the form of $b(z)$ and $A(z)$. A summary of these cases is provided in Table 9.5.

Case 1. Assume z is a one-dimensional random variable with CDF $F(\cdot)$. If $b(z) = 0$ and $A(z) = z + b(x)$, with $b(x)$ a scalar function of x, the chance constraint $\mathbb{P}(z \cdot x + b(x) \le 0) \ge 1 - \varepsilon$ can be equivalently rewritten as the following constraint:

$$F^{-1}(1 - \varepsilon)x + b(x) \le 0.$$

The convexity of $C(\varepsilon)$ can then be checked directly by inspecting this newly obtained function and, in particular, it always holds when $b(x)$ is a linear function of x.

Case 2. Assume that z is a one-dimensional random variable with CDF $F(\cdot)$. If $A(z) = a$ is a constant vector and $b(z) = z$, then $C(\varepsilon) = \{x \in \mathbb{R}^n \; : \; \mathbb{P}(a^\top x - z \le 0) \ge 1 - \varepsilon\}$ is a closed convex set for all $\varepsilon \in [0,1]$ and can be equivalently rewritten as the following linear constraint:

$$a^\top x \le F^{-1}(\varepsilon).$$

Case 3. Assume that z is an m-dimensional random vector with multivariate **log-concave** CDF $F(\cdot)$. If $A(z) = A$ is a constant $m \times n$ matrix and $b(z) = z$, then $C(\varepsilon) = \{x \in \mathbb{R}^n \; : \; \mathbb{P}(Ax \le z) \ge 1 - \varepsilon\}$ is a closed convex set for all $\varepsilon \in [0,1]$. Note that in this case there is no equivalent deterministic reformulation of the chance constraint that holds in general.

Case 4. Assume that z is an m-dimensional random vector with multivariate normal distribution $\mathcal{N}(\mu, \Sigma)$, $A(z) = z$, and $b(z) = b$. Let $\Phi^{-1}(\alpha)$ be the α-quantile of the standard normal distribution. Then, the chance constraint $\mathbb{P}(z^\top x \ge b) \ge 1 - \varepsilon$ can be equivalently rewritten as

$$\mu^\top x \ge b + \Phi^{-1}(1 - \varepsilon)\|\Sigma^{1/2}x\|_2.$$

Moreover, if $\varepsilon \le 1/2$, this constraint (and thus the set $C(\varepsilon)$) is convex and can be represented using the second-order cone, see Section 6.2.

Case 5. Assume z is an m-dimensional random vector with multivariate normal distribution $\mathcal{N}(\mu, \Sigma)$, $A(z) = (z_1, \ldots, z_{m-1})$, and $b(z) = z_m$. Then, the set $C(\varepsilon) = \{x \in$

Table 9.5 *Summary of the tractable examples of chance constraints.*

Case	Individual/joint	Conditions $b(z)$, $A(z)$	$\dim(z)$	Distribution type	Summary
1	Individual	$A(z) = z + b(x)$	1	Invertible CDF	Linear constraint
2	Individual	$A(z) = a$, $b(z) = z$	1	Invertible CDF	Linear constraint
3	Joint	Constant $A(z) = A$	Any	Log-concave CDF	Convex feasible set
4	Individual	$A(z) = z$	Any	Multivariate normal	SOCP constraint
5	Individual	$A(z) = (z_1, \ldots, z_{m-1})$, $b(z) = z_m$	Any	Multivariate normal	SOCP constraint

$\mathbb{R}^{m-1}$: $\mathbb{P}(A(z)x \geq b(z)) \geq 1 - \varepsilon\}$ is convex if $\varepsilon \leq 1/2$ and can be equivalently rewritten as

$$\mu^\top y \geq \Phi^{-1}(1 - \varepsilon)\|\Sigma^{1/2}y\|_2,$$

where $y_i = x_i$ for $i = 1, \ldots, m - 1$ and $y_m = -1$. As long as $\varepsilon \leq 0$, the result is convex and can be represented using the second-order cone; see Section 6.2.

To conclude this section, we revisit the two versions of the chance-constrained blending problem we introduced in (9.2) and (9.3).

Example 9.7 (Blending problem part 2) First, consider the case of individual constraints, that is, (9.2). If $x_1 > 0$ and $\varepsilon_1, \varepsilon_2 > 0$, we can use the method illustrated in Case 1 to deduce that the two individual chance constraints

$$\mathbb{P}(z_1 x_1 + x_2 \geq 7) \geq 1 - \varepsilon_1, \quad \text{and} \quad \mathbb{P}(z_2 x_1 + x_2 \geq 4) \geq 1 - \varepsilon_2,$$

can be rewritten as linear (hence convex) constraints in x_1, x_2, respectively, as

$$F_1^{-1}(\varepsilon_1)x_1 + x_2 \geq 7, \quad \text{and} \quad F_2^{-1}(\varepsilon_2)x_1 + x_2 \geq 4,$$

where F_1 and F_2 are the CDFs of z_1 and z_2.

As far as the joint constraint in (9.3) is concerned, if we assume that z_1 and z_2 are independent random variables, then we can factorize the probability as follows:

$$
\begin{aligned}
&\mathbb{P}(z_1 x_1 + x_2 \geq 7, z_2 x_1 + x_2 \geq 4) \\
&= \mathbb{P}(z_1 x_1 + x_2 \geq 7) \cdot \mathbb{P}(z_2 x_1 + x_2 \geq 4) \\
&= \begin{cases} \left(1 - F_1\left(\frac{7-x_2}{x_1}\right)\right) \cdot \left(1 - F_2\left(\frac{4-x_2}{x_1}\right)\right), & \text{if } x_1 > 0 \\ 1, & \text{if } x_1 = 0,\, x_2 \geq 7 \\ 0, & \text{if } x_1 = 0,\, 0 \leq x_2 < 7, \end{cases}
\end{aligned}
$$

which may result in a nonlinear constraint in x_1 and x_2, depending on the CDFs F_1 and F_2.

9.3.4 SAA-Like Approach: Sampling

One last, simple idea is to sample many scenarios of z and require the solution to meet the constraints in each of them. Intuitively, for a large enough sample, the solution should satisfy the constraints with high probability.

This approach is quite intuitive, but it is not clear how many scenarios should be sampled to guarantee that the obtained solution is feasible with a sufficiently high probability. Consider the general problem

$$
\begin{aligned}
\min \quad & c^\top x \\
\text{s.t.} \quad & \mathbb{P}\left((a_i + A_i z)^\top x \leq b_i + b_i^\top z,\, \forall\, i = 1, \ldots, m\right) \leq \varepsilon \\
& x \in \mathbb{R}^n,
\end{aligned}
\tag{9.5}
$$

where X is a convex and closed set.

The sampling-based approach works as follows: We draw N i.i.d. samples $z^1, \ldots, z^N$ from the true distribution $\mathbb{P}$ and define the **randomized sampled problem** as

$$
\begin{aligned}
\min \quad & c^\top x & (9.6) \\
\text{s.t.} \quad & (a_i + A_i z^j)^\top x \le b_i + b_i^\top z^j, \quad \forall i = 1, \ldots, m, \quad \forall j = 1, \ldots, N \\
& x \in \mathbb{R}^n.
\end{aligned}
$$

There exist theoretical results [DV04] that provide the exact N necessary for (9.6) to meet the constraints of the original problem with a sufficiently high probability. Such a theoretically guaranteed N tends, however, to be rather large, and therefore, in a realistic application, a fine-tuning of the number of samples would be required.

9.4 Risk Measures

So far, we have been considering two approaches – expectations and probabilities of constraint violations – that represent either the "reward" or "risk" side of a stochastic optimization problem. However, there exists a tool that tries to combine these two into a single measure to optimize: a **risk measure**. We will now briefly illustrate some possible variants and their implications, starting from the seafood cold warehouse problem.

Example 9.8 (Stock optimization for seafood distribution center part 3) In Example 9.2, the seafood distribution center aims to optimize its profit *on average*. However, for a particular realization of the demand z, on a particular day, the profit $f(x^*, z)$ could be very different from the corresponding expected value. This may very well happen if z has a large variability, which then makes also the profit $f(x^*, z)$ highly variable. We can measure the extent of this variability, for example, considering the variance of the profit $\text{Var}[f(x,z)]$. If the seafood distribution center wants to hedge against such variability, she can consider a slightly different optimization problem, namely

$$
\max_{x \ge 0} \mathbb{E}_z f(x,z) - \kappa \text{Var}(f(x,z)),
$$

where κ is the **risk aversion parameter**: The larger κ, the more the above optimization problem tries to find a solution with minimal profit variance. However, note that the additional variance term in the objective function destroys the convexity of the optimization problem. Other drawbacks for this choice are: (a) the variance is symmetric and thus positive deviations have the same weight as negative deviations (which in many applications is not desirable) and (b) the parameter κ is scale-dependent.

More generally, we can consider any **risk measure** ρ, that is, a function that maps a random variable/vector to a real number. A risk measure usually combines the expected value $\mathbb{E}f(x,z)$ with another scalar term, called **dispersion measure**, which measures the uncertainty of the outcome. The idea is to then use the chosen risk measure ρ and take $\rho(f(x,z))$ as the objective function to obtain a more risk-aware optimal decision. The most canonical choices for risk measures are listed next:

- $\rho(Z) = \mathbb{E}(Z)$.
- $\rho(Z) = \mathbb{E}(Z) - \kappa \text{Var}(Z)$, which is the **mean-variance model**.
- $\rho(Z) = \mathbb{E}(Z) - \kappa\sqrt{\text{Var}(Z)}$, which is the **mean-standard deviation model**.
- $\rho(Z) = \mathbb{E}(Z) - \kappa \text{VaR}_\alpha(Z)$, where $\text{VaR}_\alpha(Z)$ is the **value-at-risk (VaR)**, defined as

$$\text{VaR}_\alpha(Z) := \inf\{t \ : \ \mathbb{P}(Z \le t) \ge \alpha\}.$$

Note that $\text{VaR}_\alpha(Z) \le \varepsilon$ is just another way to write $\mathbb{P}(Z \le \varepsilon) \ge \alpha$.
- $\rho(Z) = \text{CVaR}_\alpha(Z)$, where $\text{CVaR}_\alpha(Z)$ is the **conditional value-at-risk (CVaR)**, defined as

$$\text{CVaR}_\alpha(Z) = \mathbb{E}(Z \mid Z \ge \text{VaR}_\alpha(Z)).$$

As can be seen, because CVaR is the "average" value of outcomes greater than or equal to 0, it is an upper bound on the VaR. CVaR is computationally much easier than VaR because it can be computed alternatively as

$$\text{CVaR}_\alpha(Z) = \min_\beta \beta + \frac{1}{1-\alpha}\mathbb{E}\max\left\{x^\top z - \beta, 0\right\}.$$

Therefore, an upper bound on the CVaR can be implemented as a constraint on

$$\beta + \frac{1}{1-\alpha}\mathbb{E}\max\left\{x^\top z - \beta, 0\right\},$$

with β as an additional variable. This is because the variable β will automatically adjust itself to the form that corresponds to the minimum in CVaR's definition. The above expectation lends itself to an LO reformulation as a system of linear constraints in case the SAA method is used to approximate the expectation. Because the CVaR is an upper bound on the VaR, imposing constraints on the CVaR is often used as a "proxy" method to constrain the VaR.
- $\rho(Z) = \inf_{\gamma>0} \frac{1}{\gamma}\log\mathbb{E}(\exp(\gamma Z))$, which is the **exponential risk aversion**.

Selecting the "right" function ρ depends on the modeling and optimization goals one has in mind.

Example 9.9 (Portfolio management revisited) Consider the chance-constrained portfolio problem

$$\begin{aligned} \max \quad & \mathbb{E}(x^\top z) \\ \text{s.t.} \quad & \mathbb{P}(x^\top z < 1) \le 0.2 \\ & x \in X. \end{aligned}$$

It can be written equivalently as follows:

$$\begin{aligned} \max \quad & \mathbb{E}(x^\top z) \\ \text{s.t.} \quad & \text{VaR}_{0.8}(-x^\top z) \le -1 \\ & x \in X. \end{aligned}$$

This equivalence follows from algebraic manipulations and the definition of value at risk (VaR) as an "upper" quantile of the random variable. A conservative approximation of this problem using the CVaR would then be

$$\begin{aligned} \max \quad & \mathbb{E}\big(\boldsymbol{x}^\top \boldsymbol{z}\big) \\ \text{s.t.} \quad & \mathrm{CVaR}_{0.8}\big(-\boldsymbol{x}^\top \boldsymbol{z}\big) \le -1 \\ & \boldsymbol{x} \in X. \end{aligned}$$

9.5 A Complete Example: Chance-Constrained Energy Dispatch

In this example, we will explore the applications of chance constraints to an area where high-probability guarantees on the system's functioning are required – the **economic dispatch** (ED) problem.

The problem considers the short-term determination of the optimal production of energy to meet all energy demands. Let V denote a set of nodes, each of which represents cities, industrial districts, power generators, or combinations of these. Each node $i \in V$ may have:

- a certain energy demand $d_i \ge 0$;
- a power generator whose energy production needs to be between $p_i^{\min}$ and $p_i^{\max}$ units of power. The cost of producing one unit of power at node i is given by a variable cost $c_i \ge 0$. Importantly, not all the nodes have demand and generation; more specifically, it is possible for a node to have only generation or only demand.

The goal is to determine for each node $i \in V$ the optimal production level p_i, such that:

- the total energy demand is met;
- no production limits are exceeded;
- the total energy production costs are minimized.

If we fully control the energy production and the customer demand is known, we can formulate the problem as the following MILO problem:

$$\begin{aligned} \min \quad & \sum_{i \in V} c_i p_i \\ \text{s.t.} \quad & \sum_{i \in V} p_i = \sum_{i \in V} d_i \\ & p_i^{\min} \le p_i \le p_i^{\max}, \qquad \forall i \in V. \end{aligned}$$

Now, assume that we have built several offshore wind turbines. Together, these wind turbines produce a random nonnegative amount of extra energy, denoted by ω. For a fixed value of ω, the optimization problem to be solved is thus to "fill in" to the remaining energy demand not satisfied by wind power:

$$\min \quad \sum_{i \in V} c_i p_i$$

$$\text{s.t.} \quad \omega + \sum_{i \in V} p_i = \sum_{i \in V} d_i$$

$$p_i^{\min} \le p_i \le p_i^{\max}, \qquad\qquad \forall i \in V.$$

However, the problem is that ω is a random variable and is usually not fully known before the generation levels p_i of conventional generators need to be set. Due to stochastic fluctuations in wind power generation, the ED problem is best modeled as a stochastic optimization problem. The intermittency of wind generation makes it almost impossible to perfectly balance supply and demand on a real-time basis, but in practice, there is some tolerance for error, that is, a certain degree of mismatch between supply and demand can be easily adjusted for.

To formulate the problem under this assumption, let us denote by:

- $\Delta \ge 0$ the tolerance of the absolute power mismatch between supply and demand;
- $\varepsilon \in [0,1]$ the risk level we are willing to accept for the supply to deviate from the demand more than Δ;
- ω the nonnegative random variable describing the total power production of offshore wind turbines.

In this setting, instead of requiring that supply and demand are perfectly matched, we require that the absolute difference remains below the power threshold Δ using the following chance constraint:

$$\mathbb{P}\left(\left| \omega + \sum_{i \in V} p_i - \sum_{i \in V} d_i \right| \le \Delta \right) \ge 1 - \varepsilon.$$

To formulate the problem as a MILO, we can eliminate the absolute value using two individual chance constraints – note that in this way we relax the constraint because requiring that two one-sided constraints hold with probability $1 - \varepsilon$ each is not the same as requiring that together they hold with probability $1 - \varepsilon$, but in practice we can fine-tune ε to adapt to this change. Such a split leads to the following optimization problem with chance constraints:

$$\min \quad \sum_{i \in V} c_i p_i$$

$$\text{s.t.} \quad \mathbb{P}\left(\omega + \sum_{i \in V} p_i - \sum_{i \in V} d_i \le \Delta \right) \ge 1 - \varepsilon$$

$$\mathbb{P}\left(\omega + \sum_{i \in V} p_i - \sum_{i \in V} d_i \ge -\Delta \right) \ge 1 - \varepsilon$$

$$p_i^{\min} \le p_i \le p_i^{\max}, \qquad\qquad \forall i \in V.$$

We solve this problem using the SAA approach to chance constraints, with the wind power production modeled with historical data of $N = 500$ outcomes of the total wind production. Since we have a discrete set of N possible wind production outcomes, following the method introduced in Section 9.3.2 we can reformulate the chance-constrained ED problem as a MILO problem. More specifically, we introduce a binary variable u_j for each sample j of wind production, which, thanks to the big-M technique, determines whether the constraint related to the jth sample is allowed to be violated or not, and add one constraint to ensure that the total probability that the constraint is violated is at most ε, that is,

$$\frac{1}{N}\sum_{j=1}^{N} u_j \leq \varepsilon.$$

The resulting MILO is

$$\min \quad \sum_{i\in V} c_i p_i$$

$$\text{s.t.} \quad \omega_j + \sum_{i\in V} p_i - \sum_{i\in V} d_i \leq \Delta + u_j M_j, \qquad \forall j = 1,\ldots,N$$

$$\omega_j + \sum_{i\in V} p_i - \sum_{i\in V} d_i \geq -\Delta - u_j M_j, \qquad \forall j = 1,\ldots,N$$

$$\sum_{j=1}^{N} u_j \leq \varepsilon N$$

$$p_i^{\min} \leq p_i \leq p_i^{\max}, \qquad \forall i \in V$$

$$u_j \in \mathbb{B}, \qquad \forall j = 1,\ldots,N.$$

For each sample, the supply and demand constraints are deactivated when $u_j = 1$ and $u_j = 0$ otherwise. Note that we only use one single u_j variable for each constraint. Indeed, having two separate $u_j^{(1)}$ and $u_j^{(2)}$ will yield the same objective value, but the model would be incorrect with respect to the violation of the supply and demand constraints introduced earlier.

The constant M_j here should be selected based on the data: One reasonable choice for M_j that will certainly work is to take it equal to the left-hand side minus Δ while replacing p_i for $p_i^{\max}$.

This model is implemented in Pyomo as shown below. Note that the model is formulated with ε and Δ as mutable model parameters so that we can repeatedly solve the same model instance and modify these two parameters when needed. In turn, this requires the constant M_j to be an expression.

```
def economic_dispatch(nodes, samples, eps, Delta):
    model = pyo.ConcreteModel("Economic Dispatch")

    model.n = pyo.Param(mutable=False, initialize=len(samples))
    model.eps = pyo.Param(mutable=True, initialize=eps)
    model.Delta = pyo.Param(mutable=True, initialize=Delta)

    model.N = pyo.Set(initialize=range(model.n()))
    model.nodes = pyo.Set(initialize=nodes.keys())
```

```python
    @model.Expression(model.N)
    def M(model, j):
        return (
            samples[j]
            + sum(data["p_max"] for _, data in nodes.items())
            + sum(data["d"] for _, data in nodes.items())
            - model.Delta
        )

    model.p = pyo.Var(
        nodes,
        domain=pyo.NonNegativeReals,
        bounds=lambda m, i: (nodes[i]["p_min"], nodes[i]["p_max"]),
    )
    model.u = pyo.Var(model.N, domain=pyo.Binary)

    @model.Objective(sense=pyo.minimize)
    def objective(m):
        return sum(data["c_var"] * m.p[i] for i, data in nodes.items())

    @model.Constraint(model.N)
    def supply_demand_leq(m, j):
        wind = samples[j]
        supply = sum(m.p[i] for i, data in nodes.items())
        demand = sum(data["d"] for _, data in nodes.items())
        return wind + supply - demand <= model.Delta + model.M[j] * m.u[j]

    @model.Constraint(model.N)
    def supply_demand_geq(m, j):
        wind = samples[j]
        supply = sum(m.p[i] for i, data in nodes.items())
        demand = sum(data["d"] for _, data in nodes.items())
        return wind + supply - demand >= -model.Delta - model.M[j] * m.u[j]

    @model.Constraint()
    def success_probability(m):
        return sum(m.u[j] for j in model.N) <= model.eps * model.n

    return model


# Assumes the N=500 wind power samples have already been imported,
# solve the model and report the results
eps = 0.20
Delta = 1000
model = economic_dispatch(nodes, wind_production_samples, eps, Delta)
SOLVER.solve(model)
```

For the given wind energy sample and the specified parameters, we see that the optimal energy production is 1733.39, with a cost of 7850.60.

Exercises

Exercise 9.1 A transportation company has n depots among which they send cargo. The demand for transportation between depot i and depot $j \neq i$ is modeled as a random variable z_{ij}. The total capacity of vehicles currently available at depot i is denoted by s_i, $i = 1, \ldots, n$. The company is considering repositioning its fleet to better prepare for the uncertain demand. It costs c_{ij} to move a unit of capacity from location i to location j. After repositioning, the realization of the random vector z is observed and demand is served up to the limit determined by the transportation capacity available at each location. The profit from transporting a unit of cargo from location i to location j is equal to q_{ij}. If the total demand at location i exceeds the capacity available at location i, the excess demand is lost. It is up to the company to decide how much of each demand z_{ij} will be served and which part will remain unsatisfied. For simplicity, we consider all capacity and transportation quantities as continuous variables. Formulate the problem of maximizing the expected profit as a two-stage linear stochastic optimization problem.

Exercise 9.2 This exercise illustrates that sometimes taking the average over multiple scenarios for the objective function's form can have a "convexifying" effect.

Consider an optimization problem of the form

$$\min \quad \mathbb{E}f(x, z)$$
$$\text{s.t.} \quad -1 \leq x \leq 1.$$

Assume that the uncertain parameter z has only two equiprobable scenarios:

$$\mathbb{P}(z = -1) = \mathbb{P}(z = 1) = 0.5.$$

Construct a function $f(x, z)$ that is not convex in x for any of the possible z such that the expected value function in the objective is a convex function of x.

Is the opposite possible, that is, for both $f(x, -1)$ and $f(x, 1)$ to be convex but for the expectation to not be convex?

Exercise 9.3 Consider the constraint on CVaR

$$\beta + \frac{1}{1 - \alpha} \mathbb{E} \max \left\{ x^\top z - \beta, 0 \right\} \leq t,$$

where x and β are decision variables, and z is an uncertain random vector with a known probability distribution. How would an LO reformulation of this constraint look if you use the SAA approach?

Exercise 9.4 Bonferroni's inequality is a result from probability theory frequently used in simplifying joint chance constraints. It states that the probability of a union of events is less than or equal to the sum of their individual probabilities:

$$\mathbb{P}(E_1 \cup E_2 \cup \ldots, \cup E_m) \leq \sum_{i=1}^{m} \mathbb{P}(E_m).$$

Assume that you are dealing with a joint chance constraint system:

$$\mathbb{P}\left(a_i(z)^\top x - b_i(z) \leq 0, \, i = 1, \ldots m \right) \geq 1 - \epsilon,$$

where z is known to follow a discrete probability distribution $\mathbb{P}(z = z^j) = p_j$, for $j = 1, \ldots, N$. Construct a (conservative) reformulation of this joint chance constraint leveraging the MILO reformulation of chance constraints with discrete distributions and Bonferroni's inequality. *Hint*: Introduce an extra decision variable for each inequality.

Exercise 9.5 Consider the following optimization problem:

$$\min \quad x_2 - x_1$$
$$\text{s.t.} \quad \mathbb{P}(z \geq x_1, z \leq x_2) \geq 0.8$$
$$x_1, x_2 \geq 0,$$

where z follows a probability distribution with a triangular density function $f(z)$ given by

$$f(z) = \begin{cases} 0 & \text{for } z < 0 \\ z & \text{for } 0 \leq z \leq 1 \\ 2 - z & \text{for } 1 \leq z \leq 2 \\ 0 & \text{otherwise.} \end{cases}$$

Reformulate this problem to an equivalent convex optimization problem. What is the optimal solution and optimal value? This exercise illustrates the idea of confidence intervals as the "smallest possible sets" to which the value of an unknown parameter will belong with a high probability.

Exercise 9.6 Consider a portfolio optimization problem with n assets and N possible scenarios for their returns so that the return on the ith asset in the jth scenario is denoted by r_{ij}. Each scenario has a known probability $p_j \geq 0$.

The goal is to divide the available capital using weights $x_1, \ldots, x_n \geq 0$ where $\sum_i x_i = 1$, which will yield a portfolio with a return

$$\sum_{i=1}^{N} x_i r_{ij}$$

in the jth scenario.

Formulate the optimization problem in which the goal is to minimize the portfolio return's variance. Is this problem convex?

Exercise 9.7 Consider the optimization problem

$$\min \quad \mathbb{E}(zx)$$
$$\text{s.t.} \quad -2 \leq x \leq 2,$$

where the probability distribution of z is given by

$$\mathbb{P}(z = -1) = \mathbb{P}(z = 0.5) = 0.5.$$

What is the expected performance of a solution obtained using the SAA method using sample size $N = 1$? What is the expected performance of the optimal solution obtained, taking into account the exact value of the objective function?

Exercise 9.8 Consider the linear model with random parameters inspired by the blending problem:

$$v(\boldsymbol{z}) := \min \quad x_1 + x_2$$
$$\text{s.t.} \quad z_1 x_1 + x_2 \geq 7$$
$$z_2 x_1 + x_2 \geq 4$$
$$x_1, x_2 \geq 0,$$

where $z_1 \sim \mathcal{U}[1,4]$ and $z_2 \sim \mathcal{U}[1/3,1]$.

(a) **[Explicit wait-and-see solution]** Suppose that it is possible to optimize $\boldsymbol{x}$ after observing the random vector $\boldsymbol{z} = (z_1, z_2)$. *Treating the random parameters as deterministic,* calculate the optimal solution $x_1^*(\boldsymbol{z})$ and $x_2^*(\boldsymbol{z})$ as a function of $\boldsymbol{z} = (z_1, z_2)$ and the corresponding optimal value of the problem $v(\boldsymbol{z})$. *Hint*: Use the fact that $\boldsymbol{z} \in [1,4] \times [1/3,1]$.

(b) **[Pessimistic/worst-case solution]** Using the function $v(\boldsymbol{z})$ that you calculated in (a), describe the worst case scenario, that is, the one corresponding to the value of $\boldsymbol{z}$ that returns the highest cost $v(\boldsymbol{z})$. What is the solution $\boldsymbol{x}^*(\boldsymbol{z})$ in this case?

(c) **[Optimistic/best-case solution]** Using the function $v(\boldsymbol{z})$ that you calculated in (a), describe the best-case scenario, that is, the one corresponding to the value of $\boldsymbol{z}$ that returns the highest cost $v(\boldsymbol{z})$. What is the solution $\boldsymbol{x}^*(\boldsymbol{z})$ in this case?

(d) Using the function $v(\boldsymbol{z})$ calculated in (a) and the fact that the random parameters vary uniformly in the intervals $[1,4]$ and $[1/3,1]$, calculate the expected value $\mathbb{E} v(\boldsymbol{z})$.

Exercise 9.9 Consider a normal distribution with mean μ and standard deviation σ, which has probability distribution function

$$f(x) = \frac{1}{\sigma\sqrt{2\pi}} \exp\left(-\frac{(x-\mu)^2}{2\sigma^2}\right), \qquad x \in \mathbb{R}.$$

Show that this density is log-concave in the following two ways:

(a) by using directly the definition of log-concavity and proving the corresponding inequality;
(b) by showing that $\log f(x)$ is concave. *Hint*: Check the sign of the second derivative.

Consider a Pareto distribution with exponent $\alpha > 0$ on the interval $[1, +\infty)$, which has cumulative distribution function (CDF) given by $F_\alpha(x) := 1 - x^{-\alpha}$.

(c) Show that the CDF is log-concave.
(d) Calculate the probability distribution function of the distribution by differentiating the CDF with respect to x and then prove that the probability distribution function is not log-concave. *Hint*: Check the sign of the second derivative.

Exercise 9.10

(a) Consider a semidefinite positive $n \times n$ matrix Σ and show that $\sqrt{\boldsymbol{x}^T \Sigma \boldsymbol{x}}$ is a convex function in $\boldsymbol{x}$. *Hint*: Try to rewrite the function at hand as a norm, using the fact that any semidefinite positive matrix Σ can be uniquely decomposed as $\Sigma = L^T L$, where L is also $n \times n$.

(b) Assume that $\boldsymbol{z}$ follows the multivariate normal distribution $\mathcal{N}(\mu, \Sigma)$ and let $\Phi^{-1}(\varepsilon)$ be the ε-quantile of the standard normal distribution $\mathcal{N}(0, I)$. From the theory, we know that for any $h \in \mathbb{R}$, the chance constraint $\mathbb{P}(z^T x \geq b) \geq 1 - \varepsilon$ can be equivalently rewritten as

$$b - \mu^T x + \Phi^{-1}(1 - \varepsilon)\sqrt{x^T \Sigma x} \leq 0.$$

Using (a), show that this chance constraint is convex if and only if $\varepsilon \leq 1/2$. *Hint*: Remember that the standard multivariate normal distribution is centered around zero and ask yourself what happens at $\Phi^{-1}(1 - \varepsilon)$ at $\varepsilon = 1/2$. Think about the one-dimensional case first.

10

Two-Stage Problems

We now consider problems in which the situation is not as simple as "first we make the decisions, then we observe the uncertainty and compute the costs":

$$\underset{\text{decision}}{\boldsymbol{x}} \quad \longrightarrow \quad \underset{\text{uncertain parameters observed}}{\boldsymbol{z}}.$$

Instead, we will take into account the possibility that once uncertainty is observed, some decisions can be made or adjusted. The simplest is a two-stage sequence:

$$\underset{\text{decision}}{\boldsymbol{x}} \quad \longrightarrow \quad \underset{\text{uncertain parameters observed}}{\boldsymbol{z}} \quad \longrightarrow \quad \underset{\text{decision}}{\boldsymbol{y}}.$$

It is possible that for a given $\boldsymbol{x}$, different optimal $\boldsymbol{y}$s can correspond to different outcomes of $\boldsymbol{z}$. Therefore, ideally, we optimize $\boldsymbol{x}$, taking into account all possible outcomes of $\boldsymbol{z}$ *and* the corresponding optimal decisions $\boldsymbol{y}$ that would be responses to $\boldsymbol{z}$.

The last statement is a bit vague and we need to formulate it more precisely. Sticking with the world of MILO problems outlined already in Chapter 7, we formulate a two-stage optimization problem as

$$\begin{aligned} \min \quad & \boldsymbol{c}^\top \boldsymbol{x} + \Box\, Q(\boldsymbol{x}, \boldsymbol{z}) & (10.1) \\ \text{s.t.} \quad & A\boldsymbol{x} \leq \boldsymbol{b} \\ & \boldsymbol{x} \in \mathbb{R}^{n_x}, \end{aligned}$$

known as the first-stage problem, where the **second-stage cost function** or **recourse function** $Q(\boldsymbol{x}, \boldsymbol{z})$ is formulated as the optimal value of the optimization problem solved after $\boldsymbol{z}$ becomes known:

$$\begin{aligned} Q(\boldsymbol{x}, \boldsymbol{z}) = \min \quad & \boldsymbol{q}^\top \boldsymbol{y} & (10.2) \\ \text{s.t.} \quad & R(\boldsymbol{z})\boldsymbol{x} + S(\boldsymbol{z})\boldsymbol{y} \leq \boldsymbol{t}(\boldsymbol{z}) \\ & \boldsymbol{y} \in \mathbb{R}^{n_y}. \end{aligned}$$

By convention, we define $Q(\boldsymbol{x}, \boldsymbol{z}) = +\infty$ if the second-stage problem is not feasible. In this way, if such a solution $\boldsymbol{x}$ exists, the first-stage problem is incentivized to select a solution $\boldsymbol{x}$ for which there exists a feasible second-stage decision $\boldsymbol{y}$ regardless of the outcome of $\boldsymbol{z}$. We assume that the constraints in the second-stage problem consist of m linear constraints

$$\boldsymbol{r}_j(\boldsymbol{z})^\top \boldsymbol{x} + \boldsymbol{s}_j(\boldsymbol{z})^\top \boldsymbol{y} \leq t_j(\boldsymbol{z}), \quad j = 1, \ldots, m,$$

where $\boldsymbol{r}_j(\boldsymbol{z}), \boldsymbol{s}_j(\boldsymbol{z}), t_j(\boldsymbol{z})$ depend linearly on $\boldsymbol{z}$ as in Chapter 7.

Two important remarks are in place regarding formulations (10.1)–(10.2):

- There is a $\square$ next to the second-stage outcome. The role of this is to give us a free choice as to which mindset (out of the two presented in Chapters 8 and 9) we want to apply:

  - *Robust optimization* where we care about the worst-case realization of the second-stage cost function over uncertainty set Z. In that case, the objective function of the first-stage problem becomes

  $$c^\top x + \sup_{z \in Z} Q(x, z).$$

  - *Stochastic optimization* where we care about the expected value of the second-stage cost across all possible realizations. In that case, the objective function of the first-stage problem becomes

  $$c^\top x + \mathbb{E}Q(x, z).$$

  For this, we need, of course, some information on the probability distribution of z. Differently from Chapter 9, we do not consider here the option of a "probability-like" expression, that is, a probability that the second-stage cost is above a certain value or a "risk-measure-like" expression. Although, in theory, these are all valid choices, in practice, they would make the problem too difficult to solve, and therefore, expected values are the most common choice.

- In the first-stage problem, we keep the constraints $Ax \leq b$ free of uncertainty – we assume here that even if A and b depend on uncertainty z, that has been resolved in a satisfactory way using the techniques of Chapters 8 and 9.

We will now consider a classic example of airline seat allocation by [BL11] that will help us understand the different modeling choices between single- and two-stage optimization under different types of uncertainty. Although not everything we shall discuss in this example falls exactly into the "pure" forms of two-stage optimization problems as above, it will help us to make the point that accommodating for uncertainty will typically help to make better first-stage decisions.

Example 10.1 (Airline seat allocation with finitely many demand scenarios) The following problem is adapted from an exercise presented by [BL11]. The adaptations include a change in parameters for consistency among reformulations of the problem and additional treatments for the **sample average approximation (SAA)** with chance constraints.

An airline is deciding how to allocate seats on a new plane for the Amsterdam–Buenos Aires route. This plane can seat 200 economy-class passengers. A section can be created for first-class seats, but each of these seats takes the space of two economy-class seats. A business class section can also be created, but each of these takes the space of 1.5 economy class seats. The profit for a first-class ticket is three times the profit of an economy ticket, while a business-class ticket has a profit of two times an economy ticket's profit. Once the plane is partitioned into these seating classes, it cannot be changed.

The airline knows that the plane will not always be full in every section. The airline has initially identified three scenarios to consider with about equal frequency:

1. weekday morning and evening traffic;
2. weekend traffic;
3. weekday midday traffic.

In Scenario 1, the airline thinks they can sell as many as 20 first-class tickets, 50 business-class tickets, and 200 economy tickets. Under Scenario 2, these figures are 10, 24, and 175, while under Scenario 3, they are 6, 10, and 150, respectively. Table 10.1 summarizes the forecast demand for these three scenarios.

Table 10.1 *Seat demand by scenario.*

Scenario	First-class seats	Business-class seats	Economy seats
(1) weekday morning/evening	20	50	200
(2) weekend	10	24	175
(3) weekday midday	6	10	150
Averages	**12**	**28**	**175**

The goal of the airline is to maximize ticket revenue. For marketing purposes, the airline will not sell more tickets than seats in each section (hence, no overbooking strategy). We further assume that customers seeking a first-class or business-class seat will not be downgraded if those seats are unavailable.

Model 1. Deterministic Solution for the Average Demand Scenario
A common starting point in stochastic optimization is to solve the deterministic problem where future demands are fixed at their mean values and compute the corresponding optimal solution. The resulting value of the objective has been called the *expectation of the expected value problem* (EEV) by Birge in [BL11] or the *expected value of the mean* (EVM) solution by others.

Let us introduce the set C of the three possible classes, that is, $C = \{F, B, E\}$. The objective function is to maximize ticket revenue:

$$\max_{s_c, t_c} \sum_{c \in C} r_c t_c,$$

where r_c is the revenue from selling a ticket for a seat in class $c \in C$.

Let s_c denote the number of seats of class $c \in C$ installed in the new plane. Let f_c be the scale factor that denotes the number of economy seats displaced by one seat in class $c \in C$. Then, since there is a total of 200 economy-class seats that could fit in the plane, the capacity constraint reads as

$$\sum_{c \in C} f_c s_c \leq 200.$$

Let μ_c be the mean demand for seats of class $c \in C$, and let t_c be the number of tickets sold in class $c \in C$. To ensure we do not sell more tickets than available seats nor more than demand, we need to add two more constraints:

$$t_c \leq s_c, \qquad \forall c \in C$$
$$t_c \leq \mu_c, \qquad \forall c \in C.$$

Finally, both ticket and seat variables need to be nonnegative integers, so we add the constraints $t, s \in \mathbb{Z}_+$. The model can be implemented in Pyomo as follows:

```python
# scenario data
demand = pd.DataFrame(
    {
        "morning and evening": {"F": 20, "B": 50, "E": 200},
        "weekend": {"F": 10, "B": 24, "E": 175},
        "midday": {"F": 6, "B": 10, "E": 150},
    }
).T


# global revenue and seat factor data
capacity = 200
revenue_factor = pd.Series({"F": 3.0, "B": 2.0, "E": 1.0})
seat_factor = pd.Series({"F": 2.0, "B": 1.5, "E": 1.0})



def aircraft_deterministic(demand):
    m = pyo.ConcreteModel("Aircraft seat allocation - Deterministic")

    m.CLASSES = pyo.Set(initialize=demand.columns)

    # first stage variables and constraints
    m.seats = pyo.Var(m.CLASSES, domain=pyo.NonNegativeIntegers)

    @m.Constraint(m.CLASSES)
    def plane_seats(m, c):
        return sum(m.seats[c] * seat_factor[c] for c in m.CLASSES) <=
        ↪    capacity

    # second stage variable and constraints
    m.tickets = pyo.Var(m.CLASSES, domain=pyo.NonNegativeIntegers)

    @m.Constraint(m.CLASSES)
    def demand_limits(m, c):
        return m.tickets[c] <= demand[c].mean()

    @m.Constraint(m.CLASSES)
    def seat_limits(m, c):
        return m.tickets[c] <= m.seats[c]

    # objective
    @m.Objective(sense=pyo.maximize)
    def revenue(m):
        return sum(m.tickets[c] * revenue_factor[c] for c in m.CLASSES)
```

```
    return m
```

```
model_eev = aircraft_deterministic(demand)
SOLVER.solve(model_eev)
```

For this model, the solution is illustrated in Figure 10.1 and the optimal revenue (in units of economy ticket price) is equal to 203.33.

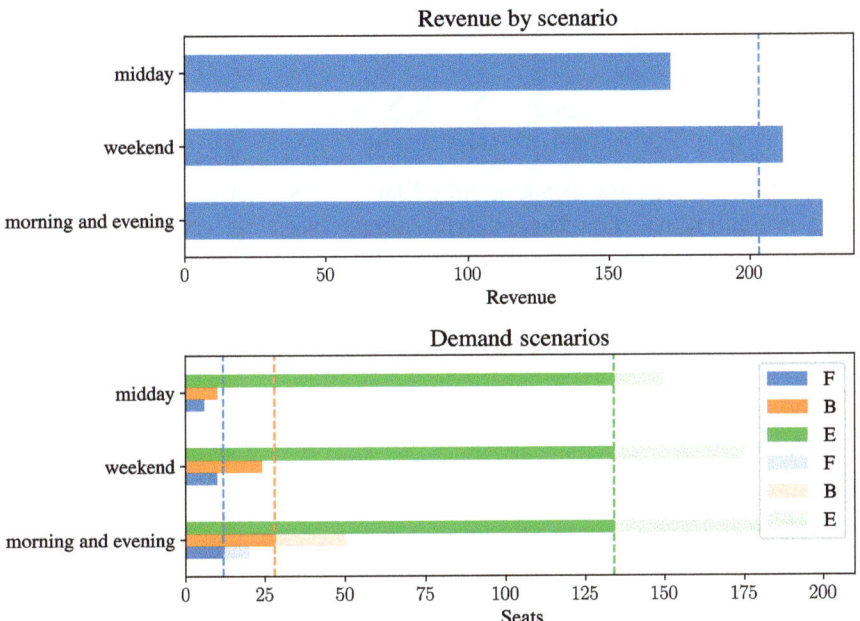

Figure 10.1 Fixed solution for the average demand scenario.

Model 2. Two-Stage SO Formulation and its Extensive Form
If we assume demand is not certain, we can formulate a two-stage SO problem. The first-stage or here-and-now variables s_c are related to the seat allocations. Due to their dependence on the realized demand z, the variables t_c describing the number of tickets sold are second-stage or recourse decision variables. The full problem formulation is as follows: the first-stage problem is

$$\max \quad \mathbb{E}_z Q(s, z)$$
$$\text{s.t.} \quad \sum_{c \in C} f_c s_c \leq 200$$
$$s \in \mathbb{Z}_+,$$

where $Q(s, z)$ is the value of the second-stage problem, defined as

$$Q(\boldsymbol{s}, \boldsymbol{z}) := \max \quad \sum_{c \in C} r_c t_c$$

$$\text{s.t.} \quad t_c \leq s_c, \qquad\qquad \forall c \in C$$

$$t_c \leq z_c, \qquad\qquad \forall c \in C$$

$$\boldsymbol{t} \in \mathbb{Z}_+.$$

In view of the assumption that there is only a finite number $N = |S|$ of scenarios for ticket demand, we can write the extensive form of the two-stage SO problem above and solve it exactly. To do so, we modify the second-stage variables $t_{c,s}$ to index them by both class c and scenario s. Therefore, the expectation can be replaced with the average revenue over the N scenarios, that is,

$$\max \quad \sum_{s \in S} \frac{1}{N} \sum_{c \in C} r_c t_{c,s},$$

where the fraction $\frac{1}{N}$ appears, as we assume that all N scenarios are equally likely. The first-stage constraint remains unchanged, while the second-stage constraints are duplicated for each scenario $s \in S$, namely

$$t_{c,s} \leq s_c, \qquad\qquad \forall c \in C$$

$$t_{c,s} \leq z_{c,s}, \qquad\qquad \forall (c,s) \in C \times S.$$

The Pyomo implementation of this model is as follows:

```python
def aircraft_stochastic(demand):
    m = pyo.ConcreteModel("Aircraft seat allocation - Demand scenarios")

    m.CLASSES = pyo.Set(initialize=demand.columns)
    m.SCENARIOS = pyo.Set(initialize=demand.index)

    # first stage variables and constraints
    m.seats = pyo.Var(m.CLASSES, domain=pyo.NonNegativeIntegers)

    @m.Constraint(m.CLASSES)
    def plane_seats(m, c):
        return sum(m.seats[c] * seat_factor[c] for c in m.CLASSES) <= capacity

    # second stage variable and constraints
    m.tickets = pyo.Var(m.CLASSES, m.SCENARIOS,
        domain=pyo.NonNegativeIntegers)

    @m.Constraint(m.CLASSES, m.SCENARIOS)
    def demand_limits(m, c, s):
        return m.tickets[c, s] <= demand[c][s]

    @m.Constraint(m.CLASSES, m.SCENARIOS)
    def seat_limits(m, c, s):
```

```
        return m.tickets[c, s] <= m.seats[c]

    # objective
    @m.Objective(sense=pyo.maximize)
    def revenue(m):
        return sum(
            m.tickets[c, s] * revenue_factor[c] for c in m.CLASSES for s in
            ↪  m.SCENARIOS
        )

    return m


model_stochastic = aircraft_stochastic(demand)
SOLVER.solve(model_stochastic)
```

For this model, the solution is illustrated in Figure 10.2 and the optimal expected revenue (in units of economy ticket price) is equal to 209.33.

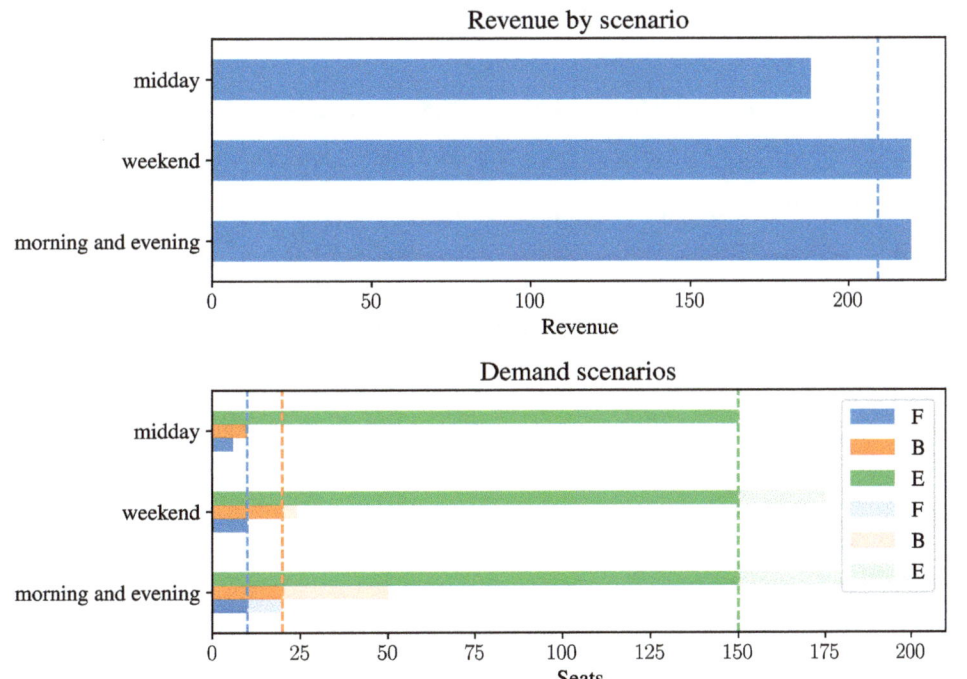

Figure 10.2 Solution of the two-stage SO problem for the $N = 3$ equiprobable scenario.

Model 3. Adding Chance Constraints

The airline wishes to provide a special guarantee for its clients enrolled in its loyalty program. In particular, it wants 98% probability to cover the demand for first-class seats and 95% probability to cover the demand for business-class seats (by clients of the loyalty program). First-class passengers are covered if they can purchase a first-class seat. Business-class passengers are covered if they buy a business-class seat or upgrade to a first-class seat.

Assume the demand of loyalty-program passengers is normally distributed as $z_F \sim \mathcal{N}(\mu_F, \sigma_F^2)$ and $z_B \sim \mathcal{N}(\mu_B, \sigma_B^2)$ for first-class and business, respectively, where the parameters are given in Table 10.2. For completeness, we also include the parameters for economy-class passengers.

Table 10.2 *Parameters for demand distributions.*

	μ	σ
F	12	4
B	28	8
E	175	20

We further assume that the demands for first-class and business-class seats are independent of each other and of the scenario (time of the week).

Let s_F be the number of first-class seats and s_B the number of business seats. The probabilistic constraints are

$$\mathbb{P}(s_F \geq z_F) \geq 0.98, \qquad \text{and} \qquad \mathbb{P}(s_F + s_B \geq z_F + z_B) \geq 0.95.$$

These can be written equivalently as linear constraints, specifically

$$\frac{s_F - \mu_F}{\sigma_F} \geq 2.054 \qquad \text{and} \qquad \frac{(s_F + s_B) - (\mu_F + \mu_B)}{\sqrt{\sigma_F^2 + \sigma_B^2}} \geq 1.645.$$

For the second constraint, we use the fact that the sum of the two independent random variables z_F and z_B is normally distributed with mean $\mu_F + \mu_B$ and variance $\sigma_F^2 + \sigma_B^2$.

We add these equivalent linear counterparts of the two chance constraints to the stochastic optimization model. Rather than writing a function to create a whole new model, we can use the prior function to create and add the two chance constraints using decorators.

```python
from scipy.stats import norm

mu = demand.mean()
sigma = {"F": 4, "B": 16, "E": 20}


def aircraft_cc(demand, QoSF=0.98, QoSFB=0.95):
    # create two-stage stochastic model as before
    m = aircraft_stochastic(demand)

    # add equivalent counterparts of the chance constraints to the 1st
    ↪    stage problem
    # the coefficients related the inverse CDF of the standard normal
    # are computed using the norm.ppf function
```

```
    @m.Constraint()
    def first_class(m):
        return (m.seats["F"] - mu["F"]) >= norm.ppf(QoSF) * sigma["F"]


    @m.Constraint()
    def business_class(m):
        return m.seats["F"] + m.seats["B"] - (mu["F"] + mu["B"]) >=
        ↪    norm.ppf(
            QoSFB
        ) * np.sqrt(sigma["F"] ** 2 + sigma["B"] ** 2)

    return m


model_cc = aircraft_cc(demand)
SOLVER.solve(model_cc)
```

For this model, the solution is illustrated in Figure 10.3 and the optimal expected revenue (in units of economy ticket price) is equal to 177.00. This is naturally lower than the revenue obtained using Model 2 since we reserve more seats for first-class and business-class passengers, aiming to meet high quality-of-service levels for the passengers in the loyalty program.

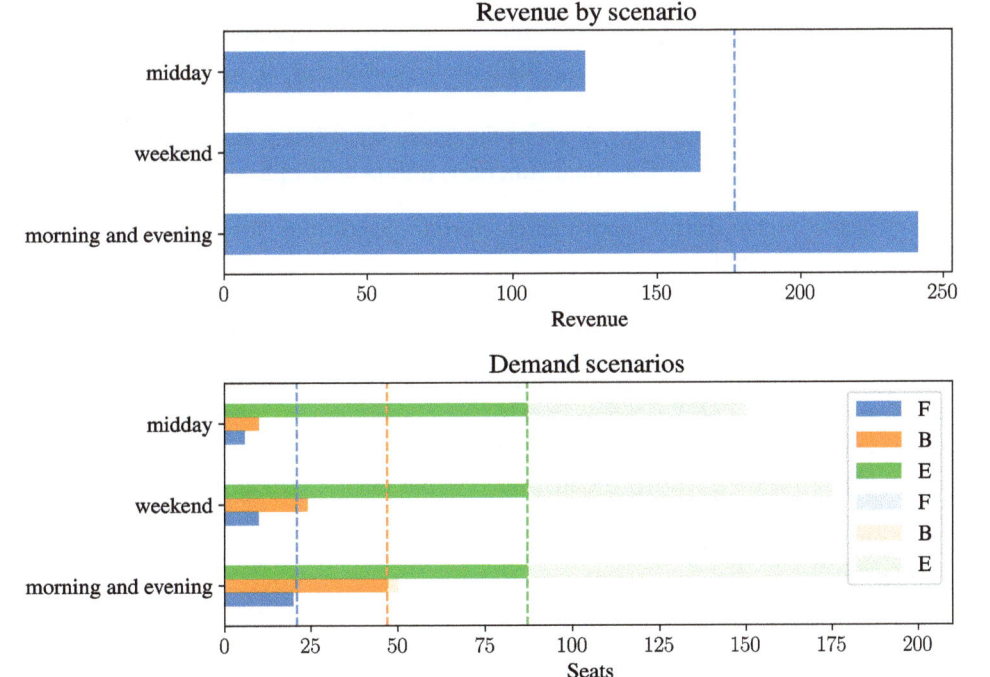

Figure 10.3 Solution of the two-stage SO problem for the $N = 3$ equiprobable scenario, including the chance constraints related to the loyalty program.

Example 10.1 illustrates the main point of two-stage modeling: Accounting for different possible zs and ys at the planning stage allows one to make better x decisions. But there is a price to pay for this – the problem becomes more difficult, especially when the set Z of possible outcomes is large or difficult to model (consider the correlated case). As visible in (10.1) it is hard because it involves a worst-case value/expected value of something that is not even a simple formula, but instead is defined for each z as a solution to a separate optimization problem.

In what follows, we illustrate common and general approaches for solving two-stage problems with both robust and stochastic optimization mindsets. Compared to Chapters 8 and 9, the problems considered here are much more difficult, and for that reason, every time a solution method needs to be carefully crafted for a specific application. For that reason, we will be brief, focused on "how to think" when designing an ad-hoc method, and only include methods that have been widely applied.

10.1 Two-Stage Robust Optimization

Solving a two-stage problem in the RO mindset is hard because it tries to combine two things: (i) covering all possible outcomes of Z, and (ii) being as flexible as possible in y when facing various outcomes z. Therefore, two approaches that have been developed make a compromise on one or the other:

- The first approach, Section 10.1.1, is similar to the adversarial approach of Section 8.4.1: It gradually builds up a finite set of z realizations, along with optimizing separate decisions for each of them and x. The hope is that the more scenarios are included, the higher fraction of Z is "covered."
- The second approach, Section 10.1.2, keeps the entire set Z intact, but restricts y to become a simple function of z (in fact, a linear function) that is later optimized along with x.

For both of these "simplifying" approaches, certain conditions have to hold and we will introduce those accordingly.

10.1.1 Column and Constraint Generation (CCG)

Suppose that instead of considering the entire set Z of scenarios we initially only care about a finite set of scenarios $\bar{Z}(K) = \{z^1, \ldots, z^K\} \subset Z$ and want to optimize for the first-stage decisions x and for the best possible second-stage decisions y^K given the scenarios in $\bar{Z}(K)$. The way to do so is, for each k, to construct a separate y^k, which makes the problem to solve be:

$$\begin{aligned}
\min \quad & c^\top x + \tau & & (10.3)\\
\text{s.t.} \quad & q^\top y^k \le \tau, & & k = 1, \ldots, K\\
& Ax \le b\\
& R(z^k)x + S(z^k)y^k \le t(z^k), & & k = 1, \ldots, K\\
& x \in \mathbb{R}^{n_x}\\
& \tau \in \mathbb{R}\\
& y^k \in \mathbb{R}^{n_y} & & k = 1, \ldots, K.
\end{aligned}$$

Now, call the optimal solution to (10.3) $\bar{x}, \bar{y}^k, \bar{\tau}$ and suppose we want to determine if there exists a scenario $z \in Z$ that is "not accounted for," that is, a realization $z \in Z$ such that none of the existing y^ks, $k = 1, \ldots, K$ is feasible for it. To do so, we can solve the pessimization problem:

$$\max \quad \theta \qquad (10.4)$$

$$\text{s.t.} \quad \theta \leq \max \left\{ \begin{array}{l} r_1(z)^\top x + s_1(z)^\top y^k - t_1(z), \\ \vdots \\ r_m(z)^\top x + s_m(z)^\top y^k - t_m(z), \end{array} \right\} \qquad \forall k = 1, \ldots, K$$

$$z \in Z$$

$$\theta \in \mathbb{R}.$$

If the optimal value of (10.4) is positive, it means that a scenario z found in this way is such that for each of the already existing recourse actions y, one of the constraints is violated.

However, the constraints of (10.4) do not admit a simple and exact MILO reformulation. There are two ways to handle this situation: (i) use a finite (but large) set Z so that it is possible to look for a violating z simply using a loop, or (ii) for a general Z, use a big-M reformulation of the above problem, namely:

$$\max \quad \theta$$

$$\text{s.t.} \quad \theta \leq r_j(z)^\top x + s_j(z)^\top y^k - t_j(z) + M u_{jk}, \qquad \forall j, k$$

$$\sum_{j=1}^{m} u_{jk} \leq m - 1, \qquad \forall k$$

$$z \in Z$$

$$\theta \in \mathbb{R}$$

$$u_{jk} \in \mathbb{B}, \qquad \forall j, k,$$

where the purpose of the last constraint is to make sure that for at least one constraint per each k the inequality is not "relaxed" with M so that the original inequality is violated for sure.

The CCG algorithm consists of alternating steps of optimizing for a fixed and finite set of z and finding a new pessimistic z:

1. Start with $k = 1$ and a set $\bar{Z}(1) = \{z^1\}$ with arbitrary $z^1 \in Z$.
2. Solve the kth iteration problem (10.3) and denote its optimal solution by $\bar{x}, \bar{y}^k$.
3. Solve the kth iteration pessimization problem (10.4) and denote its optimal value by $\bar{\theta}$. If a solution is found for which $\theta > 0$, add the corresponding $\bar{z}$ to $\bar{Z}$ and go back to Step 2. Otherwise, stop.

The benefit of this algorithm is its conceptual simplicity, but for it to work we need a procedure to solve (10.4). Second, we hope that the number of scenarios need not be that big (which is quite often the case). Luckily, even if one has to stop the algorithm earlier due to memory limits, the solution still provides some robustness. In Section 10.4 we shall demonstrate an implementation of the CCG approach.

10.1.2 Decision Rules

We now introduce an alternative approach in which we consider all $z \in Z$ by definition, but we stay away from the "maximum adaptability" of y to z by formulating y explicitly as a specific function of z. It makes sense to keep such functions simple – in particular, linear functions are a natural choice:

$$\boldsymbol{y} = P\boldsymbol{z} + \boldsymbol{p}.$$

The "shape" of such function defined by $\boldsymbol{p}$ and P is then optimized together with the decisions $\boldsymbol{x}$ so that, effectively, the problem becomes a single-stage one. For this approach to make sense, we will make the following assumptions:

- all second-stage variables $\boldsymbol{y}$ are continuous – indeed, it would not be possible to model general continuous decisions as linear functions of a parameter z that could take arbitrary values (unless in some very specific cases);
- $\boldsymbol{s}_j(\boldsymbol{z}) := \boldsymbol{s}_j$ for every $j = 1, \ldots, m$ – indeed, otherwise, in the second-stage problem's constraints we would obtain terms of the form $\boldsymbol{s}_j(\boldsymbol{z})^\top (P\boldsymbol{z} + \boldsymbol{p})$, which would make the problem difficult – we did not learn how to "robustify" constraints that are quadratic in the uncertain parameter.

Under these assumptions, the two-stage problem can be written as a single-stage one:

$$\min \quad \sup_{\boldsymbol{z} \in Z} \boldsymbol{c}^\top \boldsymbol{x} + \boldsymbol{q}^\top (P\boldsymbol{z} + \boldsymbol{p}) \tag{10.5}$$

$$\text{s.t.} \quad A\boldsymbol{x} \leq \boldsymbol{b},$$

$$\boldsymbol{r}_j(\boldsymbol{z})^\top \boldsymbol{x} + \boldsymbol{s}_j^\top (P\boldsymbol{z} + \boldsymbol{p}) \leq t_j(\boldsymbol{z}), \qquad \forall j = 1, \ldots, m, \quad \boldsymbol{z} \in Z$$

$$\boldsymbol{x} \in \mathbb{R}^{n_x}$$

$$P \in \mathbb{R}^{n_y \times l}$$

$$\boldsymbol{p} \in \mathbb{R}^{n_y}.$$

The objective function and all constraints are (i) linear in decision variables $\boldsymbol{x}, P, \boldsymbol{p}$, and (ii) linear in $\boldsymbol{z}$. Therefore, the problem can be reformulated/solved using techniques from Chapter 8.

Although formulating second-stage decisions explicitly as linear functions of $\boldsymbol{z}$ might seem restrictive, they are surprisingly powerful, and there are even problem instances in which linear decision rules are actually optimal.

10.2 Two-Stage Stochastic Optimization

Just as we can solve a two-stage problem with a worst-case mindset (see Section 10.1), we can also use knowledge about the distribution of the uncertain parameter. In Chapter 9, we considered three different ways to account for distributional knowledge: expectations, probabilities, and risk measures. Given the difficulty of solving a two-stage problem on its own, considering probabilities or risk measures has never really lifted off. For that reason, the only way to aggregate the second-stage outcomes we will introduce theoretically is the expectation, although we will also demonstrate the chance-constrained approach on an extension of Example 10.1.

When considering the expected cost of the second-stage cost function, the objective function of the first-stage problem will include a term

$$\mathbb{E}Q(\boldsymbol{x}, \boldsymbol{z}),$$

which is not easy to optimize if the probability distribution allows too many different realizations (unless we have a simple expression for $Q(x, z)$ and a reasonably simple distribution for z – this would effectively place us in the setting of Section 9.2 though).

Suppose that randomness is described by a random vector z consisting of d independent random components, each of which has only three possible realizations. In this case, the total number of scenarios would be 3^d, thus growing exponentially with d, challenging the numerical solution. The situation becomes even worse when the vector z (or only some of its components) has a continuous distribution.

In both cases (finite but very large set of scenarios or continuous distribution), a common approach is to reduce the scenario set to a manageable size through simulation. If z had a continuous distribution, we are essentially approximating its distribution by a sufficiently large number of scenarios sampled at random. Assume that we generate N independent samples $z^1, z^2, \ldots, z^N$ of the random vector z (or that we have historical data for z of the same size). We can then "pretend" that these are all the scenarios available for z and solve the problem

$$\min \quad c^\top x + \frac{1}{N} \sum_{j=1}^{N} q^\top y^j \tag{10.6}$$

$$\text{s.t.} \quad Ax \le b$$
$$R(z^j)x + S(z^j)y^j \le t(z^j), \qquad \forall j = 1, \ldots, N$$
$$x \in \mathbb{R}^{n_x}$$
$$y^j \in \mathbb{R}^{n_y}, \qquad \forall j = 1, \ldots, N,$$

where we consider these "sampled scenarios" to be equiprobable, in the sense that we take their likelihoods to be $p_i = \frac{1}{N}$ for $i = 1, \ldots, N$ (this is not necessarily the case when we have a fixed number of scenarios). This solution strategy is called the SAA method (see Section 9.2.2).

Note that the extensive form (10.6) is "easy" to solve in theory, being a MILO. For large problems, however, the size of the sample N required to make it both accurate and representative of the true distribution $\mathbb{P}$ might be large and solving (10.6) directly can be prohibitively difficult to "choke at a time." For that reason, people have developed clever approaches known as **decomposition algorithms** that make use of the fact that for a fixed x, the second-stage problem can be solved in parallel for each z^j. An example of such an approach is the L-shaped algorithm; see, for example, [BL11].

Using the SAA method, we are only approximating the value of the actual two-stage problem, but theoretical results (see [SDR14, chapter 5]) exist showing that as the sample size N increases, this approximation becomes more and more exact.

Because the second model presented in Example 10.1 was in essence already an implementation of the SAA method to minimize the expected objective function value in a two-stage setting, in what follows, we propose a quick look at how SAA can also be used to handle chance constraints in a two-stage setting with correlated uncertain parameters.

Example 10.2 (Airline seat allocation with correlated continuous demand) Consider again the airline seat allocation problem from Example 10.1. Let us now move past the simplifying assumption that there are only three equally likely scenarios and consider the case where demand is described by a random vector (z_F, z_B, z_E), where z_c is the demand for seats of class $c \in C$. Specifically, we assume the ticket demand for the three categories is captured by a three-dimensional multivariate normal distribution z with mean $\mu = (\mu_F, \mu_B, \mu_E)$, variances $(\sigma_F^2, \sigma_B^2, \sigma_E^2)$ as in Table 10.2, and covariance matrix

$$\Sigma = \begin{pmatrix} \sigma_F^2 & \rho_{FB}\sigma_F\sigma_B & \rho_{FE}\sigma_F\sigma_E \\ \rho_{BF}\sigma_B\sigma_F & \sigma_B^2 & \rho_{BE}\sigma_B\sigma_E \\ \rho_{EF}\sigma_E\sigma_F & \rho_{EB}\sigma_E\sigma_B & \sigma_E^2 \end{pmatrix}.$$

We assume $\rho_{FB} = 0.6$, $\rho_{BE} = 0.4$, and $\rho_{FE} = 0.2$.

Note that we model the demand for each class using a continuous random variable, which is a simplification of the real world, where the ticket demand is always a discrete nonnegative number. Therefore, we round down all obtained random numbers.

However, now that the number of scenarios is not finite anymore, we cannot solve the problem exactly. Instead, we can use the SAA method to approximate the expected value appearing in the objective function. The first step of the SAA is to generate a collection of N scenarios $(z_{F,s}, z_{B,s}, z_{E,s})$ for $s = 1, \ldots, N$. We can do this by sampling from the normal distributions with the given means and variances.

We then solve the stochastic problem using the generated $N = 1000$ scenarios. Note that we can use the previously defined function `airline_stochastic` to solve the model simply by calling it with a different dataframe as argument.

```
# sample size
N = 1000

# define the mean mu and standard deviation sigma of the demand for each
↪  class
mu = demand.mean()
sigma = {"F": 4, "B": 16, "E": 20}
classes = demand.columns

# correlation matrix
P = np.array([[1, 0.6, 0.2], [0.6, 1, 0.4], [0.2, 0.4, 1]])

# take P to be the identity matrix to obtain uncorrelated demand
# P = np.eye(3)

# build covariance matrix from covariances and correlations
s = np.array(list(sigma.values()))
S = np.diag(s) @ P @ np.diag(s)

# generate N samples, round each demand entry to nearest integer,
# and correct non-negative values
seed = 0
```

```
rng = np.random.default_rng(seed)
samples = rng.multivariate_normal(list(mu), S, N).round()
demand_saa = pd.DataFrame(samples, columns=classes)
demand_saa[demand_saa < 0] = 0

# find an approximate solution of the stochastic model using the SAA method
↪
model_saa = aircraft_stochastic(demand_saa)
SOLVER.solve(model_saa)
```

For this model, the statistical properties and correlation of the sample are illustrated in Figure 10.4 and the optimal expected revenue (in units of economy ticket price) is equal to 210.43.

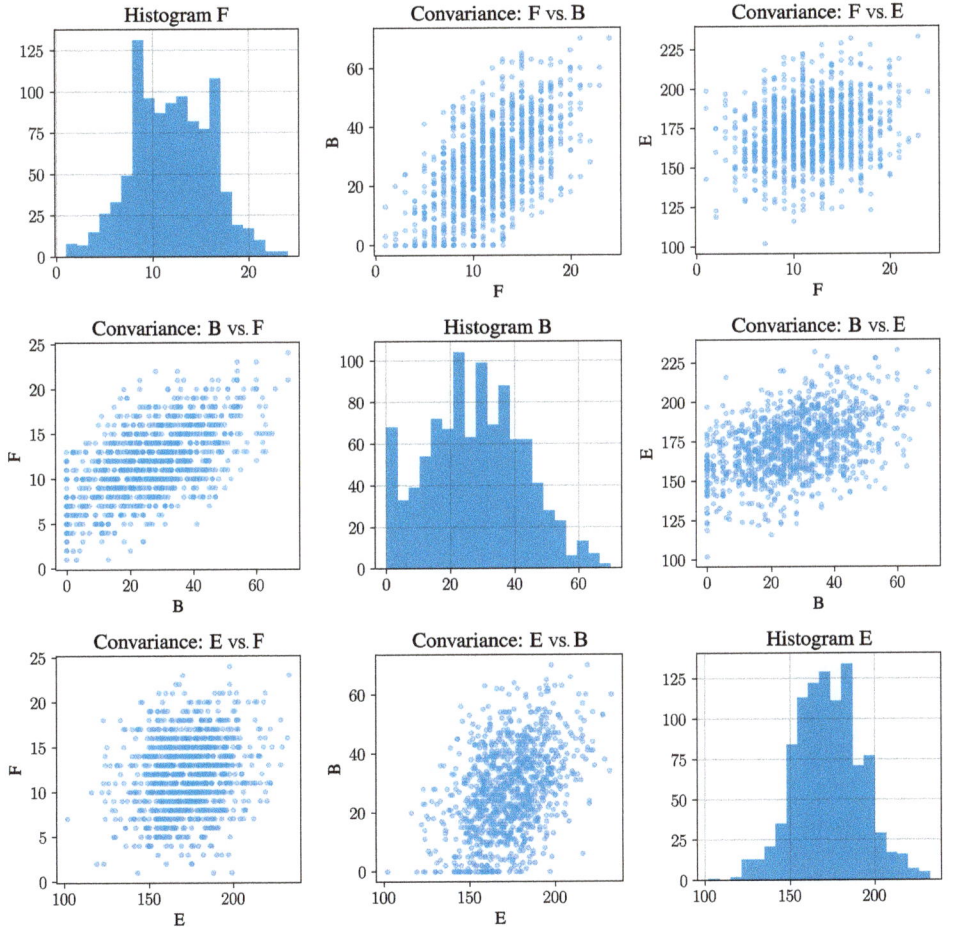

Figure 10.4 Histograms and correlation scatter plots of the sampled $N = 1000$ demand vectors.

Tackling Chance Constraints Using SAA in the Case of Correlated Demand

The linear counterparts of the chance constraints used in Model 3 of Example 10.1 have been derived under the assumption of independent normal distributions of demand for first-class and business-class travel. This assumption no longer holds for the case where demand scenarios are sampled from correlated distributions.

This final model replaces the chance constraints by approximating them using two linear constraints that explicitly track unsatisfied demand. In doing so, we introduce two new sets of integer variables y_s and w_s and a big-M constant, and use the method explained in Section 3.2 approximating the true multivariate distribution with the empirical one obtained from the sample.

The first stage remains unchanged, and so does the objective value of the second stage. The adjusted second-stage constraints are:

$$t_c \leq s_c, \qquad\qquad\qquad \forall c \in C$$
$$t_c \leq z_{c,s}, \qquad\qquad\qquad \forall (c,s) \in C \times S$$
$$s_F + M y_s \geq z_{F,s}, \qquad\qquad\qquad \forall s \in S$$
$$s_F + s_B + M w_s \geq z_{F,s} + z_{B,s}, \qquad\qquad\qquad \forall s \in S$$
$$\frac{1}{N} \sum_{s \in S} y_s \leq 1 - 0.98$$
$$\frac{1}{N} \sum_{s \in S} w_s \leq 1 - 0.95,$$

where y_s and w_s are binary variables indicating the scenarios that do not satisfy the requirements of the airline's loyalty programs for first-class and business-class passengers.

In the following code, we implement and solve this new model:

```python
# define big M constant for chance constraints MILO counterparts
bigM = 100


def aircraft_final(demand):
    m = pyo.ConcreteModel("Aircraft seat allocation - Correlated demand")

    m.CLASSES = pyo.Set(initialize=demand.columns)
    m.SCENARIOS = pyo.Set(initialize=demand.index)

    # first stage variables and constraints
    m.seats = pyo.Var(m.CLASSES, domain=pyo.NonNegativeIntegers)

    @m.Constraint(m.CLASSES)
    def plane_seats(m, c):
        return sum(m.seats[c] * seat_factor[c] for c in m.CLASSES) <=
        ↪ capacity

    # second stage variable and constraints
    m.tickets = pyo.Var(m.CLASSES, m.SCENARIOS,
    ↪ domain=pyo.NonNegativeIntegers)
```

```python
    m.first_class = pyo.Var(m.SCENARIOS, domain=pyo.Binary)
    m.business_class = pyo.Var(m.SCENARIOS, domain=pyo.Binary)

    @m.Constraint(m.CLASSES, m.SCENARIOS)
    def demand_limits(m, c, s):
        return m.tickets[c, s] <= demand[c][s]

    @m.Constraint(m.CLASSES, m.SCENARIOS)
    def seat_limits(m, c, s):
        return m.tickets[c, s] <= m.seats[c]

    @m.Constraint(m.SCENARIOS)
    def first_class_loyality(m, s):
        return m.seats["F"] + bigM * m.first_class[s] >= demand["F"][s]

    @m.Constraint()
    def first_class_loyality_rate(m):
        return sum(m.first_class[s] for s in m.SCENARIOS) <= 0.02 *
        ↪  len(m.SCENARIOS)

    @m.Constraint(m.SCENARIOS)
    def business_class_loyality(m, s):
        return (
            m.seats["F"] + m.seats["B"] + bigM * m.business_class[s]
            >= demand["B"][s] + demand["F"][s]
        )

    @m.Constraint()
    def business_class_loyality_rate(m):
        return sum(m.business_class[s] for s in m.SCENARIOS) <= 0.05 *
        ↪  len(m.SCENARIOS)

    # objective
    @m.Objective(sense=pyo.maximize)
    def revenue(m):
        return sum(
            m.tickets[c, s] * revenue_factor[c] for c in m.CLASSES for s in
            ↪  m.SCENARIOS
        )

    return m


model = aircraft_final(demand_saa)
SOLVER.solve(model)
```

For this model, the expected revenue (in units of economy ticket price) is equal to 177.05. This is naturally lower than the revenue obtained earlier, since we reserve more seats for first-class and business-class passengers aiming to meet high quality-of-service levels for the passengers in the loyalty program.

10.3 Extensions to More Time Stages

In general, multi-stage optimization can span as many time stages as one wishes:

$$\underset{\text{decision}}{\boldsymbol{x}_1} \quad \rightarrow \quad \underset{\text{uncertainty known}}{\boldsymbol{z}_1} \quad \rightarrow \quad \underset{\text{decision}}{\boldsymbol{x}_2} \quad \rightarrow \quad \underset{\text{uncertainty known}}{\boldsymbol{z}_2} \quad \rightarrow \quad \cdots.$$

However, for many practical purposes, it is sufficient to split the planning horizon only into two stages: the present stage, called *first stage*, with the here-and-now decisions implemented immediately, and the future stage, called *second stage*, with wait-and-see decisions that will be implemented after some uncertainties become known.

Why is a two-stage framework sufficient? In practice, once the here-and-now decisions are implemented, and we move to the next time stage where some information becomes known, we can set up a new optimization problem. In this problem, the decisions from time period 1 become the "here and now" decisions and all the remaining decisions become "wait-and-see" decisions. This approach is a standard practice that goes by the name **moving/rolling horizon optimization** and is illustrated in Figure 10.5 where the variables x_i describe the decisions at stage i and the z_is are the uncertain parameters revealed after stage i.

Above the axis, we depict a "dream optimization problem" to be solved had all the information across the entire time horizon been known in advance. However, solving a problem like this is difficult for two reasons:

- information is normally more uncertain the further we look into the future;
- a problem spanning the entire horizon would be too large in size.

For that reason, what is normally done is depicted below the time axis. At each time period, only a limited-size two-stage problem is solved, which does not look beyond the one-step ahead of the current step.

After the first-stage decisions are implemented and the uncertainties are observed, a new version of the small two-stage problem is set up (different versions are denoted with different

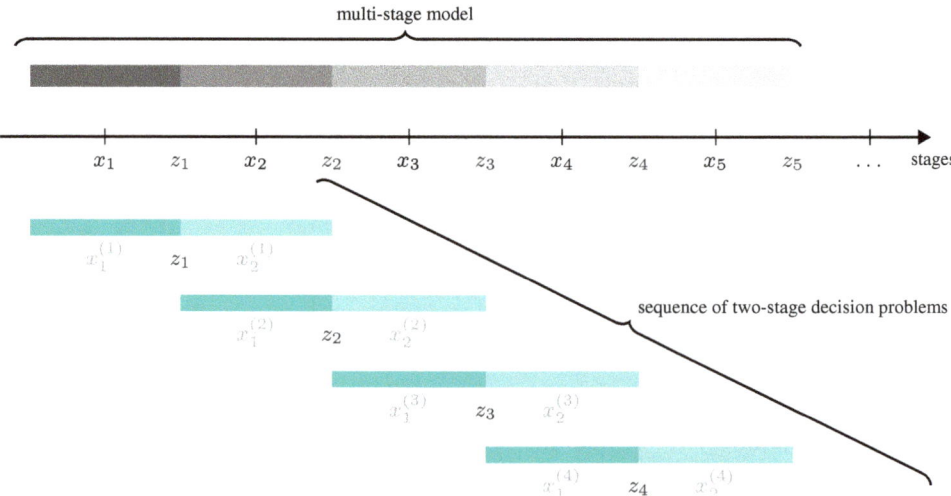

Figure 10.5 Solving a sequence of two-stage problems in a moving horizon fashion.

bracketed superscripts next to the decision variables such as $x_1^{(1)}$), and the process repeats. In this way, the size of the problem solved at each time stage is kept small, while each of the newly implemented decisions is adapted to the just-observed information.

10.4 A Complete Example: Two-Stage Production Planning

Consider again the production planning problem from Section 1.1 where a small start-up company has announced the initial production of two models, U and V. For completeness, we first recall the problem setting and then introduce the additional layer of uncertainty:

- Model U is the higher-priced version requiring 10 units of raw material. The labor requirement is estimated to be 1 hour of labor A and 2 hours of labor B. The U model will sell for $270 per unit.
- Model V requires 9 units of the same raw materials and is estimated to require 1 hour of labor A and 1 hour of labor B. Model V sells for $210 per unit with unlimited demand.
- A preproduction marketing campaign offers a guaranteed delivery of model U in return for a down payment. Initially, it is expected that the number of such orders will be 20 units. Due to the higher price, the demand for model U is limited, so no more than 40 units will be produced.
- The raw materials cost $10 per unit. The raw materials have a long lead time and must be ordered now. Unused raw materials have no waste value.
- Given current staffing, there are 80 hours of labor A available at a cost of $50/hour, and 100 hours of labor B available at a cost of $40/hour.
- The company will not have accurate knowledge of the labor required to produce each product until production starts. The estimated amount of labor A and labor B needed for each unit produced could be off by 15% and 25%, respectively. Additionally, there is uncertainty about the amount of down-payment demand for model U – it can differ by up to 25%, up or down, from the nominal value of 20. It is not expected, however, that more than two of those parameters will be perturbed by their maximum amount.

The company's chief operating officer (COO) must decide how much raw material to order now without complete information about the labor needed to manufacture the models or knowledge about the outcome of the preproduction marketing campaign. If the COO orders too much now, then raw material will be left over and wasted. If the COO orders too little, then a profit opportunity will be lost. In either case, the profit realized will be less than the potential. In what way can the best decision be made?

Problem Description

The problem has a clear two-stage structure: The ordering of the raw material has to be done before the unknown demand is observed and the labor requirements are known. The decisions about the amount of two models to be produced can be, however, made with full information about the uncertainty.

As for the uncertainty itself, no information is given other than the set of values the respective parameters can take:

- z_A: relative uncertainty in the amount of labor A required to produce each unit of product;
- z_B: relative uncertainty is the amount of labor B required to produce each unit of product;
- z_D: uncertainty in the number of initial orders for product U,

where:

$$|z_A| \le 0.15 \qquad \pm 15\% \text{ uncertainty in labor A}$$
$$|z_B| \le 0.25 \qquad \pm 25\% \text{ uncertainty in labor B}$$
$$|z_D| \le 0.25 \qquad 25\% \text{ uncertainty in initial orders for } U$$

and where at most two of them attain their maximum deviation, which can be formulated as the following budget constraint:

$$\frac{|z_A|}{0.15} + \frac{|z_A|}{0.25} + \frac{|z_D|}{0.25} \le 2.$$

Let us see how these parameters z_A and z_B affect the problem. After subtracting the unknown labor cost from the selling price, the unit profit for each device is given by:

$$P_U = 270 - 50(1 + z_A) - 80(1 + z_B) - 100 \implies P_U \in [2.5, 57.5]$$
$$P_V = 210 - 50(1 + z_A) - 40(1 + z_B) - 90 \implies P_V \in [12.5, 47.5].$$

It is not clear in advance which of the products would be more profitable. Taking into account the additional uncertainty about the preorder demand for product U, it is not possible to determine in advance what the worst-case outcome of the uncertainty z_A, z_B, z_D is for this problem.

Since we have no information about the underlying probability distribution of the uncertain parameters, a natural approach is to solve this problem with a worst-case formulation in mind. However, for sake of completeness, we also investigate how the optimal solution would look if each realization of (z_A, z_B, z_D) was equally probable.

Two-Stage Problem Formulation

We formulate the robust optimization problem where the objective is to maximize the worst-case profit subject to constraints and uncertainty as follows:

$$\max \quad -10x + \inf_{z \in Z} Q(x, z)$$
$$\text{s.t.} \quad x \ge 0,$$

where x is the amount of raw material to order, and where the uncertainty $z = (z_A, z_B, z_D)$ is given by

$$Z = \left\{ z \in \mathbb{R}^3 : |z_A| \le 0.15, |z_B|, |z_D| \le 0.25, \frac{|z_A|}{0.15} + \frac{|z_A|}{0.25} + \frac{|z_D|}{0.25} \le 2 \right\}.$$

$Q(x, z)$ is the second-stage profit defined as the optimal value of

$$Q(x, z) = \min \quad y_3$$
$$\text{s.t.} \quad (140 - 50z_A - 80z_B)y_1 + (120 - 50z_A - 40z_B)y_2 \ge y_3$$
$$y_1 \ge 20(1 + z_D)$$
$$y_2 \le 40$$

$$(1 + z_A)y_1 + (1 + z_B)y_2 \le 80$$
$$2(1 + z_A)y_1 + (1 + z_B)y_2 \le 100$$
$$-x + 10y_1 + 9y_2 \le 0$$
$$y_1, y_2, y_3 \ge 0.$$

The extra decision variable y_3 serves the purpose of keeping the objective of the second-stage problem free of uncertainty. Clearly, x plays the role of the first-stage variable here, whereas y_1, y_2, y_3 are the second-stage variables. Aligning with our textbook notation, we can formulate the problem data as

$$\boldsymbol{x} = \begin{pmatrix} x \end{pmatrix}, \, \boldsymbol{y} = \begin{pmatrix} y_1 \\ y_2 \\ y_3 \end{pmatrix}, \, \boldsymbol{z} = \begin{pmatrix} z_A \\ z_B \\ z_D \end{pmatrix}.$$

Finally, we define the data matrices as functions of uncertainties using the same notation we used at the beginning of the chapter:

$$\boldsymbol{c} = \begin{pmatrix} -10 \end{pmatrix}, \, \boldsymbol{q} = \begin{pmatrix} 0 \\ 0 \\ 1 \end{pmatrix}, \, R(\boldsymbol{z}) = \begin{pmatrix} 0 \\ 0 \\ 0 \\ 0 \\ -1 \end{pmatrix}, \, \boldsymbol{t}(\boldsymbol{z}) = \begin{pmatrix} 0 \\ -20(1 + z_D) \\ 80 \\ 100 \\ x \end{pmatrix}$$

$$S(\boldsymbol{z}) = \begin{pmatrix} -(140 - 50z_A - 80z_B) & -(120 - 50z_A - 40z_B) & 1 \\ 1 & 0 & 0 \\ 1 + z_A & 1 + z_A & 0 \\ 2(1 + z_B) & 1 + z_B & 0 \\ 10 & 9 & 0 \end{pmatrix}.$$

Solving the Robust Problem Using Sampled Scenarios

How can we solve such a two-stage problem if it is indeed the worst-case profit that we are interested in? Note that the matrix $S(\boldsymbol{z})$ depends on $\boldsymbol{z}$ and, therefore, we cannot easily formulate the second-stage decisions as linear functions of $\boldsymbol{z}$ (linear decision rules). However, we can generate a finite list of N scenarios and solve a problem that will maximize the worst-case profit across such a set of scenarios:

$$\begin{aligned} \max \quad & \boldsymbol{c}^\top \boldsymbol{x} + \tau \\ \text{s.t.} \quad & \boldsymbol{q}^\top \boldsymbol{y}^j \ge \tau, & j = 1, \dots, N \\ & R(\boldsymbol{z}^j)\boldsymbol{x} + S(\boldsymbol{z}^j)\boldsymbol{y}^j \le \boldsymbol{t}(\boldsymbol{z}^j), & j = 1, \dots, N \\ & \boldsymbol{x}, \boldsymbol{y}^1, \dots, \boldsymbol{y}^N, \tau \ge 0. \end{aligned}$$

We then generate $N = 1000$ scenarios sampling uniformly at random from the given uncertainty set. We then solve both the nominal problem (giving as input a single scenario with the nominal values for the parameters) and the robust problem using $N = 1000$ samples.

```python
# Sample the random variables z_A, z_B, z_D uniformly in the given set
def z_sample(seed):
    rng = np.random.default_rng(seed)
    while True:
        sample = {
            "z_A": 0.15 * rng.uniform(low=-1, high=1),
            "z_B": 0.25 * rng.uniform(low=-1, high=1),
            "z_D": 0.25 * rng.uniform(low=-1, high=1),
        }
        if (
            abs(sample["z_A"]) / 0.15
            + abs(sample["z_B"]) / 0.25
            + abs(sample["z_D"]) / 0.25
            <= 2
        ):
            break
    return sample


# Get a sample of 1000 realizations of the random variables z_A, z_B, z_D
N = 1000
Z = [z_sample(j) for j in range(N)]


# Function to solve the robust problem using the sampled realizations
def max_min_profit(model_params, Z):
    m = pyo.ConcreteModel("Worst case problem")

    # first stage variables
    c, q, *_ = model_params()

    m.I = pyo.Set(initialize=c.keys())
    m.J = pyo.Set(initialize=q.keys())

    m.x = pyo.Var(m.I, domain=pyo.NonNegativeReals)
    m.tau = pyo.Var()

    m.SCENARIOS = pyo.Set(initialize=range(len(Z)))

    @m.Block(m.SCENARIOS)
    def scenario(b, s):
        # get model parameters for the scenario
        _, q, R, S, t = model_params(**Z[s])

        # second stage variables
        b.y = pyo.Var(b.model().J, domain=pyo.NonNegativeReals)

        @b.Constraint()
        def stage_net_profit(b):
            return b.y["y3"] >= b.model().tau
```

```
        @b.Constraint(S.keys())
        def model_constraints(b, k):
            return (
                sum(R[k][i] * b.model().x[i] for i in b.model().I)
                + sum(S[k][j] * b.y[j] for j in b.model().J)
                <= t[k]
            )

    # worst case profit
    @m.Objective(sense=pyo.maximize)
    def worst_case_profit(m):
        return sum(c[i] * m.x[i] for i in m.I) + m.tau

    return m


# Solve the nominal problem
m = max_min_profit(model_params, [{"z_A": 0, "z_B": 0, "z_D": 0}])
SOLVER.solve(m)

# Solve the robust problem using the sampled realizations
m = max_min_profit(model_params, Z)
SOLVER.solve(m)
```

Using the nominal value for the uncertain parameters, the optimal solution is $x^* = 740.00$ and the optimal profit is 2600.00. For the worst-case profit optimization calculated over 1000 sampled scenarios, the optimal quantity of raw material to order is equal to $x^* = 547.81$ and the worst-case profit is equal to 883.04.

The Average-Case Optimization Using SAA

Instead of looking at the worst-case profit, we could be interested in optimizing for the average-case profit, assuming that every scenario within the uncertainty set is equally likely. In this case, we can approximate the optimal solution using the SAA method with the same $N = 1000$ samples as before. The corresponding optimization model that we are implementing is

$$\max \quad c^\top x + \frac{1}{N} \sum_{j=1}^{N} q^\top y^j$$

$$\text{s.t.} \quad Ax \le b$$
$$R(z^j)x + S(z^j)y^j \le t(z^j), \qquad\qquad j = 1, \ldots, N$$
$$x, y^1, \ldots, y^N, \tau \ge 0.$$

```
def max_avg_profit(model_params, Z):
    m = pyo.ConcreteModel("Average case problem (using SAA)")

    # first stage variables
    c, *_ = model_params()
```

```python
    m.I = pyo.Set(initialize=c.keys())
    m.x = pyo.Var(m.I, domain=pyo.NonNegativeReals)

    m.SCENARIOS = pyo.Set(initialize=range(len(Z)))

    @m.Block(m.SCENARIOS)
    def scenario(b, s):
        # get model parameters for the scenario
        _, q, R, S, t = model_params(**Z[s])

        # second stage variables
        b.y = pyo.Var(q.keys(), domain=pyo.NonNegativeReals)

        @b.Constraint(S.keys())
        def model_constraints(b, k):
            return (
                sum(R[k][i] * b.model().x[i] for i in m.I)
                + sum(S[k][j] * b.y[j] for j in q.keys())
                <= t[k]
            )

    # average profit
    @m.Objective(sense=pyo.maximize)
    def avg_profit(m):
        return sum(c[i] * m.x[i] for i in m.I) + sum(
            m.scenario[s].y["y3"] for s in m.SCENARIOS
        ) / len(m.SCENARIOS)

    return m


m = max_avg_profit(model_params, Z)
SOLVER.solve(m)

# Store the per-scenario profit realizations into a numpy array
avg_case_ps = np.zeros(len(Z))

for s in range(len(Z)):
    _, q, R, S, t = model_params(**Z[s])
    avg_case_ps[s] = sum(c[i] * m.x[i]() for i in m.I) +
    ↪    m.scenario[s].y["y3"]()
```

For this model, the optimal to-order quantity of the raw material changes to $x^* = 637.08$, with the average profit that is equal to 2305.93.

Of course, this is only the comparison of the two approaches by looking at the first-stage decision. One can also be interested in the following two quantities:

- The "average-case performance" of the optimal ordering decision of the robust solution – to find this one, one would need to fix this first-stage decision and for each of the sampled

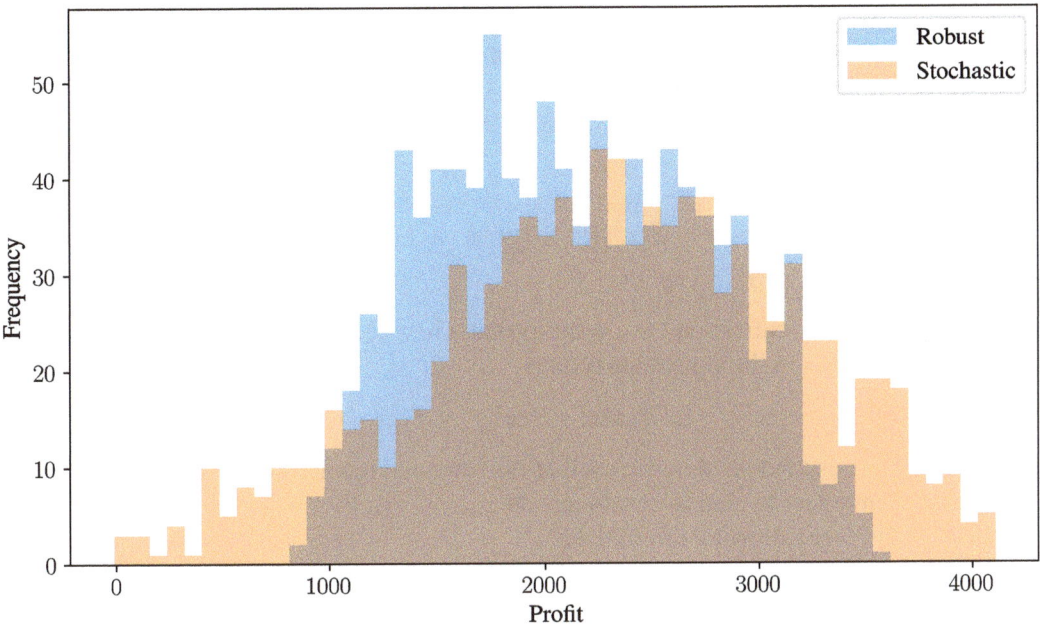

Figure 10.6 Fixed solution for the average demand scenario.

scenarios, one would need to optimize the second-stage decisions, taking the average of their costs in the end.

- The "worst-case performance" of the optimal ordering decision of the robust solution – to find this, one would need to use a function like the one we used to solve the robust solution, but fixing the first-stage decision to be the one obtained from the stochastic solution.

As it turns out, the robust-minded first-stage solution $x^* = 547.81$ has a worst-case performance of 883.04 and an average performance of 2140.78. On the other hand, the average-minded first-stage solution $x^* = 637.08$ has a worst-case performance of -9.63 and an average performance of 2305.93. There is thus a *trade-off*: Some solutions are good on average and underperform when things go very bad, and vice versa; see the histogram in Figure 10.6.

What we observe is a very typical pattern: The robust solution's "range" of values is narrower, both in the best- and worst-situation sense, and it gives thus stabler profit relations. This comes, however, at the expense of having worse performance in an "average" scenario. You can expect to observe this kind of phenomenon very often whenever you need to solve a problem under uncertainty and are unsure whether the worst-case or average-case performance should be optimized. The degree of trade-off can help you make the right decision.

Column and Constraint Generation for the Robust Solution

We now come back to trying to solve for the worst-case solution. The downside of the approach of sampling many scenarios and solving the max–min optimization problem just as before is that it requires many scenarios to be reasonably sure that we cover most of the extreme outcomes of z. This is exactly an issue that the CCG algorithm is supposed to mitigate against – the idea is to gradually build a list of scenarios that are "bad" and, hopefully, end up solving a problem with far less than 1000 problem copies included.

In the pessimization step of the CCG, having solved the problem

$$
\begin{aligned}
\max \quad & \boldsymbol{c}^\top \boldsymbol{x} + \tau \\
\text{s.t.} \quad & \boldsymbol{q}^\top \boldsymbol{y}^j \geq \tau, && j = 1, \ldots, k \\
& R(\boldsymbol{z}^j)\boldsymbol{x} + S(\boldsymbol{z}^j)\boldsymbol{y}^j \leq \boldsymbol{t}(\boldsymbol{z}^j), && j = 1, \ldots, k \\
& \boldsymbol{x}, \boldsymbol{y}^1, \ldots, \boldsymbol{y}^k, \tau \geq 0
\end{aligned}
$$

for a collection of k realizations, we want to find a new realization z where, for each j, at least one of the rows in the constraint system

$$
R(\boldsymbol{z})\bar{\boldsymbol{x}} + S(\boldsymbol{z})\bar{\boldsymbol{y}}^j \leq \boldsymbol{t}(\boldsymbol{z})
$$

is violated. If we expand the terms in each of the rows and rewrite the constraints into a form that makes them easier to read as constraints on z_A, z_B, z_D, then we obtain that for each i, at least one of the four following has to hold:

$$
\begin{aligned}
(50\bar{y}_1^i + 50\bar{y}_2^i)z_A + (80\bar{y}_1^i + 40\bar{y}_2^i)z_B &> -\bar{y}_3^i + 140\bar{y}_1^i + 120\bar{y}_2^i && \text{(profit)} \\
20z_D &> \bar{y}_1^i - 20 && \text{(demand)} \\
(y_1^i + y_2^i)z_A &> 80 - \bar{y}_1^i - \bar{y}_2^i && \text{(labor A)} \\
(2y_1^i + y_2^i)z_B &> 100 - 2\bar{y}_1^i - \bar{y}_2^i && \text{(labor B)}.
\end{aligned}
$$

Using a sufficiently large number M and binary variables $u_{ik} \in \mathbb{B}$ for each of them, we can formulate the problem of searching for an infeasible scenario as

$$
\begin{aligned}
\max \quad & \theta \\
\text{s.t.} \quad & (50\bar{y}_1^i + 50y_2)z_A + (80\bar{y}_1^i + 40\bar{y}_2^i)z_B - \theta \\
& \qquad \geq -\bar{y}_3^i + 140\bar{y}_1^i + 120\bar{y}_2^i - Mu_{i1}, && \forall j \\
& 20z_D - \theta \geq \bar{y}_1^i - 20 - Mu_{i2}, && \forall j \\
& (\bar{y}_1^i + \bar{y}_2^i)z_A - \theta \geq 80 - \bar{y}_1^i - \bar{y}_2^i - Mu_{i3}, && \forall j \\
& (2\bar{y}_1^i + \bar{y}_2^i)z_B 100 - 2\bar{y}_1^i - \bar{y}_2^i - Mu_{i4}, && \forall j \\
& \sum_{k=1}^4 u_{ik} \leq 3, && \forall j \\
& \theta \geq 0 \\
& z \in Z \\
& u_{jk} \in \mathbb{B}, && \forall j, k.
\end{aligned}
$$

This can be implemented in Pyomo as detailed below. Note that for extra flexibility the pessimization problem includes the maximum values of the parameter deviations and an uncertainty budget Γ.

Inspecting the implementation, one can notice that we look for scenarios z that will make only the "feasibility" constraints of the second-stage problem violated, without accounting for the "profit related" proxy constraint. The reason for doing this is to (possibly) accelerate the CCG convergence – the profit constraint will be taken care of anyway as the variable y_3 can adapt freely for each newly generated z^i and each newly generated solution to the master problem, and at the same time, we eliminate the risk of iteratively building a list of scenarios z that violate only the profit constraints across all i, without actually hurting the feasibility of the production plans made.

We put a hard limit of 50 iterations and decide to stop the algorithm when the optimal value of the pessimization subproblem is less than 0.1 (a constraint violation of 0.1 seems to be "acceptably small" given the fact that the labor availability numbers are in the magnitude of 80 and 100, and the amount of raw material purchased is also expected to be in the hundreds, just like the optimal profit).

```python
# Auxiliary function to create the necessary data structures
def subproblem_params(dec):
    Left = {
        "demand": {"z_A": 0, "z_B": 0, "z_D": 20},
        "profit": {
            "z_A": (50 * dec["y1"] + 50 * dec["y2"]),
            "z_B": (80 * dec["y1"] + 40 * dec["y2"]),
            "z_D": 0,
        },
        "labor A": {"z_A": dec["y1"] + dec["y2"], "z_B": 0, "z_D": 0},
        "labor B": {"z_A": 0, "z_B": 2 * dec["y1"] + dec["y2"], "z_D": 0},
    }

    Right = {
        "demand": dec["y1"] - 20,
        "profit": 140 * dec["y1"] + 120 * dec["y2"] - dec["y3"],
        "labor A": 80 - dec["y1"] - dec["y2"],
        "labor B": 100 - 2 * dec["y1"] - dec["y2"],
    }

    return Left, Right


# pessimization problem to solve at every step of the CCG algorithm
def pessimization_problem(
    subproblem_params,
    master_solution,
    z_A_max=0.15,
    z_B_max=0.25,
    z_D_max=0.5,
    Gamma=2,
):

    # Initial definitions
    m = pyo.ConcreteModel()
    big_M = 1000

    L, R = subproblem_params(master_solution[0])
```

```python
# indices for the variables
m.Z_INDICES = pyo.Set(initialize=["z_A", "z_B", "z_D"])

# indices for the scenarios
m.SCENARIOS = pyo.Set(initialize=range(len(master_solution)))

m.z = pyo.Var(m.Z_INDICES, domain=pyo.Reals)
m.z_abs = pyo.Var(m.Z_INDICES, domain=pyo.NonNegativeReals)

m.theta = pyo.Var(domain=pyo.Reals)

# Constraints on z itself
m.A_ub = pyo.Constraint(expr=m.z["z_A"] <= z_A_max)
m.A_lb = pyo.Constraint(expr=-m.z["z_A"] <= z_A_max)
m.A_abs = pyo.Constraint(expr=m.z["z_A"] <= m.z_abs["z_A"])

m.B_lb = pyo.Constraint(expr=m.z["z_B"] <= z_B_max)
m.B_ub = pyo.Constraint(expr=-m.z["z_B"] <= z_B_max)
m.B_abs = pyo.Constraint(expr=m.z["z_B"] <= m.z_abs["z_B"])

m.D_lb = pyo.Constraint(expr=m.z["z_D"] <= z_D_max)
m.D_ub = pyo.Constraint(expr=-m.z["z_D"] <= z_D_max)
m.D_abs = pyo.Constraint(expr=m.z["z_D"] <= m.z_abs["z_D"])

m.z_budget = pyo.Constraint(
    expr=m.z_abs["z_A"] / z_A_max
    + m.z_abs["z_B"] / z_B_max
    + m.z_abs["z_D"] / z_D_max
    <= Gamma
)

# bounds = {"z_A": (-0.15, 0.15),"z_B": (-0.25, 0.25), "z_D": (-0.5, 0.5)}

# Constraint violation: blockwise
@m.Block(m.SCENARIOS)
def scenario(b, s):
    # get model parameters for the scenario
    L, R = subproblem_params(master_solution[s])

    # second stage variables
    b.u = pyo.Var(L.keys(), domain=pyo.Binary)

    # hack for faster CCG convergence
    @b.Constraint()
    def at_least_one_violated(b):
        # only care for violations related to the `actual' constraints,
        # not the profit constraint
        return sum(b.u[k] for k in L.keys() if k != "profit") <=
        ↪  len(L.keys()) - 2
```

```python
        @b.Constraint(L.keys())
        def model_constraints(b, k):
            return (
                sum(L[k][i] * b.model().z[i] for i in b.model().Z_INDICES)
                - b.model().theta
                >= R[k] - b.u[k] * big_M
            )

    # worst case profit
    @m.Objective(sense=pyo.maximize)
    def max_violation(m):
        return m.theta

    return m


# Set the parameters for the CCG procedure
ccg_converged = False
stopping_precision = 0.1
max_iterations = 50
ccg_iterations = 0

# Initialize the null scenario - no perturbation
Z = [{"z_A": 0, "z_B": 0, "z_D": 0}]

while (not ccg_converged) and (ccg_iterations < max_iterations):
    # Building and solving the master problem
    m = max_min_profit(model_params, Z)
    SOLVER.solve(m)

    # Exporting the data from the master problem into a list of dictionaries
    c, q, *_ = model_params()

    master_solution = []

    print(f"\nIteration #{ccg_iterations}")

    for s in range(len(Z)):
        single_solution = {}

        for x_key in c.keys():
            single_solution[x_key] = m.x[x_key]()

        for y_key in q.keys():
            single_solution[y_key] = m.scenario[s].y[y_key]()

        master_solution.append(single_solution)

    print(f'Current solution: x = {master_solution[0]["x"]:.2f}')
```

```
# Pessimization
m = pessimization_problem(subproblem_params, master_solution, 0.15, 0.25,
↪   0.25, 2)
SOLVER.solve(m)
theta_opt, z_A, z_B, z_D = m.theta(), m.z["z_A"](), m.z["z_B"](),
↪   m.z["z_D"]()


# If pessimization yields no violation, stop the procedure,
# otherwise add a scenario and repeat
if theta_opt < stopping_precision:
    print("No violation found. Stopping the procedure.")
    ccg_converged = True
else:
    print(f"Violation found: z_A = {z_A:4.2f}, z_B = {z_B:4.2f}, z_D =
↪   {z_D:4.2f}")
    print(f"Constraint violation: {theta_opt:6.2f}")
    Z.append({"z_A": z_A, "z_B": z_B, "z_D": z_D})


ccg_iterations += 1
```

Table 10.3 shows that it takes only four iterations, instead of 1000 scenarios, to arrive at an "almost" robust solution that has essentially the same first-stage decisions as in the first case.

Thus, it pays off to generate scenarios only when they are needed and if they are needed. This principle of "delayed generation of scenarios" has its applications in many fields of large-scale optimization, not just in the context of robustness but even for deterministic problems, where it is known as **column generation** but which is not discussed in this book.

10.5 A Complete Example: Optimal Power Flow Problem With Recourse Actions

In this example, we illustrate an application of the idea of linear decision rules to a two-stage optimal power flow (OPF) problem in which the power of conventional generators has to adapt automatically to balance out the energy surplus/shortage due to changes in the input of renewable resources.

We consider a variant of the OPF problem in which each conventional generator i commits in advance to produce a specific amount p_i of energy as determined by the OPF problem, assuming that the renewable energy production from all solar panels and wind turbines will

Table 10.3 *Summary of the iterations of the CCG procedure.*

Iteration #	Current solution	Violation found	Constraint violation
0	740.00	$z_A = 0.00$, $z_B = 0.25$, $z_D = -0.25$	25.00
1	600.00	$z_A = 0.00$, $z_B = 0.10$, $z_D = 0.25$	10.00
2	600.00	$z_A = 0.02$, $z_B = 0.21$, $z_D = 0.25$	10.00
3	547.14	$z_A = -0.15$, $z_B = 0.25$, $z_D = 0.17$	3.31
4	547.14	No violation found	–

be equal to the forecast one, also denoted as $\bar{p}_j$. However, the renewable energy output of generator $j \in \mathcal{G}^{\text{wind}} \cup \mathcal{G}^{\text{solar}}$ deviates from its forecast values by an amount Δ_j, and results in a power production of

$$p_j = \bar{p}_j + \Delta_j, \quad j \in \mathcal{G}^{\text{wind}} \cup \mathcal{G}^{\text{solar}}.$$

Then conventional generators need to take a **recourse action** to ensure that the network is balanced, that is, the total energy production equals the total energy demand. This means that the problem has a two-stage structure:

- first, the "nominal" energy generation levels are set for the coal and gas units;
- second, the actual renewable energy output of the wind/solar generators is observed and power generation levels of the coal and gas units need to be adapted.

If we were optimizing for the average-case total cost, then a proper two-stage formulation of our problem would be

$$\min \quad \sum_{i \in \mathcal{G}^{\text{coal}} \cup \mathcal{G}^{\text{gas}}} c_i(\bar{p}_i) + \mathbb{E}Q(\bar{p}, \Delta)$$

$$\text{s.t.} \quad p_i^{\min} \leq \bar{p}_i \leq p_i^{\max}, \qquad \qquad \forall i \in \mathcal{G}^{\text{coal}} \cup \mathcal{G}^{\text{gas}},$$

where the recourse function $Q(\bar{p}, \Delta)$ is defined as the optimal value of the following problem:

$$\min \quad \sum_{i \in \mathcal{G}^{\text{coal}} \cup \mathcal{G}^{\text{gas}}} \hat{c}_i(r_i)$$

$$\text{s.t.} \quad p_i = \bar{p}_i + r_i \qquad \qquad \forall i \in \mathcal{G}^{\text{coal}} \cup \mathcal{G}^{\text{gas}}$$

$$p_i^{\min} \leq p_i \leq p_i^{\max}, \qquad \qquad \forall i \in \mathcal{G}^{\text{coal}} \cup \mathcal{G}^{\text{gas}}$$

$$\sum_{j:(i,j)\in E} f_{ij} - \sum_{j:(j,i)\in E} f_{ji} = p_i - d_i, \qquad \forall i \in \mathcal{G}^{\text{coal}} \cup \mathcal{G}^{\text{gas}}$$

$$\sum_{j:(i,j)\in E} f_{ij} - \sum_{j:(j,i)\in E} f_{ji} = \bar{p}_i + \Delta_i - d_i, \qquad \forall i \in \mathcal{G}^{\text{wind}} \cup \mathcal{G}^{\text{solar}}$$

$$\sum_{j:(i,j)\in E} f_{ij} - \sum_{j:(j,i)\in E} f_{ji} = \bar{p}_i - d_i, \qquad \forall i \in V \setminus \mathcal{G}^{\text{gen}}$$

$$f_{ij} = b_{ij}(\theta_i - \theta_j), \qquad \qquad \forall (i,j) \in E$$

$$-f_{ij}^{\max} \leq f_{ij} \leq f_{ij}^{\max}, \qquad \qquad \forall (i,j) \in E$$

$$\theta_i \in \mathbb{R}, \qquad \qquad \forall i \in V$$

$$f_{ij} \in \mathbb{R}, \qquad \qquad \forall (i,j) \in E$$

$$r_i \in \mathbb{R}, \qquad \qquad \forall i \in \mathcal{G}^{\text{coal}} \cup \mathcal{G}^{\text{gas}},$$

where

$$\mathcal{G}^{\text{gen}} = \mathcal{G}^{\text{wind}} \cup \mathcal{G}^{\text{solar}} \cup \mathcal{G}^{\text{coal}} \cup \mathcal{G}^{\text{gas}}$$

and $c_i(.)$ and $\hat{c}_i(.)$ are the cost functions related to the precommitted $\bar{p}_i$ and changed r_i amounts of energy, remembering that among the $\bar{p}_i$s, only those with $i \in \mathcal{G}^{\text{coal}} \cup \mathcal{G}^{\text{gas}}$ are actual decision variables while $\bar{p}_i$, $i \in \mathcal{G}^{\text{solar}} \cup \mathcal{G}^{\text{wind}}$ are known parameters (think of them as wind and solar power forecasts).

Now, we are going to implement a specific type of linear decision rule. For each conventional generator, we set a recourse action, that is real-time adjustment of its power production, based on the realization of the renewable energy. More specifically, each conventional generator $i \in \mathcal{G}^{\text{coal}} \cup \mathcal{G}^{\text{gas}}$ has a *participation factor* $\alpha_i \geq 0$ that determines to what extent the generator responds to the total imbalance $\sum_j \Delta_j$. Specifically, the power production after the recourse action at the conventional generator i is denoted by p_i and is given by

$$p_i := \bar{p}_i - \alpha_i \sum_{j \in \mathcal{G}^{\text{wind}} \cup \mathcal{G}^{\text{solar}}} \Delta_j, \qquad i \in \mathcal{G}^{\text{coal}} \cup \mathcal{G}^{\text{gas}}.$$

In the notation of the problem above, we picked the recourse action r_i to be a function of the uncertain parameters Δ rather than a decision variable, since

$$r_i = -\alpha_i \sum_{j \in \mathcal{G}^{\text{wind}} \cup \mathcal{G}^{\text{solar}}} \Delta_j.$$

The participation factor $\alpha_i \in [0,1]$ indicates the fraction of the power imbalance that generator i needs to compensate for. To ensure that the power balance is satisfied, we need to have $\sum_{i \in \mathcal{G}^{\text{coal}} \cup \mathcal{G}^{\text{gas}}} \alpha_i = 1$. Indeed, in this case, assuming that the power was balanced in the first stage, that is, $\sum_{i \in \mathcal{G}} p_i - \sum_{i \in V} d_i = 0$, then the net power balance after the second stage is

$$\sum_{i \in \mathcal{G}} p_i - \sum_{i \in V} d_i$$

$$= \sum_{j \in \mathcal{G}^{\text{wind}} \cup \mathcal{G}^{\text{solar}}} (\bar{p}_j + \Delta_j) + \sum_{i \in \mathcal{G}^{\text{coal}} \cup \mathcal{G}^{\text{gas}}} \left(\bar{p}_i - \alpha_i \sum_{j \in \mathcal{G}^{\text{wind}} \cup \mathcal{G}^{\text{solar}}} \Delta_j \right) - \sum_{i \in V} d_i$$

$$= \sum_{j \in \mathcal{G}^{\text{wind}} \cup \mathcal{G}^{\text{solar}}} \Delta_j - \sum_{j \in \mathcal{G}^{\text{wind}} \cup \mathcal{G}^{\text{solar}}} \left(\sum_{i \in \mathcal{G}^{\text{coal}} \cup \mathcal{G}^{\text{gas}}} \alpha_i \right) \Delta_j + \sum_{i \in \mathcal{G}} \bar{p}_i - \sum_{i \in V} d_i$$

$$= \sum_{i \in \mathcal{G}} \bar{p}_i - \sum_{i \in V} d_i = 0.$$

The participation factors α_i do not have to be equal for all generators and in fact, they can be optimized jointly together with the initial power levels p_i. Since the energy produced as recourse action is more expensive, we account for this by adding to the objective function the cost term $\sum_{i \in \mathcal{G}^{\text{coal}} \cup \mathcal{G}^{\text{gas}}} \hat{c}_i(\alpha_i \sum_{j \in \mathcal{G}^{\text{wind}} \cup \mathcal{G}^{\text{solar}}} \Delta_j)$ for some cost functions $\hat{c}_i(.)$.

The resulting two-stage SO problem is

$$\min \quad \sum_{i \in \mathcal{G}^{\text{coal}} \cup \mathcal{G}^{\text{gas}}} c_i(\bar{p}_i) + \mathbb{E}Q(\bar{p}, \alpha, \Delta)$$

$$\text{s.t.} \quad \sum_{i \in \mathcal{G}^{\text{coal}} \cup \mathcal{G}^{\text{gas}}} \alpha_i = 1$$

$$p_i^{\min} \leq \bar{p}_i \leq p_i^{\max}, \qquad\qquad\qquad \forall\, i \in \mathcal{G}^{\text{coal}} \cup \mathcal{G}^{\text{gas}}$$

$$\alpha_i \geq 0, \qquad\qquad\qquad\qquad\qquad \forall\, i \in \mathcal{G}^{\text{coal}} \cup \mathcal{G}^{\text{gas}},$$

where the recourse function $Q(\bar{p}, \alpha, \Delta)$ is defined as the optimal value of

$$\min \quad \sum_{i \in \mathcal{G}^{\mathrm{coal}} \cup \mathcal{G}^{\mathrm{gas}}} \hat{c}_i(\alpha_i \Delta^{\mathrm{tot}})$$

$$\begin{aligned}
\text{s.t.} \quad & \Delta_{\mathrm{tot}} = \sum_{j \in \mathcal{G}^{\mathrm{wind}} \cup \mathcal{G}^{\mathrm{solar}}} \Delta_j \\
& p_i^{\min} \leq p_i - \alpha_i \Delta^{\mathrm{tot}} \leq p_i^{\max}, && \forall\, i \in \mathcal{G}^{\mathrm{coal}} \cup \mathcal{G}^{\mathrm{gas}} \\
& \sum_{j\,:\,(i,j)\in E} f_{ij} - \sum_{j\,:\,(j,i)\in E} f_{ji} = \bar{p}_i - \alpha_i \Delta_{\mathrm{tot}} - d_i, && \forall\, i \in \mathcal{G}^{\mathrm{coal}} \cup \mathcal{G}^{\mathrm{gas}} \\
& \sum_{j\,:\,(i,j)\in E} f_{ij} - \sum_{j\,:\,(j,i)\in E} f_{ji} = \bar{p}_i + \Delta_i - d_i, && \forall\, i \in \mathcal{G}^{\mathrm{wind}} \cup \mathcal{G}^{\mathrm{solar}} \\
& \sum_{j\,:\,(i,j)\in E} f_{ij} - \sum_{j\,:\,(j,i)\in E} f_{ji} = \bar{p}_i - d_i, && \forall\, i \in V \setminus \mathcal{G}^{\mathrm{gen}} \\
& f_{ij} = b_{ij}(\theta_i - \theta_j), && \forall\, (i,j) \in E \\
& -f_{ij}^{\max} \leq f_{ij} \leq f_{ij}^{\max}, && \forall\, (i,j) \in E \\
& \theta_i \in \mathbb{R}, && \forall\, i \in V \\
& f_{ij} \in \mathbb{R}, && \forall\, (i,j) \in E.
\end{aligned}$$

The only remaining question is where to take the values for Δ_j from. One way of doing things would be to construct an uncertainty set for Δ_j and make sure that the inequality constraints hold for all realizations of Δ_j using the technique of robust counterparts/adversarial approach from Chapter 8. This would be a solid approach if we were optimizing for the worst-case value of the objective function, instead of an expectation.

However, since in this particular application we optimize for the expected value, it makes sense for us to resort to the SAA method. More specifically, we sample T realizations of the renewable fluctuations Δ, indexed as $\{\Delta^s\}_{s \in S}$ with $S = \{1, \ldots, T\}$, and we approximate the expectation through an empirical average across all those samples while enforcing that the constraints hold for every such realization. In this way, the resulting problem we actually solve is

$$\min \quad \sum_{i \in \mathcal{G}^{\mathrm{coal}} \cup \mathcal{G}^{\mathrm{gas}}} c_i(\bar{p}_i) + \frac{1}{T} \sum_{s=1}^{T} \sum_{i \in \mathcal{G}^{\mathrm{coal}} \cup \mathcal{G}^{\mathrm{gas}}} \hat{c}_i(\alpha_i \Delta_{\mathrm{tot}}^s)$$

$$\begin{aligned}
\text{s.t.} \quad & \Delta_{\mathrm{tot}}^s = \sum_{j \in \mathcal{G}^{\mathrm{wind}} \cup \mathcal{G}^{\mathrm{solar}}} \Delta_j^s \\
& p_i^{\min} \leq \bar{p}_i - \alpha_i \Delta_{\mathrm{tot}}^s \leq p_i^{\max}, && \forall\, i \in \mathcal{G}^{\mathrm{coal}} \cup \mathcal{G}^{\mathrm{gas}}, \forall\, s \in S \\
& \sum_{j\,:\,(i,j)\in E} f_{ij}^s - \sum_{j\,:\,(j,i)\in E} f_{ji}^s = \bar{p}_i - \alpha_i \Delta_{\mathrm{tot}}^s - d_i, && \forall\, i \in \mathcal{G}^{\mathrm{coal}} \cup \mathcal{G}^{\mathrm{gas}}, \forall\, s \in S \\
& \sum_{j\,:\,(i,j)\in E} f_{ij}^s - \sum_{j\,:\,(j,i)\in E} f_{ji}^s = \bar{p}_i + \Delta_i^s - d_i, && \forall\, i \in \mathcal{G}^{\mathrm{wind}} \cup \mathcal{G}^{\mathrm{solar}}, \forall\, s \in S \\
& \sum_{j\,:\,(i,j)\in E} f_{ij}^s - \sum_{j\,:\,(j,i)\in E} f_{ji}^s = \bar{p}_i - d_i, && \forall\, i \in V \setminus \mathcal{G}^{\mathrm{gen}}, \forall\, s \in S \\
& f_{ij}^s = b_{ij}(\theta_i^s - \theta_j^s), && \forall\, (i,j) \in E, \forall\, s \in S
\end{aligned}$$

$$-f_{ij}^{\max} \le f_{ij}^s \le f_{ij}^{\max}, \qquad\qquad \forall\,(i,j) \in E, \forall\,s \in S$$

$$\theta_i^s \in \mathbb{R}, \qquad\qquad \forall\,i \in V, \forall\,s \in S$$

$$f_{ij}^s \in \mathbb{R}, \qquad\qquad \forall\,(i,j) \in E, \forall\,s \in S$$

$$\sum_{i \in \mathcal{G}^{\mathrm{coal}} \cup \mathcal{G}^{\mathrm{gas}}} \alpha_i = 1$$

$$p_i^{\min} \le \bar{p}_i \le p_i^{\max}, \qquad\qquad \forall\,i \in \mathcal{G}^{\mathrm{coal}} \cup \mathcal{G}^{\mathrm{gas}}$$

$$\alpha_i \ge 0, \qquad\qquad \forall\,i \in \mathcal{G}^{\mathrm{coal}} \cup \mathcal{G}^{\mathrm{gas}}.$$

To recover a linear problem, we assume that the energy generation costs are modeled as

$$c_i(x) := c_i x, \quad \hat{c}_i(x) = 2c_i|x|,$$

where c_i is the unit production cost of the i-th generator. This structure for the $\hat{c}_i(\cdot)$ functions means that we assume that any real-time adjustment in the energy dispatch of a generator is twice as costly as a prescheduled unit of energy generated there.

```python
# OPF with recourse actions for conventional generators based on participation
↪  factors
def OPF_participationfactors(
    network,
    imbalances,
    totalimbalances,
    abstotalimbalances,
    uniformparticipationfactors=False,
):
    # Define a model
    model = pyo.ConcreteModel("OPF with participation factors")

    # Define sets
    model.T = pyo.Set(initialize=range(len(imbalances)))
    model.V = pyo.Set(initialize=network["nodes"].keys())
    model.E = pyo.Set(initialize=network["edges"].keys())
    model.SWH = pyo.Set(
        initialize=[
            i
            for i, data in network["nodes"].items()
            if data["energy_type"] in ["wind", "solar", "hydro"]
        ]
    )
    model.CG = pyo.Set(
        initialize=[
            i
            for i, data in network["nodes"].items()
            if data["energy_type"] in ["coal", "gas"]
        ]
    )
    model.NG = pyo.Set(
        initialize=[
            i for i, data in network["nodes"].items() if
            ↪  pd.isna(data["energy_type"])
```

```python
        ]
    )

    # Declare decision variables
    model.p = pyo.Var(model.V, domain=pyo.NonNegativeReals)
    model.r = pyo.Var(model.V, model.T, domain=pyo.NonNegativeReals)
    model.alpha = pyo.Var(model.V, domain=pyo.NonNegativeReals)
    model.theta = pyo.Var(model.V, model.T, domain=pyo.Reals)
    model.f = pyo.Var(model.E, model.T, domain=pyo.Reals)
    model.abs_total_imbalance = pyo.Param(
        model.T, domain=pyo.NonNegativeReals, initialize=abstotalimbalances
    )
    model.total_imbalance = pyo.Param(
        model.T, domain=pyo.Reals, initialize=totalimbalances
    )

    # Declare objective function including the recourse actions
    model.objective = pyo.Objective(
        expr=sum(
            data["c_var"] * model.p[i]
            for i, data in network["nodes"].items()
            if data["is_generator"]
        )
        + 1
        / len(model.T)
        * sum(
            sum(
                2 * data["c_var"] * model.alpha[i] *
                ↪    model.abs_total_imbalance[t]
                for i, data in network["nodes"].items()
                if data["energy_type"] in ["coal", "gas"]
            )
            for t in model.T
        ),
        sense=pyo.minimize,
    )

    # Declare constraints
    # First-stage production levels must meet generator limits
    model.generation_upper_bound = pyo.Constraint(
        model.V, rule=lambda m, i: m.p[i] <= network["nodes"][i]["p_max"]
    )
    model.generation_lower_bound = pyo.Constraint(
        model.V, rule=lambda m, i: network["nodes"][i]["p_min"] <= m.p[i]
    )

    # Wind, solar and hydro generators have zero participation factors
    model.windsolarhydro_nopartecipationfactors = pyo.Constraint(
        model.SWH, rule=lambda m, i: m.alpha[i] == 0
    )
```

```python
# Load nodes have zero participation factors
model.load_nopartecipationfactors = pyo.Constraint(
    model.NG, rule=lambda m, i: m.alpha[i] == 0
)


# Participation factors must sum to one
model.sum_one = pyo.Constraint(rule=sum(model.alpha[i] for i in model.V)
↪   == 1)


if uniformparticipationfactors:
    # Participation factors must be equal
    model.equal_participationfactors = pyo.Constraint(
        model.CG, rule=lambda m, i: m.alpha[i] == 1 / len(model.CG)
    )


# Second-stage production levels must also meet generator limits
model.power_withrecourse = pyo.Constraint(
    model.V,
    model.T,
    rule=lambda m, i, t: m.r[i, t] == m.p[i] - m.alpha[i] *
    ↪   m.total_imbalance[t],
)
model.generation_upper_bound_withrecourse = pyo.Constraint(
    model.CG,
    model.T,
    rule=lambda m, i, t: m.r[i, t] <= network["nodes"][i]["p_max"],
)
model.generation_lower_bound_withrecourse = pyo.Constraint(
    model.CG,
    model.T,
    rule=lambda m, i, t: network["nodes"][i]["p_min"] <= m.r[i, t],
)


# Expressions for outgoing and incoming flows
model.outgoing_flow = pyo.Expression(
    model.V,
    model.T,
    rule=lambda m, i, t: sum(m.f[i, j, t] for j in model.V if (i, j) in
    ↪   model.E),
)
model.incoming_flow = pyo.Expression(
    model.V,
    model.T,
    rule=lambda m, i, t: sum(m.f[j, i, t] for j in model.V if (j, i) in
    ↪   model.E),
)


# Net power production at each node after recourse actions
model.flow_conservation = pyo.Constraint(
```

```
        model.V,
        model.T,
        rule=lambda m, i, t: m.incoming_flow[i, t] - m.outgoing_flow[i, t]
            == m.r[i, t] + imbalances[t][i] - nodes[i]["d"],
    )

    model.susceptance = pyo.Constraint(
        model.E,
        model.T,
        rule=lambda m, i, j, t: m.f[(i, j), t]
            == network["edges"][(i, j)]["b"] * (m.theta[i, t] - m.theta[j, t]),
    )

    model.flows_upper_bound = pyo.Constraint(
        model.E,
        model.T,
        rule=lambda m, i, j, t: m.f[(i, j), t] <= network["edges"][(i,
        ↪  j)]["f_max"],
    )
    model.flows_lower_bound = pyo.Constraint(
        model.E,
        model.T,
        rule=lambda m, i, j, t: -m.f[(i, j), t] <= network["edges"][(i,
        ↪  j)]["f_max"],
    )

    return model
```

Scenario Generation

We generate $T = 100$ scenarios in which the wind and solar production deviate from the forecast values. Such deviations, named *imbalances*, are generated uniformly at random assuming that the realized wind or solar power is between 0.5 and 1.5 times the forecast value.

For ease of calculations, for each scenario, we define a separate data structure with the total energy imbalance and the total absolute imbalance.

```
# Define the set of nodes with possible deviations from forecasts,
# i.e. those with either a wind or a solar generator
SW = {48, 53, 58, 60, 64, 65}
SW_df = nodes_df[nodes_df["node_id"].isin(SW)]

# Define the number of scenarios and the random seed
T = 100
seed = 0
rng = np.random.default_rng(seed)

# Imbalances are generated uniformly at random assuming the realized
# wind or solar power is between 0.5 and 1.5 times the forecasted value
imbalances = [
    {
```

```
            i: rng.uniform(-nodes_df["p_min"][i] / 2, nodes_df["p_min"][i] / 2)
            if i in SW
            else 0
            for i in nodes_df.index
        }
        for t in range(T)
]
totalimbalances = {t: sum(imbalances[t].values()) for t in
↪    range(len(imbalances))}
abstotalimbalances = {t: abs(totalimbalances[t]) for t in
↪    range(len(totalimbalances))}
```

Perfect Forecast Case (No Imbalances)

We first solve the optimization model in the case where the forecast for solar and wind power is perfect, which means there is no imbalance. In this case, the recourse actions are not needed, and the second-stage part of the problem is trivial.

```
# Trivial arrays for the case of perfect forecast and no need of recourse
↪    actions
zeroimbalances = [{i: 0 for i in nodes_df.index}]
zerototalimbalances = {0: sum(zeroimbalances[0].values())}
zeroabstotalimbalances = {0: abs(zerototalimbalances[0])}

# Solve the model
m = OPF_participationfactors(
    network,
    zeroimbalances,
    zerototalimbalances,
    zeroabstotalimbalances,
    uniformparticipationfactors=True,
)
SOLVER.solve(m)
firststagecost = sum(
    data["c_var"] * m.p[i].value
    for i, data in network["nodes"].items()
    if data["is_generator"]
)
```

The resulting (first-stage only) energy production cost is $40,385.16$.

It is key to understand how this solution would perform if there were perturbations in renewable energy production.

First of all, it is not guaranteed that it is possible to find a feasible solution in such a case. We can solve for the remaining variables while keeping the initial solution $\bar{p}$ fixed and check that with uniform participation factors, it is not possible to find a feasible flow in any of the scenarios we consider. If instead we allow for nonuniform participation factors, then this is not possible in 13 out of 100 scenarios.

Putting the feasibility issues aside for a moment, let us check how much extra cost would there be with uniform participation factors, which would be on average across our scenarios. We can calculate this by taking the total imbalance and computing the cost of the recourse action to cover it, assuming that every coal and gas generator adjusts its production proportionally to the optimal participation factor previously obtained. The average cost of energy production due to recourse actions is 5354.25, resulting in a total average cost of 45,739.41.

Stochastic Case (Nonzero Imbalances)

If we assume that the forecast for solar and wind power is not perfect, then the total energy imbalance in the network will be nonzero in each scenario. The resulting average total cost of energy production would be much higher than in the deterministic scenario. This is intuitive because recourse actions are needed to cover the imbalance and recourse actions are much more expensive than first-stage production-level decisions.

We now solve the two-stage SO model that accounts for the fluctuations of solar and wind power from their forecasts, using the 100 generated scenarios. In this case, the recourse actions are still needed, but we assume fixed uniform participation factors equal to 0.1 for all 10 conventional generators in $\mathcal{G}^{coal} \cup \mathcal{G}^{gas}$.

```python
m = OPF_participationfactors(
    network,
    imbalances,
    totalimbalances,
    abstotalimbalances,
    uniformparticipationfactors=True,
)
SOLVER.solve(m)
print(
    "The optimal production levels for the conventional generators are",
    [np.round(m.p[i].value, 2) for i in m.CG],
)
print(
    "The participation factors for the conventional generators are",
    [np.round(m.alpha[i].value, 2) for i in m.CG],
)

firststagecost = sum(
    data["c_var"] * m.p[i].value
    for i, data in network["nodes"].items()
    if data["is_generator"]
)
averagerecoursecost = (
    1
    / T
    * sum(
        sum(
            2 * data["c_var"] * m.alpha[i].value * m.abs_total_imbalance[t]
            for i, data in network["nodes"].items()
        )
        for t in m.T
    )
)
```

The first-stage energy production cost is $41,719.19$, and the average energy production cost due to recourse actions is 5354.25, so the total cost is $47,073.44$.

We see that the total average production cost is slightly higher than in the first "perfect forecast" nominal scenario, but the benefit of this newly obtained solution is that we are sure that we have a feasible power flow dispatch in all scenarios.

We argue that using this solution should be preferable: Even if the "nominal" solution had a slightly lower average production cost, we need to factor in how costly it will be for the network operator when the solution becomes infeasible. Having an infeasible network configuration means that there is a risk of cascading failures and/or blackout, which, in addition to potentially damaging the infrastructure, is dramatically more expensive from a financial and societal perspective, and having a 13% chance of this happening is just unaffordable.

Next, we optimize the participation factors α_i together with the initial power levels and achieve a reduction in the average total cost. We can do this using the same function as before, but changing the argument `uniformparticipationfactors` to `False`. The resulting first-stage energy production cost is $40,446.13$, and the average energy production cost due to recourse actions is 5969.05, so the total cost is $46,415.18$. This energy dispatch is about 1.4% cheaper than the solution with uniform participation factors. It might seem like a small difference, but given the high volumes of energy produced and consumed, say at the national level, this makes a huge difference.

Exercises

Exercise 10.1 Consider the following two-stage SO problem:

$$\min \quad x + \mathbb{E}Q(x,z) \tag{10.7}$$
$$\text{s.t.} \quad x \geq 0,$$

where the uncertain parameter z is one-dimensional and follows a uniform distribution on the $[0,4]$ interval, and where the second-stage problem is

$$\min \quad 2y$$
$$\text{s.t.} \quad y + x \geq z - 2$$
$$\quad y + x \geq 4 - 2z$$
$$\quad y \geq 0.$$

What is the optimal solution and optimal value of this problem? What would be the solution obtained assuming the average value of z and its corresponding average-case performance in terms of the objective function?

Exercise 10.2 Consider the following two-stage RO problem:

$$\min \quad x + \sup_{z \in Z} Q(x,z) \tag{10.8}$$
$$\text{s.t.} \quad x \geq 0,$$

where $Z = [0,3]$ and where

$$Q(x,z) = \min \quad y$$
$$\text{s.t.} \quad y \leq x + z + 1$$
$$y \geq x + z - 1$$
$$y \geq 0.$$

What is the optimal solution and the optimal value of this problem?

Exercise 10.3 Consider again the problem of the previous exercise. It is common to simplify two-stage RO problems by formulating them as a single-stage problem (in which case the second-stage decision y cannot vary depending on z).

In the above problem's case, such a formulation would look as follows:

$$\min \quad x + y$$
$$\text{s.t.} \quad y \leq x + z + 1, \qquad \forall z \in Z$$
$$y \geq x + z - 1, \qquad \forall z \in Z$$
$$x, y \geq 0.$$

What is the optimal solution and the optimal value of this problem?

Exercise 10.4 Consider the optimization problem of Exercise 10.2 and solve it using the linear decision rule approach. Is the optimal decision x and the optimal value the same as in the analytical optimal solution?

Exercise 10.5 Single-stage RO problems with polyhedral uncertainty set (box, budgeted, etc.) can be solved equivalently by enumerating all vertices (extreme points) of the uncertainty set and enforcing the constraints on them. That is, a problem involving a constraint

$$(a + Az)^\top x \leq b - b^\top z, \quad \forall z \in Z$$

is equivalent to that with a constraint

$$(a + Az)^\top x \leq b - b^\top z, \quad \forall z \in \bar{Z},$$

where $\bar{Z}$ is the set of vertices of Z. However, this intuition does not hold for two-stage problems. To show it, construct a robust two-stage optimization problem with a single-dimensional interval uncertainty set $Z = [a,b]$ in which the worst-case realization of the uncertain parameter is not one of its endpoints.

Exercise 10.6 Consider again the problem in (10.7). Solve it using the SAA method for increasing sample sizes. Do you observe convergence to the optimal solution and value obtained through mathematical analysis?

Exercise 10.7 Solve the RO version of problem (10.7), that is, solve the problem:

$$\min \quad x + \sup_{z \in Z} Q(x,z)$$
$$\text{s.t.} \quad x \geq 0,$$

where $Z = [0,4]$, with the second-stage problem remaining the same. Is the optimal solution obtained the same or is it different from that in the two-stage SO case? What is the average performance of the robust solution? What is the worst-case performance of the SO solution?

Exercise 10.8 Consider a two-stage SO problem in which all the first-stage and second-stage decision variables are continuous (there are no integrality constraints). Show that the second-stage value function $Q(x)$ corresponding to solving the problem with the SAA technique:

$$Q(\boldsymbol{x}) = \min \quad \frac{1}{N}\sum_{j=1}^{N} \boldsymbol{q}^\top \boldsymbol{y}^j$$

$$\text{s.t.} \quad R(\boldsymbol{z}^j)\boldsymbol{x} + S(\boldsymbol{z}^j)\boldsymbol{y}^j \leq \boldsymbol{t}(\boldsymbol{z}^j), \qquad j = 1,\ldots,N$$

$$\boldsymbol{y}^j \in \mathbb{R}^{n_y}, \qquad\qquad\qquad\qquad j = 1,\ldots,N,$$

is convex in $\boldsymbol{x}$.

Exercise 10.9 Construct a two-stage SO problem where a solution obtained using the SAA method is always infeasible, that is, when an optimal first-stage decision is used and the original two-stage SO problem is solved with that fixed first-stage decision, there exists no feasible solution.

Exercise 10.10 Consider a generation expansion planning problem with a set of generators $g \in \{A,B,C\}$ depicted in the following figure.

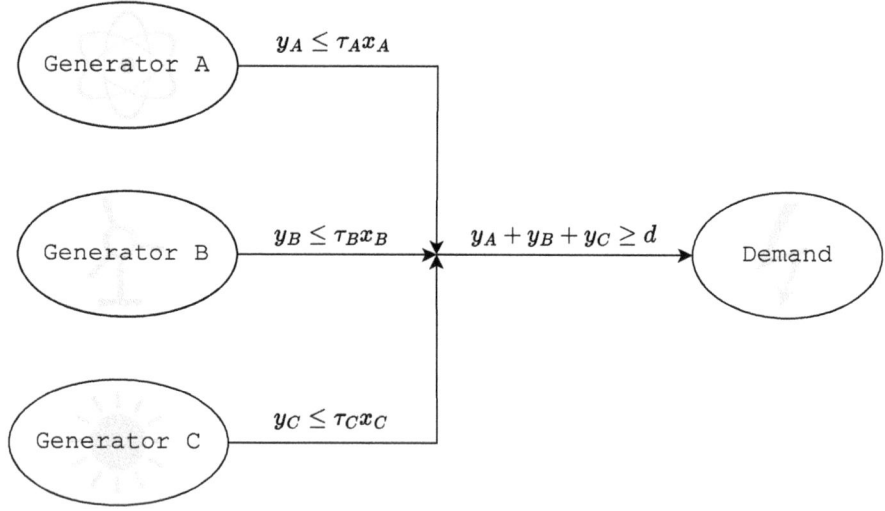

The goal is to determine which generators to "expand" and utilize to satisfy the energy demand at minimal cost.

For this problem, we have six decision variables: For each generator g, there is a *strategic* (first-stage) investment decision x_g and an *operational* (second-stage) energy flow decision y_g. The size of the investment is discrete and comes in three possible sizes: small, medium, or large (i.e., $x_g \in \{0,1,2,3\}$). The energy flows y_g are continuous and nonnegative (i.e., $y_g \geq 0$).

Our initial optimization model contains 10 parameters:

- total demand d;
- capacity factor τ_g, $g \in \{A,B,C\}$;
- fixed cost f_g, $g \in \{A,B,C\}$;
- variable cost c_g, $g \in \{A,B,C\}$.

If there had been no uncertainty in the parameters, the "nominal" problem could be formulated mathematically as the following single-stage optimization problem:

$$\min \quad \sum_g f_g x_g + \sum_g c_g y_g \tag{10.9}$$

$$\text{s.t.} \quad \sum_g y_g \geq d$$

$$y_g \leq \tau_g x_g, \qquad\qquad \forall g \in \{A,B,C\}$$

$$y_g \geq 0, \qquad\qquad \forall g \in \{A,B,C\}$$

$$x_g \in \{0,1,2,3\}, \qquad\qquad \forall g \in \{A,B,C\}.$$

Of the 10 parameters in our optimization model (10.9), three are considered uncertain. These are the demand d and capacity factors τ_B and τ_C for wind and solar, respectively. The other seven parameters are considered known/fixed and take on the following values:

$$\tau_A = 0.90, \quad \begin{pmatrix} f_A \\ f_B \\ f_C \end{pmatrix} = \begin{pmatrix} 8 \\ 4 \\ 2 \end{pmatrix}, \quad \begin{pmatrix} c_A \\ c_B \\ c_C \end{pmatrix} = \begin{pmatrix} 1.0 \\ 0.1 \\ 0.1 \end{pmatrix}.$$

For uncertain parameters, we assume that there is a discrete set of four scenarios that occur with known probabilities. For each scenario $s \in \{1,2,3,4\}$, let p^s, d^s, τ_B^s, and τ_C^s denote the probability and parametric values associated with scenario s, these are presented in the table below.

s	p^s	d^s	τ_B^s	τ_C^s
1	1/4	1.5	0.40	0.30
2	1/4	1.8	0.30	0.30
3	1/4	2.2	0.40	0.20
4	1/4	2.5	0.30	0.20

(a) *Nominal solution and its performance.* In this part of the exercise, we test the performance of a solution obtained using the common practice of optimizing for the "average scenario." For the uncertain parameters, assume a single "nominal scenario" being the average of the four scenarios provided:

$$d = 2, \quad \begin{pmatrix} \tau_B \\ \tau_C \end{pmatrix} = \begin{pmatrix} 0.35 \\ 0.25 \end{pmatrix}. \tag{10.10}$$

For this nominal scenario, determine the optimal solution to (10.9). Next, assume that the decisions x are fixed but that the decisions y can be adapted depending on the observed values of the uncertain parameters. What is the probability of infeasibility (there is no feasible decision y) for the x-decisions obtained for the nominal solution? Among the scenarios for which there exists a feasible solution, what is the average objective function value?

(b) *No-regret moves.* Identify if there are decisions that remain optimal regardless of the assumed data scenario. That is, for each of the scenarios considered, solve the nominal problem (10.9) and inspect which parts of the solution remain the same and which remain the same across the scenarios.

(c) *Stochastic optimization.* Formulate a two-stage stochastic optimization version of (10.9) under the four scenarios provided and solve this problem to optimality. What is the expected objective function value of this solution? How do the optimal x decisions differ from the best "nominal solution"?

(d) *Robust optimization.* Formulate a two-stage robust optimization version of (10.9) and solve it assuming interval uncertainty sets for each of the uncertain parameters, where $\mathcal{U}_d = \{d \colon 1.5 \leq d \leq 2.5\}$, $\mathcal{U}_{\tau_B} = \{\tau_B \colon 0.30 \leq \tau_B \leq 0.40\}$, and $\mathcal{U}_{\tau_C} = \{\tau_C \colon 0.20 \leq \tau_C \leq 0.30\}$. Use the CCG approach or linear decision rules for the second stage. How do the optimal x decisions differ from the best "nominal solution"?

Appendix A

Linear Algebra Primer

We briefly review some key concepts of linear algebra and introduce some notation that we will use throughout the book. First, we need to distinguish between scalars, vectors, and matrices.

- A scalar is a single numerical value, usually a real number $s \in \mathbb{R}$.
- A vector is an array of scalars $\boldsymbol{v} = [v_1, v_2, \ldots, v_n] \in \mathbb{R}^n$, each of which we can refer to as entry, component, or element. We refer to the ith entry of the vector $\boldsymbol{v}$ as v_i. The dimension of a vector is equal to the number of its entries, so the dimension of $\boldsymbol{v} \in \mathbb{R}^n$ is n. We will always write vectors in bold font.
- A $m \times n$ matrix is a rectangular array of scalars containing both rows and columns, that is,

$$
A = \begin{bmatrix} A_{11} & A_{12} & \ldots & A_{1n} \\ \vdots & & \ddots & \\ A_{m1} & \ldots & & A_{mn} \end{bmatrix} \in \mathbb{R}^{m \times n}.
$$

An element of a $m \times n$ matrix is accessed with the pair of indices (i, j), where i is the row index, which varies from 1 to m, and j is the column index, varying from 1 to n. If a matrix has the same number of rows as columns, it is referred to as a square matrix. We will always write matrices using capital letters.

A vector can be seen as a special case of a matrix and, similarly, a scalar can be seen as a special case of both vectors and matrices. More specifically, a scalar is a 1×1 matrix or vector. By default, vectors are taken to be column vectors. Hence, an n-dimensional vector can be viewed as a $n \times 1$ matrix.

It is possible to multiply a vector $\boldsymbol{v}$ with a scalar s by multiplying each component of the vector with the scalar, obtaining a new vector defined as

$$
s\boldsymbol{v} = [sv_1, \ldots, sv_n].
$$

Two vectors of the same dimension, say $\boldsymbol{v}, \boldsymbol{w} \in \mathbb{R}^n$, can be added or subtracted and these operations are performed component-wise:

$$
\boldsymbol{v} + \boldsymbol{w} = [v_1 + w_1, \ldots, v_n + w_n].
$$

Matrices can be added or subtracted in a similar way when they have exactly the same size.

It is often convenient to use the notation 0_n for the $n \times n$ matrix with all entries equal to zero and the notation I_n for the identity matrix of size n, that is, the $n \times n$ matrix with ones on the diagonal and zero elsewhere.

The transpose of a $m \times n$ matrix A is the $n \times m$ matrix $A^\top$ obtained by flipping its rows and columns, that is, $(A^\top)_{ij} = A_{ji}$. Vectors being a special case of a matrix, we can define the transpose of a $n \times 1$ (column) vector as its representation as a row vector $1 \times n$.

A matrix A is said to be symmetric if $A = A^\top$, that is, when its entries are symmetric with respect to the main diagonal.

The ordered multiplication between two matrices A and B is well defined only if the number of columns of the first matrix is equal to the number of rows of the second matrix. Assuming that this is the case and that A and B are respectively a $m \times n$ and a $n \times d$ matrices, then AB is the $m \times d$ matrix defined entry-wise as

$$(AB)_{ij} = \sum_{k=1}^{n} A_{ik} B_{kj}.$$

The matrix–vector multiplication is a special case of this and is possible when the number of columns of the matrix matches the number of rows of the vector. More specifically, if A is a $m \times n$ and v is a $n \times 1$ vector, then Av is the $m \times 1$ vector defined entry-wise as

$$(Av)_i = \sum_{k=1}^{n} A_{ik} v_k.$$

If we consider the multiplication between two vectors of the same dimension, then we recover the so-called dot product between two vectors. If $v, w \in \mathbb{R}^n$, their dot product is the scalar defined as

$$v \cdot w = v^\top w = \sum_{i=1}^{n} v_i w_i \in \mathbb{R}.$$

For any positive integer p, we can define the p-norm of a vector v as the nonnegative scalar

$$\|v\|_p := \left(\sum_{i=1}^{n} |v_i|^p \right)^{1/p}.$$

In the special case of $p = 2$, we recover the Euclidean norm, which can be equivalently reformulated using the dot product,

$$\|v\|_2 = \sqrt{v^T v}.$$

Given a $n \times n$ matrix A, a scalar $\lambda \in \mathbb{R}$ is said to be an eigenvalue of A if there exists a vector $v \in \mathbb{R}^n$ such that

$$Av = \lambda v.$$

In principle, the eigenvalues could be complex numbers, but a well-known spectral theorem due to Charles Hermite guarantees that all the eigenvalues of a real symmetric matrix are real numbers.

Appendix B

Solutions of Selected Exercises

Solution 1.1

(a) For each item, you must decide whether to take it or not: There is thus one decision variable per potential item to load.

(b) You should consider its weight relative to the maximum weight that you are willing to carry. The joint weight will be bound by the limit you have stipulated and will give rise to a constraint. In order to optimize, we need to value solutions in some way. One way is to measure utility per item: a number that expresses the importance of the items and that you need to decide by yourself. We already had the joint weight, raising a constraint, now we have the joint utility: the objective function that you want to maximize.

(c) Denoting by I the set of items, by g_i the weight in grams, and by u_i the utility of item i, one may think of decisions x_i indicating whether item i is taken. These variables should take the value 1 if and only if the item is selected.

Our model becomes:

$$
\begin{aligned}
\max \quad & \sum_{i \in I} u_i x_i \\
\text{s.t.} \quad & \sum_{i \in I} g_i x_i \le w \\
& x_i \in \mathbb{B}, \qquad\qquad\qquad \forall\, i \in I.
\end{aligned}
$$

We maximize total utility while ensuring that the total weight is within the stipulated bound $w = 1500$.

(d) Assuming a dataframe called `data` we can write the code below.

```python
def Luggage(data, max_weight):
    m = pyo.ConcreteModel()
    m.I = pyo.Set(initialize=data.index)
    m.g = pyo.Param(m.I, initialize=data.gram)
    m.u = pyo.Param(m.I, initialize=data.utility)
    m.w = pyo.Param(initialize=max_weight)
    m.x = pyo.Var(m.I, domain=pyo.Binary)

    @m.Objective(sense=pyo.maximize)
    def utility(m):
        return pyo.quicksum(m.u[i] * m.x[i] for i in m.I)
```

```
    @m.Constraint()
    def weight(m):
        return pyo.quicksum(m.g[i] * m.x[i] for i in m.I) <= m.w

    return m
```

For completeness, let us see a way to get from the text to the data:

```
text = """\item slippers, 150 gram, utility 10
\item pyjamas, 350 gram, utility 9
\item pair of trousers, 250 gram, utility 6
\item blue T-shirt, 100 gram, utility 6
\item red T-shirt, 120 gram, utility 7
\item thermal bottle, 500 gram, utility 6
\item dental hygiene kit, 120 gram, utility 9
\item this book, 550 gram, utility 8
\item the latest Dan Brown, 850 gram, utility 8
\item binoculars, 350 gram, utility 6
\item underwear, 100 gram,  utility 10
\item socks, 50 gram, utility 9"""


def unpack(line):
    d, g, u = line.split(",")
    return (
        d.replace(r"\item", "").strip(),
        int(g.replace(" gram", "").strip()),
        int(u.replace("utility", "").strip()),
    )


records = [unpack(line) for line in text.splitlines()]

data = pd.DataFrame.from_records(
    records, columns=["item", "gram", "utility"]
).set_index("item", drop=True)
```

We can now solve our problem with the following code:

```
lug = Luggage(data, 2000)
SOLVER.solve(lug)
sol = pd.DataFrame.from_dict(
    lug.x.extract_values(), orient="index", columns=["used"], dtype=bool
)
```

The solution provided in the table below is stored in a dataframe.

	used
slippers	True
pajamas	True
pair of trousers	True
blue T-shirt	True
red T-shirt	True

thermal bottle	False
dental hygiene kit	True
this book	False
the latest Dan Brown	False
binoculars	False
underwear	True
socks	True

In this case, the interpretation is relatively straightforward: We take the items whose indicator variable is set to 1, or to `True` as shown in the table.

Of course, we can list the solution in many other ways, such as a packing list consisting only of the selected items.

We are also interested in the weight loaded and the utility carried in all cases. A call to `lug.utility()` yields 66 while `lug.weight()` implies that we will be carrying 1240 grams.

(e) You either set the decision variable for this book to take the value 1 with an additional constraint, or you lower the allowed weight to $1500 - 550 = 950$ grams, remove the book from the model, put it in the bag, and solve the smaller problem. Ignoring the decision about this book leads to the selection of additional 890 grams to carry with a utility of 57; basically, it means that you will leave your pajamas at home, as the following table illustrates.

	used
slippers	True
pajamas	False
pair of trousers	True
blue T-shirt	True
red T-shirt	True
thermal bottle	False
dental hygiene kit	True
the latest Dan Brown	False
binoculars	False
underwear	True
socks	True

Solution 2.1

(a) The manager can decide how many *offers* and how many *boxes* to sell. We denote these two by x_1 and x_2, respectively. Each offer sold consumes one phone and two prepaid SIM cards from the stock. Each box consumes one phone, one hands-free kit, and three prepaid SIM cards.

The objective is to maximize the money earned by selling the decided amount of offers and boxes, and the constraints ensure that the stock is respected. The model can be formulated as follows:

$$\begin{aligned} \max \quad & 7x_1 + 9x_2 \\ \text{s.t.} \quad & x_1 + x_2 \leq 8 \\ & x_2 \leq 4 \\ & 2x_1 + 3x_2 \leq 19 \\ & x_1, x_2 \geq 0. \end{aligned}$$

In the following, we see the graphical resolution with an isoline of the objective function drawn on the optimal solution $x_1^* = 5$, $x_2^* = 3$.

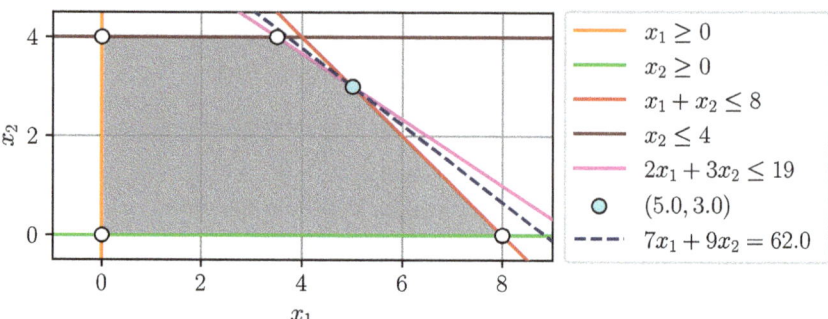

Below is a straightforward implementation of the model in Pyomo that uses indices for the decision variables.

```python
def PhoneStore_v2():
    m = pyo.ConcreteModel("Phone Store with indices")

    m.T = pyo.Set(initialize=["offer", "box"])
    m.x = pyo.Var(m.T, within=pyo.NonNegativeReals)

    @m.Objective(sense=pyo.maximize)
    def obj(m):
        return 7 * m.x["offer"] + 9 * m.x["box"]

    @m.Constraint()
    def phones(m):
        return m.x["offer"] + m.x["box"] <= 8

    @m.Constraint()
    def handfree(m):
        return m.x["box"] <= 4

    @m.Constraint()
    def sim(m):
        return 2 * m.x["offer"] + 3 * m.x["box"] <= 19

    return m
```

(b) If λ_1, λ_2, and λ_3 denote the minimum price to accept for a phone, a hands-free kit, and a prepaid SIM card, respectively, the manager wants to solve the problem below to find the lowest prices that the salesperson should pay to get extra phones, hands-free kits, and prepaid cards, respectively, while the manager does not lose money compared to the original plan:

$$\begin{aligned} \min \quad & 8\lambda_1 + 4\lambda_2 + 19\lambda_3 \\ \text{s.t.} \quad & \lambda_1 + 2\lambda_3 \geq 7 \\ & \lambda_1 + \lambda_2 + 3\lambda_3 \geq 9 \\ & \lambda_1, \lambda_2, \lambda_3 \geq 0. \end{aligned}$$

This problem is formulated to minimize the total amount that the salesperson pays to the manager while ensuring that the revenues from an offer and a box are at least met.

(c) The two problems are each other's duals.

(d) Since the manager knows that at most she may get $62 for the whole stock, the first two options are no-brainers: sell! The third option is also interesting in the event that the trouble to organize the campaign is estimated to cost at least $2. The importance of valuing the trouble of organizing a campaign as part of the decision-making process illustrates that the model is clearly not complete, but definitely useful. This is the fate of most models!

Solution 3.1

(a) The decision variables are the same as in Example 3.10:

$$x_j = \begin{cases} 1 & \text{facility } j \text{ is installed} \\ 0, & \text{otherwise} \end{cases}$$

and

$$y_{ij} = \begin{cases} 1 & \text{customer } i \text{ is served at } j \\ 0, & \text{otherwise.} \end{cases}$$

The models are as follows. First, the aggregated model:

$$\max \quad \sum_{i \in I} \sum_{j \in J} v_i y_{ij}$$

$$\text{s.t.} \quad \sum_{i \in I} y_{ij} \leq n x_j, \qquad \forall j \in J$$

$$\sum_{j \in J} y_{ij} \leq 1, \qquad \forall i \in I$$

$$\sum_{j \in J} x_j = p$$

$$y_{ij} = 0, \qquad \forall i \in I, \forall j \in J : d_{ij} > S$$

$$x_j \in \mathbb{B}, \qquad \forall j \in J$$

$$y_{ij} \in \mathbb{B}, \qquad \forall i \in I, \forall j \in J.$$

Now, the disaggregated model:

$$\max \quad \sum_{i \in I} \sum_{j \in J} v_i y_{ij}$$

$$\text{s.t.} \quad y_{ij} \leq x_j, \qquad \forall i \in I, \forall j \in J$$

$$\sum_{j \in J} y_{ij} \leq 1, \qquad \forall i \in I$$

$$\sum_{j \in J} x_j = p$$

$$y_{ij} = 0, \qquad\qquad \forall i \in I, \forall j \in J : d_{ij} > S$$
$$x_j \in \mathbb{B}, \qquad\qquad \forall j \in J$$
$$y_{ij} \in \mathbb{B}, \qquad\qquad \forall i \in I, \forall j \in J.$$

(b) For the single-index model, we use the same variables x_j for whether the facility j is built, but instead of variables y_{ij} to indicate if customer i is served at j, we now simply define variables z_i for each customer $i \in I$ to indicate if that customer can be served by a facility that is in fact installed:

$$\max \quad \sum_{i \in I} v_i z_i$$

$$\text{s.t.} \quad z_i \le \sum_{j \in J : d_{ij} \le S} x_j, \qquad\qquad \forall i \in I$$

$$\sum_{j \in J} x_j = p$$

$$x_j \in \mathbb{B}, \qquad\qquad \forall j \in J$$

$$z_i \in \mathbb{B}, \qquad\qquad \forall i \in I.$$

The first constraint effectively allows z_i to be served if at least one facility is installed within reach. For that reason, we refer to this model as a covering formulation.

We implemented all three models and generated random instances with the dimensions listed below.

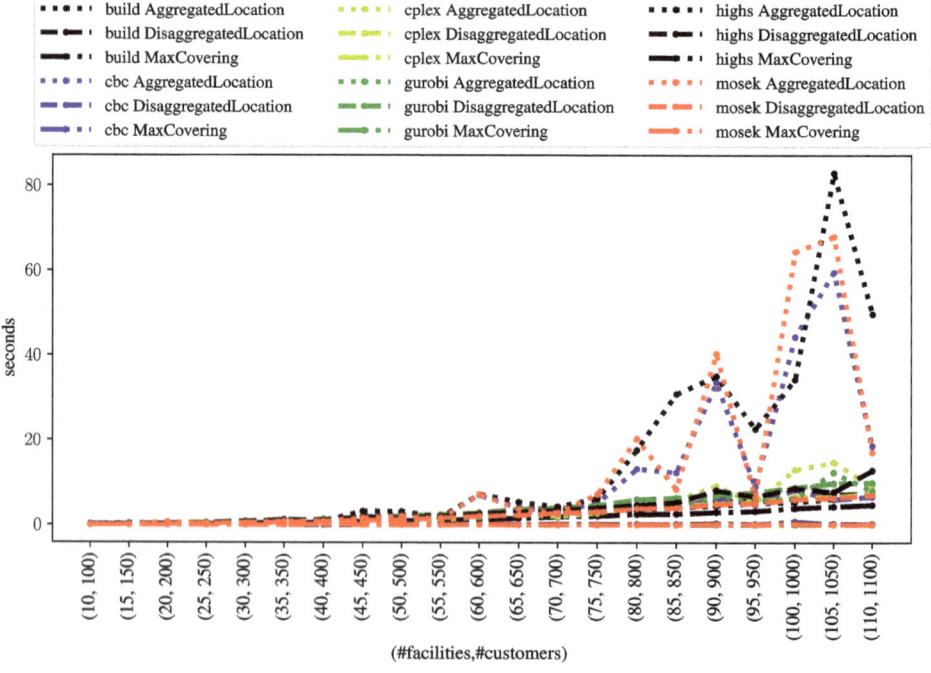

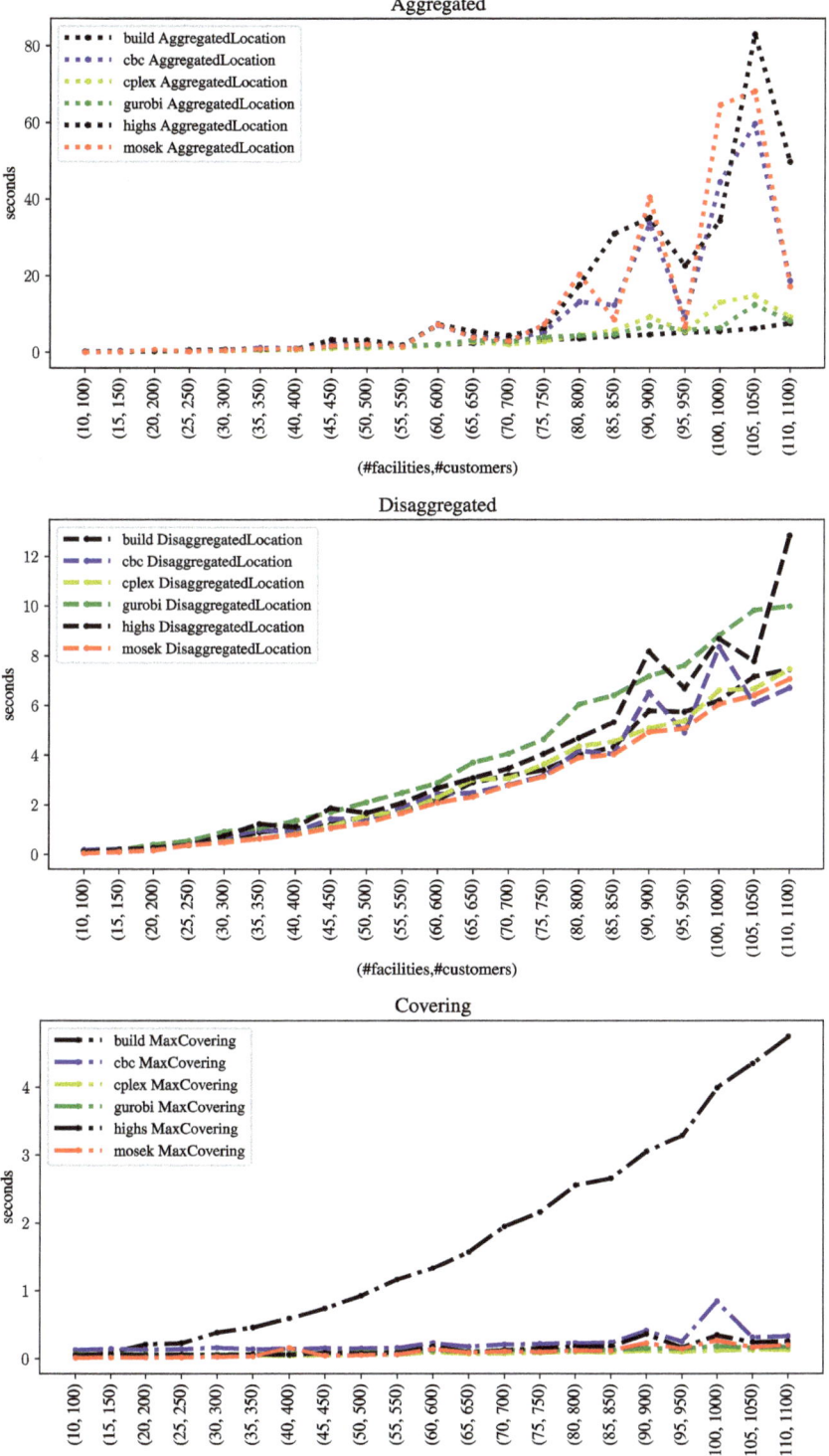

Several things may strike us from these experiments: First, the disaggregated model consistently outperforms the aggregated one in this case, even without switching cut generation off. Second, for the disaggregated models, and especially the max covering model, the choice of solver does not seem to influence the solving times significantly. Finally, and more remarkably, the covering model scales up much better with the size of the instance. Note that the time to create the Pyomo model is larger than the time to solve it. Furthermore, the time taken to model scales up in a worse way. Note that all three models are equivalent and each of them gives the same optimal solution.

(c) The idea is to have our problem be solved in such a way that one objective has a "priority" over another objective. It can be achieved in two ways: (i) sequential solving of two problems where in the second round we "lock" the objective value of the first one through a constraint, and (ii) choosing a smart penalty term for one of the objectives so that two problems can be solved in one go. We choose the second approach here.

We need to involve the x_i variables in the objective function, adding:

$$\sum_j w_j x_j.$$

The w_j should be negative to penalize while maximizing but chosen small not to obfuscate the main objective. Each additional customer served should count more than what any number of facilities maintained closed can discount. The choice

$$w_j = \frac{-\min(v_i)}{p+1}$$

is good, assuming that each v_i is positive. The last model becomes:

$$
\begin{aligned}
\max \quad & \sum_{i \in I} v_i z_i + \sum_{j \in J} w_j x_j \\
\text{s.t.} \quad & z_i \le \sum_{j \in J : \, d_{ij} \le S} x_j, && \forall i \in I \\
& \sum_{j \in J} x_j \le p \\
& x_j \in \mathbb{B}, && \forall j \in J \\
& z_i \in \mathbb{B}, && \forall i \in I.
\end{aligned}
$$

The other two models can be modified in the same way.

(d) This is a good moment to remember that, in Pyomo, parameters may be mutable. Defining parameter p as mutable allows modifying and redoing the optimization of the model for each value of p. One of our randomly generated instances gave the following Pareto frontier:

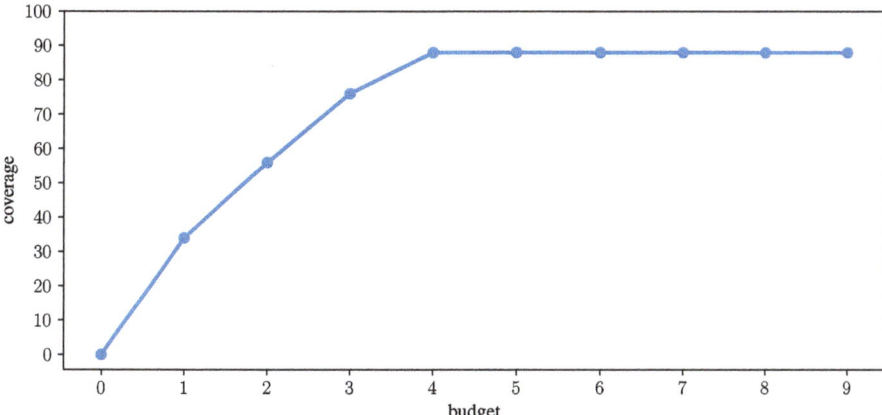

We can see that, for this example, despite the budget, we cannot cover more than about 88% of the customers.

Solution 4.1 Although small, the problem considered in this exercise is representative of many practical applications, including the assignment of aircraft and crews to already scheduled flights.

(a) Representing each event by a node, we draw an arrow from event s to event t if the start time of t is not before the end time of s plus the travel time from s to t. We obtain the following network.

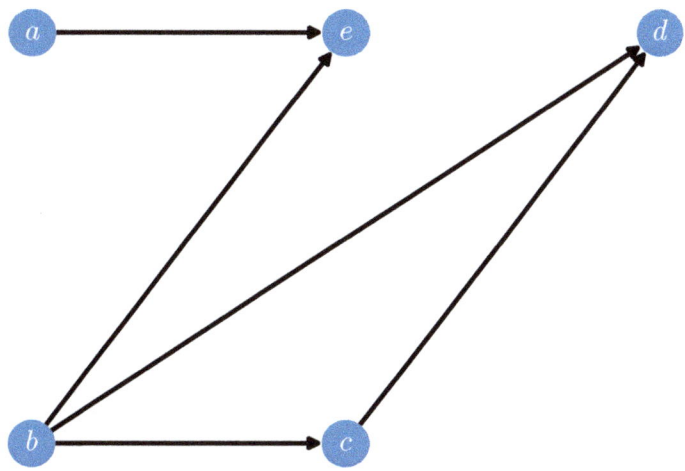

We easily see that two juries suffice: One attends a and e and the other attends b, c, and d in that order.

It is not easy to develop a general procedure for this problem, but we can at least formulate it better. In fact, we are looking for a minimal set of distinct *paths* through the graph that covers all nodes, and where no two paths share any node, see [Wik23].

As it turns out, as long as our graph does not have any cycles, one can solve any instance of such a problem by finding the max flow on a specially constructed graph with a procedure described in [BL23]. We build a network where the max-flow problem will correspond to utilizing connecting as subsequent event pairs with a single jury as possible. In that graph, each event receives two nodes, x (that event being a predecessor of another event) and y (that event being a successor of another event), and we link the x node of one event and the y node of another event if the first event can precede the latter. Furthermore, we add a source node s connected to all x nodes and a target node t connected to all y nodes. We assume that each arc has a capacity of 1.

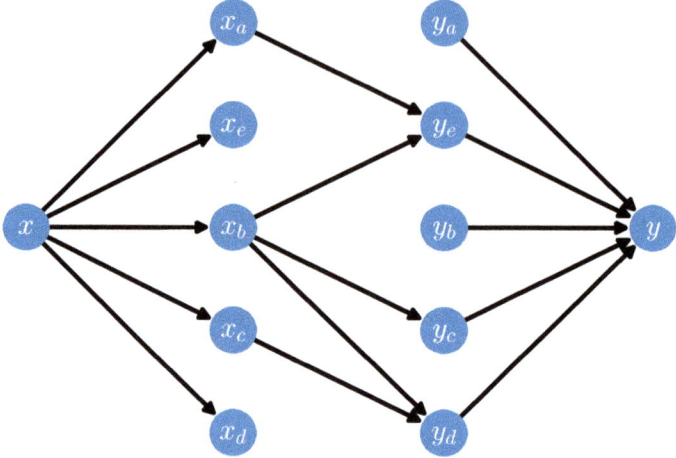

The idea is that the higher the total flow, the more pairs of events are linked with the same jury, and the fewer juries we need overall. Determining a max flow from s to t gives the following solution, from which we can easily recover the original one.

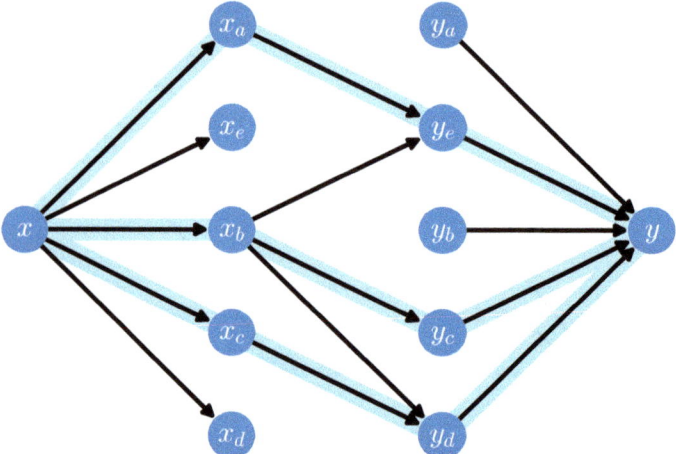

(b) Each event receives start and finish nodes and is represented by an edge connecting them. Additional edges are created from the finish node of an event to the start node of the subsequent event within reach. Just as in Example 4.1, an edge is added from the start of the network to each start of an event and another from each finish of an event to the end of the network.

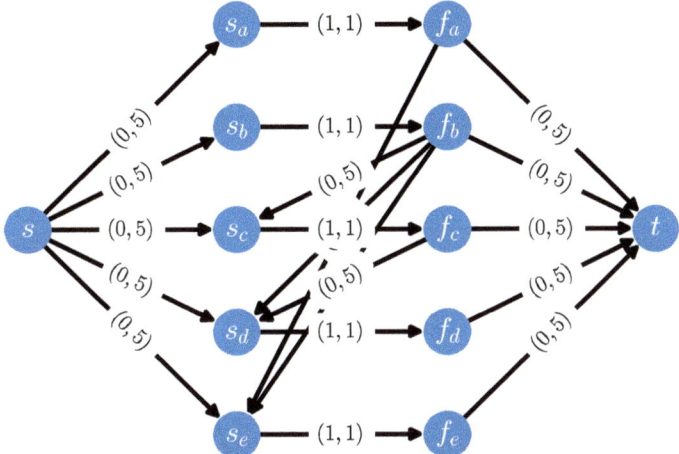

On each edge, we indicate in brackets a lower and an upper bound for the flow that should pass through it. To ensure that each event receives a jury, both the lower and upper bounds for these edges are equal to 1. All other edges allow flow from 0 to 5, which is the total number of events and hence also the maximal number of juries needed.

We can formulate this problem as a min-cost flow problem where the unit flow on each edge is priced at 1. Since the constraint structure is that of a flow problem, the problem is TU and we can use an LO solver to find that the optimal solution assigns the value 1 to the variables:

$$x_{s_e,f_e}, x_{s_d,f_d}, x_{f_b,s_c}, x_{s,s_a}, x_{f_a,s_e}, x_{s_b,f_b}, x_{s_a,f_a}, x_{f_d,t}, x_{f_e,t}, x_{s_c,f_c}, x_{f_c,s_d}, x_{s,s_b},$$

and 0 to all others. The value of v is 2, therefore two juries suffice.

The following is the corresponding Pyomo implementation:

```python
def Scheduling(G, s="s", t="t", domain=pyo.Reals):
    m = pyo.ConcreteModel("Scheduling event juries")
    m.E = G.edges
    m.N = G.nodes

    def _bounds(m, s, t):
        return G.get_edge_data(s, t).get("lower", 0), G.get_edge_data(s,
        ↪    t).get(
            "upper", None)
        )

    m.x = pyo.Var(m.E, within=domain, bounds=_bounds)
    m.v = pyo.Var(within=domain)

    @m.Objective(sense=pyo.minimize)
    def obj(m):
        return m.v

    @m.Constraint(m.N)
    def flow_balance(m, n):
```

```
        rhs = m.v if n == s else -m.v if n == t else 0
        return (
            sum(m.x[e] for e in G.out_edges(n)) - sum(m.x[e] for e in
            ↪    G.in_edges(n))
            == rhs
        )

    return m
```

The solution can also be seen on the network:

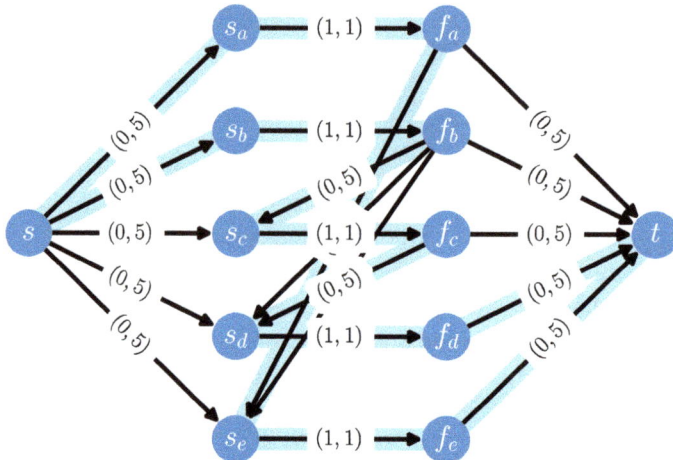

(c) Our problem becomes a min-cost flow problem where we assign a weight equal to the event duration or travel time to each arc corresponding to an event or travel. As nothing changes about the constraints, the problem remains TU and can be solved with an LO solver.

The following is the corresponding Pyomo implementation:

```
def SchedulingMinimumWorkingTime(G, s="s", t="t", domain=pyo.Reals):
    m = pyo.ConcreteModel("Scheduling event juries minimizing total
    ↪    working hours")
    m.E = G.edges
    m.N = G.nodes
    m.d = pyo.Param(m.E, initialize=nx.get_edge_attributes(G, "duration"))

    def _bounds(m, s, t):
        return G.get_edge_data(s, t).get("lower", 0), G.get_edge_data(s,
        ↪    t).get(
            "upper", None
        )

    m.x = pyo.Var(m.E, within=domain, bounds=_bounds)
    m.v = pyo.Var(within=domain)

    @m.Objective(sense=pyo.minimize)
```

```
def obj(m):
    return sum(m.d[e] * m.x[e] for e in m.E)


@m.Constraint(m.N)
def flow_balance(m, n):
    rhs = m.v if n == s else -m.v if n == t else 0
    return (
        sum(m.x[e] for e in G.out_edges(n)) - sum(m.x[e] for e in
        ↪   G.in_edges(n))
        == rhs
    )


return m
```

The solution is, not surprisingly, to use five juries, since, in that way, no traveling occurs during working time (events have to be covered anyway, so the problem boils down to a problem of minimizing the total travel time).

(d) In this case, the expression in the objective function of the previous model becomes used in a constraint. The constraint matrix now includes other numbers than 0, 1, and -1, making it obviously not TU. For that reason, the problem cannot be solved with an LO solver alone – the solution of the LO version would be a fractional solution using 2.75 juries with a total working time of 10 hours. Making it a MILO by imposing the integrality on the x variables leads to the following scheduling of three juries, with a total working time of 9 hours.

The Pyomo model is straightforward:

```
def SchedulingLimitingTotalWorkingTime(G, s="s", t="t", domain=pyo.Reals):
    m = pyo.ConcreteModel("Scheduling event juries limiting total working
    ↪   hours")
    m.E = G.edges
    m.N = G.nodes
    m.d = pyo.Param(m.E, initialize=nx.get_edge_attributes(G, "duration"))
    m.d_max = pyo.Param(initialize=10, mutable=True)

    def _bounds(m, s, t):
        return G.get_edge_data(s, t).get("lower", 0), G.get_edge_data(s,
        ↪   t).get(
            "upper", None
        )

    m.x = pyo.Var(m.E, within=domain, bounds=_bounds)
    m.v = pyo.Var(within=domain)

    @m.Objective(sense=pyo.minimize)
    def obj(m):
        return m.v

    @m.Constraint()
    def limit_total_working_time(m):
        return sum(m.d[e] * m.x[e] for e in m.E) <= m.d_max
```

```
@m.Constraint(m.N)
def flow_balance(m, n):
    rhs = m.v if n == s else -m.v if n == t else 0
    return (
        sum(m.x[e] for e in G.out_edges(n)) - sum(m.x[e] for e in
        ↪   G.in_edges(n))
        == rhs
    )

return m
```

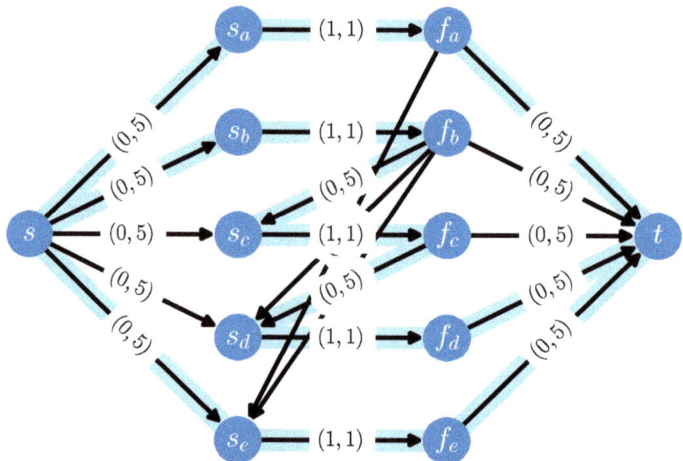

(e) Before we start devising a special approach for the problem with a new constraint added, it is always a good idea to verify if perhaps the solution without that constraint could already be feasible here (in which case it would also be optimal). If we take the solution of (a), we find that the jury covering a and e starts at 10 and ends at 18, violating the maximum of six working hours. The other jury attends three events but within the maximum. Therefore, we need another solution.

As it turns out, imposing limits *per jury* leads to the need to differentiate the flows. We end up with a problem formulation known as the *multicomodity flow problem*, which is notoriously difficult to solve in practice. The idea here is that we assume that at most K different commodities can flow through the network (the best bound on K is the total number of events, i.e., $K = 5$). Each such flow will correspond to a path through the system and we will impose the flow balance constraints separately for each of them, and also separately for each we will count the total time spent at events and in travel. The following is the mathematical model for a general graph $G = (V, E)$ with a source node s and target node t:

$$\min \quad \sum_{k \in \{1,\ldots,K\}} \sum_{j:\, (s,j) \in E} x_{sjk}$$

$$\text{s.t.} \quad \sum_{i \in V:(i,j) \in E} x_{ijk} - \sum_{i \in V:(j,i) \in E} x_{jik} = b_j, \qquad \forall j \in V, \forall k \in \{1,\ldots,K\}$$

$$\sum_{(i,j)\in E} x_{ijk} w_{ij} \leq 6, \qquad\qquad \forall k \in \{1,\dots,K\}$$

$$l_{ij} \leq \sum_{k\in\{1,\dots,K\}} x_{ijk} \leq u_{ij}, \qquad\qquad \forall (i,j) \in E$$

$$x_{ijk} \in \mathbb{B}, \qquad\qquad \forall (i,j) \in E, \forall k \in \{1,\dots,K\},$$

where $b_s = 1$, $b_t = -1$, and $b_j = 0$ otherwise. The first constraint is nothing else than a series of flow-balance constraints that force the flow corresponding to the commodity (jury) k to be a path. The second constraint imposes the per-commodity (jury) total work duration constraint. The third constraint says that each event must be covered by at least one path.

A solution of the above model implemented on our problem instance is visualized in the following figure, where the different commodities' flows have been distinguished using different colors.

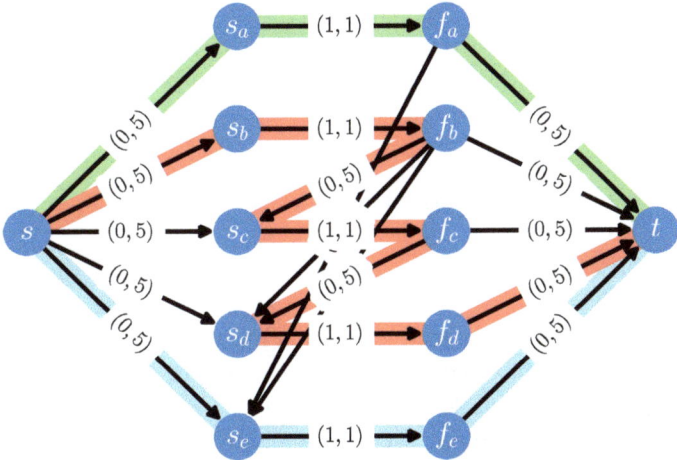

Note that we could have also used the above problem formulation to solve the original problem by removing the total work time constraint.

The Pyomo implementation is as follows:

```python
def SchedulingLimitingIndividualWorkingTime(G, s="s", t="t",
    nof_events=len(events)):
    m = pyo.ConcreteModel("Scheduling event juries limiting individual
        working hours")
    m.E = G.edges
    m.N = G.nodes
    m.J = pyo.RangeSet(nof_events)
    m.EN = pyo.Set(
        initialize=[e for e in m.E if all("_" in x for x in e) and
            e[0][-1] == e[1][-1]]
    )
    m.d = pyo.Param(m.E, initialize=nx.get_edge_attributes(G, "duration"))
    m.max_d = pyo.Param(initialize=6, mutable=True)
```

```
m.x = pyo.Var(m.E, m.J, within=pyo.Binary)
m.v = pyo.Var(m.J, within=pyo.Reals)

@m.Objective(sense=pyo.minimize)
def obj(m):
    return sum(m.v[j] for j in m.J)

@m.Constraint(m.N, m.J)
def flow_balance(m, n, j):
    rhs = m.v[j] if n == s else -m.v[j] if n == t else 0
    return (
        sum(m.x[e, j] for e in G.out_edges(n))
        - sum(m.x[e, j] for e in G.in_edges(n))
        == rhs
    )

@m.Constraint(m.J)
def limit_duration(m, j):
    return sum(m.d[e] * m.x[e, j] for e in m.E) <= m.max_d

@m.Constraint(m.EN)
def attend_all_events(m, u, v):
    return sum(m.x[(u, v), j] for j in m.J) == 1

return m
```

Solution 5.1 We can make the objective function linear by using additional variables w_i and extra constraints. The objective becomes $\sum_{i=1}^{n} w_i$, and the additional constraints to include are $w_i \geq i + \boldsymbol{a}_1^\top y$ and $w_i \geq i(1 + \boldsymbol{a}_2^\top x)$, for $i = 1, \ldots, n$.

To linearize the first constraint, we multiply both sides of the inequality by $1 + y_i$ (this is strictly positive since $y_i \geq 0$) and obtain $1 + b_i x_i \leq 1 + y_i$.

To linearize the second constraint, we can take the natural logarithm $\ln(\cdot)$ on both sides and obtain the following linear constraint:

$$1 + 3\boldsymbol{a}_2^\top \boldsymbol{x} \leq 0.$$

Since $\ln(\cdot)$ is an increasing function, the inequality sign $\leq$ remains the same. Putting all these ingredients together, the equivalent linear optimization problem is

$$
\begin{aligned}
\min \quad & \sum_{i=1}^{n} w_i \\
\text{s.t.} \quad & w_i \geq i + a_1^\top y, && \forall i \\
& w_i \geq i\left(1 + \boldsymbol{a}_2^\top \boldsymbol{x}\right), && \forall i \\
& 1 + b_i x_i \leq 1 + y_i, && \forall i \\
& 1 + 3\boldsymbol{a}_2^\top \boldsymbol{x} \leq 0 \\
& x, y \geq 0.
\end{aligned}
$$

Solution 6.1 Let x_i and y_i denote the number of hours worked, respectively, by divisions A and B in week i. Furthermore, let w_i be the number of goods produced in week i, u_i the amount of stored products at the end of week i, l_{Ai} the minimum hours that should be worked at division A in week i, and u_{Ai} the maximum hours. The notation l_{Bi} and u_{Bi} is used for division B. Using these notations we can state the problem as follows:

$$
\min \quad \sum_{i=1}^{n}(x_i + y_i)
$$

$$
\text{s.t.} \quad \sqrt{x_i}\sqrt{y_i} \geq w_i, \quad i = 1,\ldots,n
$$

$$
\sum_{j=1}^{i} w_j - u_i = \sum_{j=1}^{i} d_j, \quad i = 1,\ldots,n
$$

$$
0 \leq u_i \leq C, \quad i = 1,\ldots,n
$$

$$
w_i \geq 0, \quad i = 1,\ldots,n
$$

$$
l_{Ai} \leq x_i \leq u_{Ai}, \quad i = 1,\ldots,n
$$

$$
l_{Bi} \leq y_i \leq u_{Bi}, \quad i = 1,\ldots,n.
$$

Note that $\sqrt{x_i}\sqrt{y_i} \geq w_i$ is equivalent to $w_i^2 \leq x_i y_i$. Hence, the problem above is equivalent to

$$
\min \quad \sum_{i=1}^{n}(x_i + y_i)
$$

$$
\text{s.t.} \quad \sum_{j=1}^{i} w_j - u_i = \sum_{j=1}^{i} d_j, \quad i = 1,\ldots,n
$$

$$
(2w_i, x_i - y_i, x_i + y_i) \in \mathbf{K}_{\text{SOCO}}^3, \quad i = 1,\ldots,n
$$

$$
0 \leq u_i \leq C, \quad i = 1,\ldots,n
$$

$$
w_i \geq 0, \quad i = 1,\ldots,n
$$

$$
l_{Ai} \leq x_i \leq u_{Ai}, \quad i = 1,\ldots,n
$$

$$
l_{Bi} \leq y_i \leq u_{Bi}, \quad i = 1,\ldots,n.
$$

Solution 7.1

(a) An obvious uncertainty is the travel speed on each segment of the road.

This uncertainty has two main sources: traffic conditions and our capabilities (type of vehicle, driver skills, etc.).

Another uncertainty is the accessibility of each road segment during the journey.

Different locations may have different sources for this uncertainty: Some parts of the world may be severely affected by the weather, such as African mud roads in the rainy season.

Software giants such as Google own a wealth of data from which accurate predictions can be made.

One can only guess how the uncertainty of such predictions is accounted for during optimization.

We guess some robustness is observed, as users would be extremely annoyed by arriving much later than predicted or being often forced to reroute due to obstructions.

(b) Such an introduction is a great invitation to consider stochastic optimization.

(c) Life critical implies that errors may be fatal.

Sounds like a great case for robust optimization.

(d) Making a model robust or stochastic increases the complexity of the model.

Simple hacks are likely to be the way to go here.

Solution 8.1 A basic formulation of the shortest path problem without uncertainty would be

$$
\begin{aligned}
\min \quad & \sum_{(i,j)\in E} x_{ij} w_{ij} \\
\text{s.t.} \quad & \sum_{(s,i)\in E} x_{si} = 1 \\
& \sum_{(i,t)\in E} x_{it} = 1 \\
& \sum_{(i,j)\in E} x_{ij} - \sum_{(j,k)\in E} x_{jk} = 0, & \forall\, j \in V \setminus \{s,t\} \\
& x_{ij} \in \mathbb{B}, & \forall\, (i,j) \in E.
\end{aligned}
$$

As we can see, uncertainty will be present only in the objective function, so we need to derive its worst-case value. We can readily apply the budgeted uncertainty set formula adapted to this case, but we can also derive the dual of the pessimization problem from scratch. This pessimization problem is

$$
\begin{aligned}
\max \quad & \sum_{(i,j)\in E} x_{ij}(1 + z_{ij}) w_{ij} \\
\text{s.t.} \quad & 0 \le z_{ij} \le 1, & \forall\, (i,j) \in E \\
& \sum_{(i,j)\in E} z_{ij} \le \Gamma.
\end{aligned}
$$

We can derive its dual through the Lagrangian defined as follows:

$$
\begin{aligned}
L(\boldsymbol{x},\alpha,\beta) &= \sum_{(i,j)\in E} x_{ij}(1 + z_{ij}) w_{ij} + \sum_{(i,j)\in E} \alpha_{ij}(1 - z_{ij}) + \beta \left(\Gamma - \sum_{(i,j)\in E} z_{ij} \right) \\
&= \sum_{(i,j)\in E} (x_{ij} w_{ij} + \alpha_{ij}) + \beta\Gamma + \sum_{(i,j)\in E} (x_{ij} w_{ij} - \alpha_{ij} - \beta) z_{ij}.
\end{aligned}
$$

In this context, we can see that the dual function is

$$
\begin{aligned}
g(\alpha,\beta) &= \sup_{\alpha\ge 0, \beta\ge 0} L(\boldsymbol{x},\alpha,\beta) \\
&= \begin{cases} \displaystyle\sum_{(i,j)\in E} (x_{ij} w_{ij} + \alpha_{ij}) + \beta\Gamma, & \text{if } x_{ij} w_{ij} \le \alpha_{ij} + \beta \quad \forall\, (i,j) \in E \\ +\infty, & \text{otherwise.} \end{cases}
\end{aligned}
$$

The dual of the pessimization problem is the problem of minimizing this function. Therefore, it makes sense to embed it into our original optimization problem formulation to obtain the following robust problem formulation:

$$\min \quad \sum_{(i,j)\in E} (x_{ij}w_{ij} + \alpha_{ij}) + \beta\Gamma$$

$$\text{s.t.} \quad \sum_{(s,i)\in E} x_{si} = 1$$

$$\sum_{(i,t)\in E} x_{it} = 1$$

$$\sum_{(i,j)\in E} x_{ij} - \sum_{(j,k)\in E} x_{jk} = 0, \qquad \forall\, j \in V \setminus \{s,t\}$$

$$x_{ij}w_{ij} \le \alpha_{ij} + \beta, \qquad \forall\, (i,j) \in E$$

$$x_{ij} \in \mathbb{B}, \qquad \forall\, (i,j) \in E$$

$$\alpha_{ij} \ge 0, \qquad \forall\, (i,j) \in E$$

$$\beta \ge 0.$$

This problem is not TU since it is possible to construct problem instances where the optimal solution is fractional. For example, consider a network with vertices s, t, u_1, u_2, and four arcs with the following lengths: $w_{su_1} = w_{su_2} = 0$, $w_{u_1t} = w_{u_2t} = 1$, and $\Gamma = 1$. One can verify that, without integrality constraints, it is possible to arrive at an optimal solution $x_{su_1} = x_{su_2} = x_{u_1t} = x_{u_2t} = 0.5$ with the worst-case objective function value of 1.5. In contrast, the two available integer solutions both yield the worst-case objective function value of 2:

$$x_{su_1} = x_{u_1t} = 1, \quad x_{u_2t} = x_{su_2} = 0$$

and

$$x_{su_1} = x_{u_1t} = 0, \quad x_{su_2} = x_{u_2t} = 1.$$

Solution 9.1 Let x_{ij} denote the capacity of vehicles that we move from depot i to depot j. Note that we can always take this quantity to be nonnegative, that is, $x_{ij} \ge 0$, since if we move capacity from the depot j to depot i, we can take $x_{ij} = 0$ and its "mirror" variable to be nonnegative $x_{ji} \ge 0$. Noticing that the cost of moving vehicles before the demand is realized amounts to $\sum_{i \neq j} c_{ij}x_{ij}$, the first-stage problem for the profit maximization can be written as follows:

$$\max \quad -\sum_{i\neq j} c_{ij}x_{ij} + \mathbb{E}Q(\boldsymbol{x}, \boldsymbol{z})$$

$$\text{s.t.} \quad \boldsymbol{x} \ge 0.$$

After the first-stage problem, the new capacity at depot i, which we denote by l_i, is equal to the starting value s_i, minus the capacity that we moved away from depot i, plus the capacity that we moved into depot i, that is,

$$l_i = s_i - \sum_{j\neq i} x_{ij} + \sum_{j\neq i} x_{ji}.$$

The second-stage problem describes what happens at every depot while aiming to maximize profit *after* we have already decided how much capacity $\boldsymbol{x}$ to move between depots and *upon the realization of demand $\boldsymbol{z}$*. Let y_{ij} be the second-stage variable describing the amount of demand to move from i to j that we decide to satisfy at depot i. Clearly, $0 \leq y_{ij} \leq z_{ij}$, but at the same time, each depot cannot exceed the available vehicle capacity, hence $\sum_{j \neq i} y_{ij} \leq l_i$. Trivially, at each depot i, the profit we make is equal to the total demand that we move from that depot, that is, $\sum_{j \neq i} q_{ij} y_{ij}$. Given these considerations, the second-stage problem becomes

$$Q(\boldsymbol{x}, \boldsymbol{z}) := \max \quad \sum_{i \neq j} q_{ij} y_{ij}$$

$$\text{s.t.} \quad \sum_{j \neq i} y_{ij} \leq s_i - \sum_{j \neq i} x_{ij} + \sum_{j \neq i} x_{ji},$$

$$y_{ij} \leq z_{ij},$$

$$\boldsymbol{y} \geq 0.$$

Solution 10.1 This problem can be interpreted as follows: x is the "guess" that one makes on the value of the uncertain quantity $\max\{4 - 2z, z - 2\}$, and after z is observed, y_1 is the twice more expensive "extra" that one has to add up to the value of x to obtain a number equal at least equal to $\max\{4 - 2z, z - 2\}$. Given the interval $[0,4]$, we know that the largest value $\max\{4 - 2z, z - 2\}$ takes is 4, so it does not make sense to select an x larger than this, so we can only consider $x \in [0,4]$.

To calculate the recourse function $\mathbb{E}Q(x, z)$, we need to evaluate the expectation. We will break it into two cases depending on whether $z \leq 2$ or $z > 2$, each with probability 0.5.

If $z \leq 2$, then if $4 - 2z \leq x$ with conditional probability $x/4$, in which case we can set $y = 0$, or $x > 4 - 2z$ with conditional probability $1 - x/4$, in which case, on average, y will need to be equal to $(4 - x)/2$.

If $z > 2$, then if $x \geq 2$, we can set $y = 0$. Otherwise, with conditional probability $x/2$, we will have $x \geq z - 2$, and with conditional probability $1 - x/2$, we will have $x < z - 2$, in which case, on average, y will need to be equal to $(2 - x)/2$.

In general, we have the following formulation for the recourse function:

$$\mathbb{E}Q(x, z) = \begin{cases} \frac{1}{2} 2 \left(1 - x/4\right) \left(2 - x/2\right), & \text{if } 2 \leq x \leq 4 \\ \frac{1}{2} 2 \left((1 - x/4)(2 - x/2) + (1 - x/2)^2\right), & \text{if } 0 \leq x \leq 2. \end{cases}$$

We can minimize the function $x + \mathbb{E}Q(x, z)$ analytically, setting its derivative to 0, to obtain the optimal point $x^* = 4/3$, with expected performance equal to 2.333.

The solution obtained for the average value of $z = 2$ would have been $x = 0$, and its corresponding average performance would have been equal to 3, thus substantially worse than that of the optimal stochastic solution.

References

[ADV23]   Martin S. Andersen, Joachim Dahl, and Lieven Vandenberghe. *CVXOPT – Python Software for Convex Optimization*. 2023. URL: https://cvxopt.org/index.html [accessed November 2023].

[App+07] David L. Applegate et al. *The Traveling Salesman Problem: A Computational Study*. Princeton Series in Applied Mathematics. Princeton University Press, 2007.

[BJS10]   M. S. Bazaraa, John J. Jarvis, and Hanif D. Sherali. *Linear Programming and Network Flows*. John Wiley & Sons Inc, 2010.

[BL11]   John R Birge and Francois Louveaux. *Introduction to Stochastic Programming*. Springer Science & Business Media, 2011.

[BL23]   Michelle Bodnar and Andrew Lohr. *Introduction to Algorithms – CLRS, Chapter 26 Problems*. 2023. URL: https://walkccc.me/CLRS/Chap26/Problems/26-2/ [accessed November 2023].

[BS04]   Dimitris Bertsimas and Melvyn Sim. "The price of robustness." *Operations Research* 52.1 (2004), pp. 35–53. https//doi.org/10.1287/opre.1030.0065.

[BV04]   Stephen Boyd and Lieven Vandenberghe. *Convex Optimization*. Cambridge University Press, 2004.

[Coo]   W. Cook. *The Traveling Salesman Problem*. www.math.uwaterloo.ca/tsp/index.html.

[CR74]   Richard Church and Charles ReVelle. "The maximal covering location problem." *Papers of the Regional Science Association* 32.1 (1974), pp. 101–118. https//doi.org/10.1007/BF01942293.

[Dia+23a] Steven Diamond et al. *CVXPY – A Python Library for Convex Optimization Problems*. 2023. www.cvxpy.org/ [accessed November 2023].

[Dia+23b] Steven Diamond et al. *Disciplined Convex Programming (DCP) Rules – CVXPY Tutorial*. 2023. www.cvxpy.org/tutorial/dcp/index.html [accessed November 2023].

[Dij22]   Edsger W. Dijkstra. "A note on two problems in connexion with graphs." In: *Edsger Wybe Dijkstra: His Life, Work, and Legacy*. Morgan Claypool, 2022, pp. 287–290.

[DV04]   D. P. De Farias and B. Van Roy. "On constraint sampling in the linear programming approach to approximate dynamic programming." *Mathematics of Operations Research* 29.3 (2004), pp. 462–478.

[For+]   John Forrest et al. *COIN-OR CBC User's Manual*. https://coinor.github.io/Cbc/intro.html [accessed November 2023].

[GPS02]   Christelle Guéret, Christian Prins, and Marc Sevaux. *Applications of Optimization with Xpress-MP*. Dash Optimization, 2002.

[Hal+]   Julian Hall et al. *Highs Solver*. https://highs.dev/ [accessed November 2023].

[Har+17] William E. Hart et al. *Pyomo – Optimization Modeling in Python*, 2nd ed. Springer International Publishing, 2017. https//doi.org/10.1007/978-3-319-58821-6.

[Har15]   Ford W. Harris. "Operations and cost." *Factory Management Series* 48 (1915), p. 52.

[KIG14]   Serkan Kiranyaz, Turker Ince, and Moncef Gabbouj. "Optimization techniques: an overview." *Multidimensional Particle Swarm Optimization for Machine Learning and Pattern Recognition*. Springer, 2014, pp. 13–44. https//doi.org/10.1007/978-3-642-37846-1_2.

[Kit11]   Mitri Kitti. *History of Optimization*. 2011. www.mitrikitti.fi/opthist.html [accessed 15 October 2021].

[KY91]   Hiroshi Konno and Hiroaki Yamazaki. "Mean-absolute deviation portfolio optimization model and its applications to Tokyo Stock Market." *Management Science* 37.5 (1991), pp. 519–531.

[Loh13]   Volker Lohweg. *Banknote Authentication*. 2013. https://doi.org/10.24432/C55P57.

[Loz18]   Christian Carballo Lozano. *Modeling and Optimization of a Weekly Workforce with Python and Pyomo*. 2018. https://bit.ly/3IHmEcm [accessed 26 September 2023].

[Mos23a] Mosek. *Exponential Cone – MOSEK Modeling Cookbook*. 2023. http://docs.mosek.com/modelingcookbook/expo.html [accessed November 2023].

[Mos23b] Mosek. *Power Cone – MOSEK Modeling Cookbook*. 2023. http://docs.mosek.com/modelingcookbook/powo.html [accessed November 2023].

[NW82]   Subhash C. Narula and John F. Wellington. "The minimum sum of absolute errors regression: A state of the art survey." *International Statistical Review / Revue Internationale de Statistique* 50.3 (1982), p. 317. https//doi.org/10.2307/1402501.

[Pos+23] Krzysztof Postek et al. *Companion JupyterBook for "Hands-On Mathematical Optimization with Python"* 2023. https://mobook.github.io/MO-book/intro.html.

[Pyo23]   Pyomo Development Team. *Pyomo Documentation*. 2023.

[Pyt23]   Python. *Python 3 Documentation – Input and Output*. Python Software Foundation, 2023. https://docs.python.org/3/tutorial/inputoutput.html [accessed November 2023].

[San23]   Sandia. *Rotated Quadratic Cone – Pyomo Documentation*. Sandia National Laboratories, 2023. https://bit.ly/3x3McOi [accessed November 2023].

[Sch03]   Alexander Schrijver. *Combinatorial Optimization: Polyhedra and Efficiency*. Springer, 2003.

[SDR14]   Alexander Shapiro, Darinka Dentcheva, and Andrzej Ruszczyński. *Lectures on Stochastic Programming: Modeling and Theory*. SIAM, 2014.

[Smi+]   Kevin D. Smith et al. *PEP 318 – Decorators for Functions and Methods*. Python Software Foundation, 2023. https://peps.python.org/pep-0318/ [accessed November 2023].

[WB06]   Andreas Wächter and Lorenz T Biegler. "On the implementation of an interior-point filter line-search algorithm for large-scale nonlinear programming." *Mathematical Programming* 106.1 (2006), pp. 25–57.

[Wik23]   Wikipedia. *Path Cover*. 2023. https://en.wikipedia.org/wiki/Path_cover [accessed November 2023].

[Wil13]   H. Paul Williams. *Model Building in Mathematical Programming*. John Wiley & Sons, 2013.

Index

For EU product safety concerns, contact us at Calle de José Abascal, 56–1°,
28003 Madrid, Spain or eugpsr@cambridge.org.

www.ingramcontent.com/pod-product-compliance
Ingram Content Group UK Ltd.
Pitfield, Milton Keynes, MK11 3LW, UK
UKHW051018071225
465726UK00006BA/263